Extreme Value Theory in Engineering

This is a volume in
STATISTICAL MODELING AND DECISION SCIENCE

Gerald L. Lieberman and Ingram Olkin, editors
Stanford University, Stanford, California

Extreme Value Theory in Engineering

Enrique Castillo

Departamento de Matemática Aplicada
Escuela de Ingenieros de Caminos
Universidad de Cantabria
Santander
Spain

ACADEMIC PRESS, INC.
Harcourt Brace Jovanovich, Publishers

Boston   San Diego   New York
Berkeley   London   Sydney
Tokyo   Toronto

ACADEMIC PRESS, INC.
1250 Sixth Avenue, San Diego, CA 92101

United Kingdom Edition published by
ACADEMIC PRESS, INC. (LONDON) LTD.
24-28 Oval Road, London NW1 7DX

Library of Congress Cataloging-in-Publication Data

Castillo, Enrique, Date
  Extreme value theory in engineering.

  (Statistical modeling and decision science)
  Bibliography: p.
  Includes index.
  1. Extreme value theory.   2. Mathematical statistics.
I. Title. II. Series.
QA273.6.C36   1987   519.5   87-16792
ISBN 0-12-163475-2

To whoever does not suffer from uncertainty

Contents

vii

**IV   Asymptotic Distribution of Sequences of
         Dependent Random Variables**

V   Multivariate Case

Preface

Before initiating the task of writing a book, one needs an encouraging motivation. After several years of work with engineers in the area of extremes, several Ph. D. Theses of my students on this topic and the contact, personal or through their many publications, with mathematicians and statisticians, I discovered the lack of close communication and collaboration between practitioners and theorists. This fact is the main cause, on the one hand, for engineers to be unaware of the very many recent advances in the area of extremes, and, on the other hand, for the mathematicians and statisticians to find practical applications and motivations for their theoretical work. The exponentially increasing number of publications on any subject makes it impossible to become a specialist in a wide area of knowledge. Thus, the need for collaboration and joint work. All this has increased my motivation for writing the present book and clearly defined its goal. However, to be a bridge between engineers and mathematicians or statisticians is not an easy task, as experience demonstrates; their daily languages and main concerns are far enough apart to make it difficult. In addition, everything must be done at the risk of being criticized from both sides. In this context, the present book will be only a grain of sand in a beach still to be built.

In order to make it more readable for engineers and to avoid unnecessary repetition, most of the proofs have been intentionally omitted, and only new or very simple ones are included. With the aim of facilitating the understanding of the definitions and theorems, many clarifying examples have been inserted. A collection of 14 sets of data has also been included and many techniques described throughout the book have been illustrated by their direct application to them. Some computer codes, which appear in appendix B, were prepared with the aim of facilitating the use of the above mentioned techniques to daily practice problems. The language Basic was selected to make this possibility of a wider range. Fortran or Pascal users will not find any trouble in making a translation of codes.

The book can be used as a textbook or as a consulting book. In the first case the examples and computer codes will facilitate the work of students, individually or in groups, by its application to fictitious or real problems taken from imagination or daily practice. The extensive bibliography, included in the book, can also serve as the basis for some student additional work.

The book covers a wide range of the available material to date. Engineers can find useful results on extremes and other order statistics and motivating examples and ideas. Mathematicians and statisticians can find some motivating practical ideas and examples to illustrate their theoretical aspects.

Some theoretical parts of the book are completely original, as is chapter 9 and parts of chapters 3, 4 and 6. The rest is a recapitulation of existing material with added personal views of the author. However, throughout the book, a very large percentage of the included examples are original and many of them refer to engineering problems.

One of the main contribution of the book is the emphasis on the tail behaviour as the basis for the analysis. This is indicated mainly in chapters 3, 4, 5 and 6, and leads to a change in the recommended techniques of analysis, as probability paper, estimation of parameters or selection of domains of attraction techniques, for example.

I want to publicly thank Professor J. Galambos for his extremely valuable help. He made possible my sabbatical period at Temple University, during which a great deal of the work included in this book was completed. He also contributed the section on characterization of distributions and many suggestions and corrections to this book. This was initially planned to be a joint book, but previous engagements with the editor of the second edition of his book on extreme order statistics made it impossible.

I want also to thank the US-Spain Joint Committee for Research and the University of Santander, who financed my sabbatical leave, and the Computer Center and the Department of Mathematics of Temple University, who provided me the use of computers and the library and gave me room space.

Thanks also to my family for being generous in allowing me to dedicate part of its time to make the book possible, to my wife for her collaboration in preparing the bibliography and index, to Alberto Luceño for some valuable comments on the manuscript, to Agustin Manrigue for drawing all figures, and to Deborah Beaumont for her careful proofreading of the manuscript.

Finally, I want to mention the scientific community, mainly those included in the Bibliography and those who were, surely unintentionally, omitted. They, through their life work, made this book a possibility. To all of them my sincere thanks.

I Introduction

Chapter 1

Introduction and Motivation

1.1. Introduction

Experienced engineers know how successful design depends on the adequate selection of the method of analysis. The modeling or idealization of the problem under consideration (water supply system, structure, road, harbour, dam, etc.) should be sufficiently simple, logically irrefutable, admit of mathematical solution, and, at the same time, reproduce sufficiently well the actual problem. As in any other branch of knowledge, the selection of the idealized model should be achieved by detecting and reproducing the essential first-order factors, and discarding or neglecting the inessential second-order factors. The new elements under design are represented by the simplest geometrical, structural, social, etc. elements and the relationships between them are idealized.

In many situations, the final design is a compromise between the capacity of the element (strength, service capacity, production capacity, etc.) and the actual operating conditions (actual loads, service, traffic, demand, etc.). Classical methods of analysis lie quite far from the operating conditions of the actual elements. The most important reason for this is that the properties of the elements and external conditions are assumed to be known with complete certainty. However, all these properties are under the influence of such amount of incontrollable external factors that they become random variables. Two extreme results of this classical design are those related to under- or over-design which lead to failure of the designed element or to a large waste of resources.

The aim of the engineering design of an element is to guarantee that none of the catastrophic limiting states occur during its lifetime. To be more specific, the element shall satisfy the following requirement

(1.1) $O_t \leq C_t$ for any t

where O_t and C_t are the actual operating conditions (loads, demands, traffic, etc.) acting on the element and the real capacity of the element at time t, respectively.

Both components are clearly random variables whose distribution functions may be established only by systematically analyzing the available history for similar elements and conditions. Fortunately, since operating conditions, O_t, and capacities, C_t, are statistically independent random variables, their distributions can be examined separately. This avoids a great deal of difficulty.

In some cases, condition (1.1) is replaced by

(1.2) $O_{max} \leq C_{min}$

where O_{max} and C_{min} are some upper and lower bounds, respectively, that can under no condition be exceeded. However, conditions (1.1) and (1.2) are not equivalent.

The random nature of the fundamental design parameters is taken into account by the so-called safety factor. This factor summarizes, in some simple way, the random character of design parameters. Consequently, capacities are lessened (diminished) and operating conditions strengthened (increased). This means that under classical design techniques expression (1.1) is replaced by

(1.3) $S_o O \leq \dfrac{C}{S_c}$

where S_o and S_c are two safety factors or coefficients whose values are larger than unity and O and C are values given by design codes.

However, the safety factor technique is so simple that it becomes completely insufficient for the analysis of many problems whose solution is possible only on the basis of Probability Theory and Mathematical Statistical methods.

In order to achieve satisfactory behaviour of the element, condition (1.1) must hold for the entire lifetime of the element and because many values of O_t and C_t occur, we must have

(1.4) $\text{Max}(O_t - C_t) \leq 0$

As stated before, in some occasions, this condition is substituted by

$$(1.5) \qquad\qquad \text{Max } O_t < \text{Min } C_t$$

where $\text{Max}(O_t - C_t)$, $\text{Max } O_t$ and $\text{Min } C_t$ are also random variables.

That is why the problem of the distribution of extreme values is of very great interest. In fact, the design engineer is not interested in the distribution of $(O_t - C_t)$, O_t or C_t but in the distribution of their extremes instead. This is so, because those values are the only ones affecting the failure of any given system. For example, the strength of a series of elements is always the strength of its weakest element. The strength of any other element in no way affects the strength of the system. This is the reason why the optimal design criterion for series systems is that of equal strength for all the elements if they are subjected to the same demands.

In other situations, equation (1.1) need not hold for the whole lifetime and some events leading to failure of (1.1) are allowed. Thus, a system is said to operate satisfactorily if only $1, 2, \ldots, k$ times in its lifetime, condition (1.1) is violated. In this case the main role, before played by the maxima and minima, is now played by the k-th largest and the k-th smallest values of O_t and C_t, respectively. This is just one reason for the importance of other order statistics.

1.2. Some Engineering Examples Where Extreme Value Theory Is of Significance

There exists a very long list of engineering areas where the extreme value theory plays a decisive role. Most of them have problems that can be fitted into the general framework given in the preceding section. However, in order to be more specific, we have given here some concrete examples.

STRUCTURAL ENGINEERING.   Modern building codes and standards provide information on extreme winds in the form of wind speeds corresponding to various specified mean recurrence intervals (return periods). These speeds are estimates of extreme winds that can occur at the place where the building or engineering work is to be located and have a large influence on their design characteristics and final costs. Precise estimates of the probabilities of occurrence of extreme winds are required in order to allow for realistic margins of safety in structural design, on one hand, and for economical solutions, on the other hand.

For a complete analysis of this problem the reader is referred to Court (1953), Davenport (1968a,b,1972,1978), Duchêne Marullaz (1972), Grigoriu

(1984a,b), Hasofer (1972), Hasofer and Sharpe (1969), Sachs (1972), Shellard (1958), Simiu and Filliben (1975,1976), Simiu et al (1977,1978,1979,1982), Thom (1967,1968a,b), Wilson (1966), Zidek et al. (1979), etc.

Building codes also give the design loads to be used in calculations. These loads are closely related to the largest loads acting on the structure during its lifetime. Small design loads can lead to collapse of the structure and many induced damages and their consequences. On the contrary, large design loads lead to uneconomical structures and large waste of money. A correct design is possible only if the statistical properties of largest loads are well known. For some works in this area see Ang (1973), Lévy (1949), Mistéth (1973), Moses (1974), Murzewski (1972), Prot (1949a,b,1950), etc.

Another important factor to be taken into account in structural design is the possibility of earthquakes. Some material related to this problem can be found in Bogdanoft and Schiff (1972). Most of the building codes provide information about seismic incidence in the form of areas of equal risk. A building or engineering work will survive a given period if it is designed in order to withstand the most severe earthquake occurring in that period. Thus, the maximum earthquake intensity plays an important role though not the only one. The probabilistic risk assessment of seismic events has notable importance in nuclear power plants where the losses are due not only to material damage of structures, but to the very dangerous indirect consequences that can follow.

OCEAN ENGINEERING.   In the area of Ocean engineering, every one knows that wave height is the main factor to be considered for design purposes. Thus, the design of offshore platforms, breakwaters, dikes and other harbour works rely upon the knowledge of the probability distribution of the largest waves. Another problem of crucial interest in this area is to find the joint distribution of the heights and periods of sea waves. More precisely, the engineer is interested in the periods associated with the largest waves. This is clearly a problem, which in the extreme value field, is known under the name of concomitants of order statistics. Some of the publications dealing with these problems are those of Battjes (1977), Borgman (1963,1970,1973), Bretschneider (1959), Castillo et al. (1977), Cavanie et al. (1976), Chakrabarti and Cooley (1977), Corniero (1981), Court (1953), Dattatri (1973), Draper (1963), Earle et al. (1974), Giménez Curto (1979), Goodknight and Russel (1963), Günbak (1978), Houmb and Overvik (1977), Longuet-Higgins (1952,1975), Losada et al. (1978), Putz (1952), Sellars (1975), Suárez Bores (1964), Thom (1967,1971,1973), Thrasher and Aagaard (1970), Tucker (1963), Wiegel (1964), Wilson (1966), Yang et al. (1974), etc.

HYDRAULICS ENGINEERING.   Stream discharge and flood flow have long been measured and used by engineers in the design of hydraulic structures (dams, canals, etc.) and flood protection works and in planning for flood plain use. A flood frequency analysis is the basis for the engineering design of many projects and the economic analysis of flood-control projects. High losses in human lives and property, due to damages caused by floods, have recently pointed out the need for precise estimates of probabilities and return periods of these extreme events. However, hydraulic structures and flood protection works are affected not only by the intensity of floods but by their frequency, as occurs with a levee, for example. Thus, we can conclude that knowledge of the extreme statistical behaviour of stream discharge is of essential importance in hydraulic engineering. Some works related to these problems are those of Beard (1962), Benson (1968), Chow (1951,1964), Dalrymple (1960), Gumbel and Goldstein (1964), Gupta et al. (1976), Hershfield (1962), Karr (1976), Kirby (1969), Matalas and Wallis (1973), Mistéth (1974), North (1980), Reich (1970), Shane et al (1964), Todorovic (1970,1971,1978,1979), Zelenhasic (1970), etc.

POLLUTION STUDIES.   With the existence of large concentrations of people (producing smoke, human wastes, etc.) or the appearance of new industries (chemical, nuclear, etc.), the pollution of air, rivers and coasts has become a common problem of industrialized countries. The pollutant concentration, expressed as the amount of pollutant per unit volume (of air or water), is forced, by city codes, to remain below a given critical level. Thus, the regulations are satisfied if, and only if, the largest pollution concentration during the period of interest is less than the critical level. Here then, the largest value plays the fundamental role in design. For some applications see Barlow and Singpurwalla (1974), Larsen (1969), Leadbetter et al. (1983), Roberts (1979a,b), Singpurwalla (1972b), etc.

METEOROLOGY.   Extreme meteorological conditions are known to influence many aspects of human life as in the flourishing of agriculture and animals, the behaviour of some machines, the lifetime of certain materials, etc. In all these cases the engineers, instead of centering interest on the mean values (temperature, rainfall, etc.), are concerned only with the occurrence of extreme events (very high or very low temperature, rainfall, etc.). Accurate prediction of the probabilities of those rare events thus becomes the aim of the analysis. See Leadbetter (1983) and Sneyers (1984) for some examples.

MATERIAL STRENGTH.   One interesting application of extreme value theory to material strength is the analysis of size effect. In many engineering problems, the strength of actual structures has to be inferred from the strength of small elements or reduced size prototypes or models, which are tested under laboratory conditions. In such cases, extrapolation from small to much larger sizes is needed. In this context, extreme value theory becomes very useful in order to analyze the size effect and make extrapolations possible and reliable. If the strength of a piece is determined, or largely affected, by the strength of its weakest (real or hypothetical) subpiece into which the piece can be subdivided, as it usually occurs, the minimum strength of the weakest subpiece determines the strength of the piece. Thus, large pieces are statistically weaker than small pieces. For a complete list of references before 1978 the reader is referred to Harter (1977,1978a,b).

FATIGUE STRENGTH.   Modern Fracture Mechanics Theory reveals that fatigue failure is due to propagation of cracks when elements are under the action of repetitive loads. The fatigue strength of a piece is governed by the largest crack in the piece. If the size and the shape of the crack were known, the lifetime, measured in number of cycles to failure, could be deterministically obtained. However, the presence of cracks in pieces is random in number, size and shape, and thus, the random character of fatigue strength. Assume a longitudinal piece hypothetically subdivided in subpieces of the same length and being subjected to a fatigue test. Then, all the subpieces are subjected to the same loads and the lifetime of the piece is that of the weakest subpiece. Thus, the minimum lifetime of the subpieces determines the lifetime of the piece. Some references related to fatigue are Andrä and Saul (1974,1979), Batdorf (1982), Batdorf and Ghaffanian (1982), Birkenmaier and Narayanan (1982), Birnbaum and Saunders (1958), Bühler and Schreiber (1957), Castillo et al. (1983,1985), Castillo and Galambos (1985), Coleman (1956,1957a,b,1958a,b,c), Dengel (1971), Düebelbeiss (1979), Epstein (1954), Epstein and Sobel (1954), Fernández-Canteli (1982), Fernández-Canteli et al (1984), Freudenthal (1975), Gabriel (1979), Grover (1966), Hajdin (1976), Helgason and Hanson (1976), Maennig (1967,1970), Phoenix (1978), Phoenix and Smith (1983), Phoenix and Tierney (1983), Phoenix and Wu (1983), Smith (1980,1981), Spindel et al (1979), Tide and van-Horn (1966), Tierney (1982), Tilly and Moss (1982), Tradinick et al. (1981), Warner (1982), Warner and Hulsbos (1966), Weibull (1959), Yang and Trapp (1974), etc.

ELECTRICAL STRENGTH OF MATERIALS.   The lifetime of some electrical devices depends not only on their random quality but on the random voltage levels acting on them too. The device survives a given period if the maximum voltage level does not surpasses a critical value. Thus, the maximum voltage in the period is one of the governing variables in this problem. For some examples see Endicott and Weber (1956,1957), Hill and Schmidt (1948), Lawless (1982), Nelson (1982), Weber and Endicott (1956,1957), etc.

HIGHWAY TRAFFIC. Due to economical considerations, many highways are designed in such a manner that traffic collapse is assumed to take place a limited number, k, of times during a given period of time. Thus, the design traffic is that associated with the k-th largest traffic intensity during that period. Obtaining accurate estimates of the cdf of the k-th order statistic pertains to the theory of extreme order statistics and allows a reliable design to be performed.

CORROSION RESISTANCE. Corrosion failure takes place by the progressive size increase and penetration of initially small pits through the thickness of an element, due to the action of chemical agents. It is clear that the corrosion resistance of an element is determined by the largest pits and largest concentrations of chemical agents and that small and intermediate pits and concentrations do not have any affect on the corrosion strength of the element. Some references related to this area are Aziz (1956), Eldredge (1957), Logan (1936), Logan and Godsky (1931), Thiruvengadam (1972), etc.

A similar model explains the leakage failure of batteries, which gives another example where extremes are the design values.

1.3. New Developments in Extreme Value Theory

Since the publication of the book of Gumbel (1958) *Statistics of Extremes*, which caused a large impact on the engineering community, many new advances in the area of Extreme Value Theory have taken place. This was due mainly to applied engineers and scientists who began to state new problems arising from their daily practice, and to the work of mathematicians and statisticians who incorporated many of the extreme value problems into their working areas.

Among these advances, we can mention the following:

(a) Several theorems allowing the characterization of limit distributions, domains of attractions, etc.
(b) The unified von-Mises form to the classical, three fold, limit distributions for the i.i.d. case.

(c) New formulas for obtaining bounds are approximating probabilities of events, with applications to order statistics.

(d) New methods of estimation based on order statistics.

(e) New algorithms for estimation and simulation, mainly with censored samples.

(f) The identification of feasible limit distributions for k-th order statistics in the i.i.d. case with separated analyses for upper or lower, moderately upper or lower and central order statistics.

(g) Identification of necessary and sufficient conditions under which the limit distributions for the i.i.d. case still hold for dependent sequences.

(h) The study of many types of dependent sequences as exchangeable, stationary, m-dependent, those satisfying different mixing conditions, etc.

(i) Many results for important models, as Gaussian or moving average, for example.

(j) The analysis of multivariate extreme value distributions.

(k) Some important results for concomitants of order statistics.

(l) The appearance of new regression models related to extremes.

and

(m) very many other results.

1.4. Aim of the Book

Many of the above advances have been published in mathematical and statistical journals which are not written in a language many engineers and scientists can easily understand. The thrust of some of the above studies was mainly of theoretical interest and very mathematical and rigorous statements and proofs were given, such that even the practical meaning of their statements can be understood only by true specialists in the area. This fact, which is common to many other areas, has been the reason why many engineers and applied scientists are unaware of a large list of these advances and, consequently, they have not incorporated them into their daily procedures.

The aim of this book is to bring to engineers and applied scientists some selected results of extreme value theory which can be useful in application. We are aware that it is not an easy task, and that any effort to bring together mathematicians and engineers, as experience shows, has many associated difficulties. We have, intentionally, omitted most of the proofs of theorems, in order to make the text more readable to engineers.

With the purpose of facilitating the understanding and comprehension of the main theoretical results, many examples have been inserted. In spite of these efforts, an unavoidable (if one wants to include important results) minimum of theoretical concepts and theorems still remains.

Finally, the book tries to clarify some aspects and to point out some of the common errors encountered in the past.

1.5. Organization of the Book

The present book is mainly divided in four parts. The first part (chapter 2) is devoted to the general theory of order statistics. A study of the asymptotic properties of order statistics is not conceivable without a previous understanding of the behaviour and properties of order statistics from finite samples.

The second part (chapters 3 to 7) is concerned with the asymptotic distribution of order statistics for sequences of independent random variables. This part includes most of the classical problems in the area and additional ones coming from modern extreme value theory.

The third part (chapter 8) analyzes the case of sequences of dependent random variables.

Finally, the fourth and last part (chapters 9 and 10) addresses the problem of multivariate extremes and some regression and multivariate models related to extremes.

As mentioned before, a fairly large collection of examples has been inserted in order to illustrate some of the possible applications of the theoretical results just explained. In addition, a collection of 14 sets of data, which have been included and explained in Appendix C, are used to facilitate the understanding of the different algorithms, estimation methods, tests of hypotheses, etc. Many methods are systematically applied to these sets and the results are analyzed and discussed in the light of their physical meaning.

Finally, a collection of computer codes, in BASIC, are also given in order to facilitate the use of the material included in the book.

1.6. Some Classical Statistical Concepts

By way of introduction before starting with the core of the book, we define in this section some classical and very important statistical concepts related to extremes as exceedances, return period and characteristic and record values. We also include some distribution free results and formulas that allow the solution of some simple engineering design problems, as illustrated in the examples.

1.6.1.   Exceedances.

In many situations, the engineer or scientist, instead of dealing with maxima and minima, is interested in the events associated with the exceedances of certain values of the random variable under study. If a sea wave destroys a breakwater, or a flood or wind causes serious damages it does not matter whether the wave height was 10 or 12 m., the flood covered the soil with 20 or 80 cm. of water, or the wind has a speed of 100 or 150 km/hour. The engineer knows which are the critical values of sea wave height, flood amount or wind speed that lead to damages and his only interest is centered in the frequencies of exceedance of such values. This allows to deal with the frequencies instead of the values on the random variable itself.

So, we have the following definition.

Definition 1.1. (Exceedance). Let X be a random variable and u a real number. We say that the event $X = x$ is an exceedance of the level u if $x > u$.

∎

In the following unless the contrary being indicated we deal with absolutely continuous random variables.

One important practical problem is the following: Assuming independent and equally distributed trials, determine the probability of r exceedances in the next n trials.

Due to these assumptions, we deal with the same Bernouilli experiment (only two outcomes are possible: "exceedance" or "not exceedance") repeated n times. Thus, the number of exceedances is a binomial variable $B(n, p(x))$ with parameters n and $p(x)$, where $p(x)$ is the probability of exceedance of the level x of the variable under study.

The probability of r exceedances is

$$(1.6) \qquad P[\underline{m}_n(x) = r] = \binom{n}{r} p^r(x)[1 - p(x)]^{n-r}; \qquad 0 \le r \le n$$

where $\underline{m}_n(x)$ is the number of exceedances of the value x in a series of n trials.

But we can write $p(x)$ as

$$(1.7) \qquad\qquad\qquad p(x) = P[X > x] = 1 - F(x)$$

where $F(x)$ is the cdf of X.

Hence, expression (1.6) becomes

$$(1.8) \qquad\qquad P[\underline{m}_n(x) = r] = \binom{n}{r}[1 - F(x)]^r F^{n-r}(x)$$

Assume now that we make the level x dependent on n, x_n say, and that we have

(1.9)
$$\lim_{n \to \infty} n[1 - F(x_n)] = \tau; \qquad 0 \le \tau \le \infty$$

Then, the probabilities in (1.8) can be approximated by those of a Poisson process and we have the following theorem (see Leadbetter et al. (1983)).

Theorem 1.1. (Exceedances as a Poisson process). *Let $\{X_n\}$ be a sequence of independent and identically distributed (i.i.d.) random variables with cdf $F(x)$. Assume that the sequence $\{x_n\}$ of real numbers satisfies the condition*

(1.10)
$$\lim_{n \to \infty} n[1 - F(x_n)] = \tau; \qquad 0 \le \tau \le \infty$$

then

(1.11)
$$\lim_{n \to \infty} P[\underline{m}_n(x_n) = r] = \frac{\exp(-\tau)\tau^r}{r!}; \qquad r \ge 0$$

where $\underline{m}_n(x_n)$ is the number of X_i ($i = 1, 2, \ldots, n$) exceeding x_n, and the right hand side of (1.11) must be taken as 1 or zero depending on whether $\tau = 0$ or $\tau = \infty$, respectively.

The proof of this theorem is just the well known Poisson limit for the binomial distribution if $0 < \tau < \infty$ and some trivial considerations for the two extreme cases $\tau = 0$ and $\tau = \infty$.

The practical importance of this theorem lies in the fact that exceedances approximately follow a Poisson distribution, in the sense of (1.11) if (1.10) holds.

Another interesting problem related to exceedances is the following: Assuming independent and equally distributed trials, determine the probabilities of r exceedances, in the next N trials, of the m-th largest observation in the past n trials.

In this case, if p_m is the probability of exceedance of the m-th largest observation in the past n trials, the probability of r exceedances in the N future trials is given by

(1.12)
$$w(n, m, N, r/p_m) = \binom{N}{r} p_m^r [1 - p_m]^{N-r}$$

But

(1.13)
$$p_m = 1 - F(X_{n-m+1:n}) = F(X_{m:n}) = U_{m:n}$$

where $X_{n-m+1:n}$ is the $(n - m + 1)$-th order statistic of the sample of n past observations and $U_{m:n}$ is the m-th order statistic in a sample of size n drawn from a uniform $U(0, 1)$ population. Note that (1.13) is independent of the underlying cdf and has a clear intuitive meaning. Note also that p_m is random (it depends on sample values) with pdf (see (2.40))

$$(1.14) \qquad f(p_m) = \frac{p_m^{m-1}[1 - p_m]^{n-m}}{B(m, n - m + 1)}; \qquad 0 \le p_m \le 1$$

and then the total probability rule with (1.12) and (1.14) leads to

$$w(n, m, N, r) = \int_0^1 w(n, m, N, r/p_m) f(p_m) \, dp_m$$

$$= \binom{N}{r} \int_0^1 \frac{p_m^{r+m-1}[1 - p_m]^{N-r+n-m}}{B(m, n - m + 1)} \, dp_m$$

$$(1.15) \qquad = \binom{N}{r} \frac{B(r + m, N - r + n - m + 1)}{B(m, n - m + 1)}$$

The mean number of exceedances, $\bar{r}(n, m, N)$, taking into account that the mean of the binomial variable in (1.12) is Np_m, the total probability rule and that the mean of the m-th order statistic $U_{m:n}$ is $m/(n + 1)$ becomes

$$(1.16) \qquad \bar{r}(n, m, N) = \int_0^1 Np_m f(p_m) \, dp_m = N\mu_{U_{m:n}} = \frac{Nm}{n + 1}$$

By a similar process with the second moment we get for the variance of the number of exceedances

$$(1.17) \qquad \sigma^2(n, m, N) = \frac{Nm(n - m + 1)(N + n + 1)}{(n + 1)^2(n + 2)}$$

The variance takes a minimum value for $m = (n + 1)/2$. However, the coefficient of variation decreases with m, as should be expected.

Note that these results are distribution free in the sense that they are independent of $F(x)$.

Example 1.1. (Temperatures). The yearly maximum temperature during the last 40 years in a given region was $42°$ C. Determine the mean value and variance of the number of exceedances of $42°$ C of the yearly maximum temperature during the next 30 years.

Solution: From (1.16) and (1.17) we get

$$\bar{r}(40, 1, 30) = 30/41 = 0.732$$

$$\sigma^2(40, 1, 30) = \frac{30 \times 40 \times 71}{41^2 \times 42} = 1.207 \qquad \blacksquare$$

Example 1.2. (Floods).   The yearly maximum of flood discharges, in cubic meters per second, at a given location of a river, during the last 60 years are shown in Table C2 of appendix C. Choose a flood design value in order to have a mean value of 4 exceedances in the next 20 years.

Solution: According to (1.16), we have

$$\bar{r}(60, m, 20) = \frac{20m}{61} \approx 4 \quad \Rightarrow \quad m \approx 12$$

which shows that the value to be chosen is the 12th largest order statistic in the series, i.e. 50.17 cubic meters per second.   ∎

1.6.2.   Return periods.

Assume now an event (flood, dam failure, exceedance of a given temperature, etc.) such that its probability of occurrence in a unit period of time (normally one year) is p. Assume also that occurrences of such an event in different periods are independent. Then, as time passes, we have a sequence of equally likely Bernouilli experiments (only two outcomes: (a) occurrence or (b) not occurrence, are possible). Thus, the time (measured in unit periods) to the first occurrence is a geometric random variable $Ge(p)$, with mean value $1/p$. This motivates the following definition.

Definition 1.2. (Return period).   Let A be an event, and T the random time between consecutive occurrences of A events. The mean value, τ, of the random variable T is called the return period of the event A.   ∎

Note that if $F(x)$ is the cdf of the yearly maximum of a random variable, the return period of that random variable to exceed the value x is $1/[1 - F(x)]$ years. Similarly, if $F(x)$ is the cdf of the yearly minimum of a random variable, the return period of the variable to go below the value x is $1/F(x)$ years.

Also note that if a given engineering work fails when, and only when, the event A occurs, its mean lifetime coincides with the return period of A. The importance of return periods in engineering is due to the fact that many design criteria are defined in terms of return periods.

The probability of occurrence of the event A before the return period is (see the geometric distribution)

(1.18) $$F(\tau) = 1 - (1 - p)^{\tau} = 1 - (1 - p)^{1/p}$$

which for $\tau \to \infty$ $(p \to 0)$ tends to the value 0.63212.

Example 1.3. (Flood return period). The cdf of the yearly maximum discharge, in cubic meters per second (cms), of a river at a given location is given by

$$F(x) = \exp\left[-\exp\left(-\frac{x - 38.5}{7.8} \right) \right]$$

The return periods of yearly maximum discharges of 60 and 70 cms. are

$$\tau_{60} = \frac{1}{1 - F(60)} = 16.25 \text{ years, and}$$

$$\tau_{70} = \frac{1}{1 - F(70)} = 57.24 \text{ years,}$$

This means that discharges of 60 and 70 cms occur, in mean, once every 16.25 and 57.24 years, respectively. ∎

Example 1.4. (Breakwater design wave height). If the design wave height of a breakwater is defined as that value with a return period of 50 years, and the yearly maximum wave height, in feet, is known, from past experience, to have a limiting Gumbel distribution with cdf

$$F(x) = \exp\left[-\exp\left(-\frac{x - 15}{4} \right) \right]$$

the design wave height, h, must satisfy the equation

$$\frac{1}{1 - F(h)} = 50$$

from which we get $h = 30.61$ feet. ∎

1.6.3. Characteristic values.

If now the event A consists of the exceedances of the level x of a random variable X, we can write $A(x)$ instead of A, and then p becomes $p(x) = 1 - F(x)$ and

(1.19) $$\tau(x) = \frac{1}{1 - F(x)}$$

In this case, the expected value of the number of events $A(x)$ in n periods is given by (see the binomial random variable)

$$(1.20) \qquad\qquad n[1 - F(x)]$$

Of special interest is the value of the level x leading to an expected value of 1. This value is called the characteristic largest value for that period. More precisely, we have the following definition.

Definition 1.3. (Characteristic values). A certain value x of a random variable X is said to be the characteristic largest value for a period of duration n units if the mean value of the number of exceedances of that value in such a period is unity. Because of its dependence on n, the characteristic largest value will be denoted by u_n. ∎

Change of X into $-X$ allows us to define, in a similar manner, the characteristic smallest value, v_n.

Note that the characteristic largest value satisfies the equation

$$(1.21) \qquad\qquad n[1 - F(u_n)] = 1 \quad \Rightarrow \quad F(u_n) = 1 - \frac{1}{n}$$

Similarly, the characteristic smallest value satisfies

$$(1.22) \qquad\qquad nF(v_n) = 1 \quad \Rightarrow \quad F(v_n) = \frac{1}{n}$$

Note the symmetry of expressions (1.21) and (1.22).

The probability of exceeding the characteristic largest value in the period is

$$(1.23) \qquad\qquad 1 - F^n(u_n) = 1 - \left(1 - \frac{1}{n}\right)^n$$

which for large n tends to $1 - \exp(-1) = 0.6321$.

Example 1.5. (Characteristic largest value of a flood in a lustrum). The cdf of the yearly maximum of flow discharge, in cubic meters per second, of a river is given by

$$F(x) = \exp\left[-\exp\left(-\frac{x - 38.5}{7.8}\right)\right]$$

Then, according to (1.21), the characteristic largest value of flow discharge for a lustrum is the solution of the following equation

$$F(u_5) = 1 - \frac{1}{5} = 0.8$$

from which we get $u_5 = 50.2$ cms. ∎

Example 1.6. (Characteristic largest value by extrapolation). Given the data of yearly maximum flood discharge, in cms, calculate, by extrapolation, the characteristic value of a period of 100 years.

Solution: First, we can draw the expected number of exceedances in the next 100 years for the largest sample values by means of expression (1.16). Figure 1.1. shows the expected values and the extrapolated line. From it a characteristic largest value of 80.2 cms can be found. ■

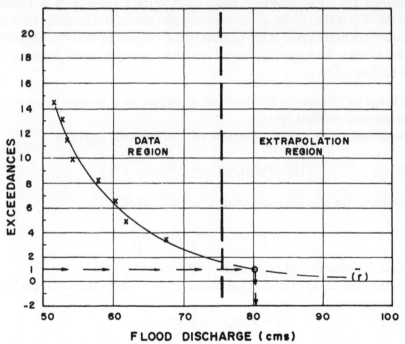

FIGURE 1.1. Expected number of exceedances of maximum flood discharge in a period of 100 years

1.6.4. Record values.

Definition 1.4. (Record value). Let $\{X_n\}$ be a sequence of random variables. X_k is said to be a record value of that sequence if $X_k > X_i$ for $i < k$. ■

Note that this definition corresponds in meaning to the same term utilized in sports and athletic competitions, i.e. the value of the random variable is said to be a record if never before that value has been neither exceeded nor equalled.

We include this term here for reasons of completeness. However, its use in engineering applications is scarce as yet.

Example 1.7. (Record snowfall values). The yearly total snowfall, in inches, registered in Philadelphia during the period 1939-1977 is shown in table 1.1. Apart from previous data, the values 22.3, 31.4, 31.7, 41.8, 49.1 and 54.9 are record values, because they all exceed previous snowfall figures.

Year	Amount	Year	Amount	Year	Amount	Year	Amount
1939	22.3	1940	31.4	1941	10.3	1942	16.3
1943	15.7	1944	21.1	1945	18.7	1946	23.7
1947	31.7	1948	19.3	1949	1.9	1950	4.6
1951	16.2	1952	16.8	1953	22.6	1954	12.1
1955	23.0	1956	7.9	1957	41.8	1958	5.1
1959	21.8	1960	49.1	1961	29.2	1962	20.5
1963	32.9	1964	26.2	1965	27.4	1966	44.3
1967	15.9	1968	23.7	1969	20.3	1970	18.3
1971	12.2	1972	7.0	1973	20.8	1974	13.6
1975	17.5	1976	18.7	1977	54.9		

TABLE 1.1. Total snowfall amount in inches during the period 1939-1977 in Philadelphia

■

II   Order Statistics

Chapter 2   Order Statistics

2.1. Introduction

As indicated in chapter one, order statistics are the basic elements upon which many engineering designs rely. Extreme value distributions cannot be completely understood if a full comprehension of order statistics is not previously established.

We start this chapter by giving a precise definition of order statistics (section 2.2) and deriving the probability density functions (pdf) and the cumulative distribution functions (cdf) of single and joint order statistics (sections 2.3.1 and 2.3.2). A full set of examples of practical applications is inserted with the aim of illustrating and making more easily understandable any of the new concepts.

Section 2.3.3 deals with the problem of samples with random size, which appears in many applications. The required modifications in order to take into account this effect are explained and discussed.

In section 2.3.4 the particular, and very important, case of a uniform parent is analyzed in detail. Some basic theorems for several simulation techniques are explained in section 2.3.5.

Section 2.3.6 is fully devoted to the problem of determining the moments of order statistics. For some populations, as uniform populations for example, exact values are given. In more complicated cases bounds or approximation methods or formulas are provided. The problem of estimation based on order statistics, which is later used in chapter 5, is studied in section 2.3.7, and some characterizations of distributions are included in section 2.3.8. Finally, the problem of order statistics from dependent samples is analyzed in section 2.4.

2.2.   Concept of Order Statistic

Definition 2.1 (Order statistic).   Let $(X_1, X_2, \ldots, X_n)$ be a sample from a given population. If the values of the sequence $X_1, X_2, \ldots, X_n$ are rearranged in an increasing order $X_{1:n} \leq X_{2:n} \leq \ldots \leq X_{n:n}$ of magnitude, then the r-th member of this new sequence is called the r-th order statistic of the sample.

∎

Note that the sample size, n, is included in the notation $X_{r:n}$ and that any order statistic must have an associated sample size. However, when this is clear from the context or the explicit reference to n is not needed, the notation $X_{(r)}$ or simply X_r will be used instead.

The two important terms $X_{1:n} = \text{Min}(X_1, X_2, \ldots, X_n)$ and $X_{n:n} = \text{Max}(X_1, X_2, \ldots, X_n)$ are called extremes and play an essential role in applications.

Example 2.1. (Grades in Mathematics).   Let $X_1, X_2, \ldots, X_n$ be the final grades in Mathematics of n students in a randomly selected classroom from the set of all of a given level in a given country. The grades of the worst, second worst and best students in Math, which are obviously random variables, are the first, second and last order statistics, respectively. Intuitively, we know that the grades are mainly concentrated in the interval 90-100 for the best and in the range 0-40 for the worst, if the number of students in the class is large enough.

∎

Example 2.2. (Rubble-mound breakwater).   As is well known, the armor course of a rubble-mound breakwater is composed of many large rock pieces, selected as to size and shape to well defined slopes, in order to protect the core. When the height of sea waves reaching the breakwater surpasses a given value, some of the rock pieces in the armor course are removed, by water energy, from their initial position and do not return to it, initiating in this way the progressive failure of the breakwater. Because m or more blocks must be removed in order for the failure to occur, and because this process requires k waves to exceed a given height H_0, the design engineer is interested in the probabilities of the k-th largest sea waves exceeding H_0, i.e. he is interested in the $(n - k + 1)$-th order statistic of the series of sea wave heights occurring at the location of the breakwater during a given period of time.

On the other hand, if the breakwater is rigid, i.e. constructed by a monolitic piece, the failure occurs the first time the sea wave height exceeds a critical value. Thus, in this case, only the last order statistic (maximum) is of interest for design purposes.   ■

Example 2.3. (Water levels of a river).   If $X_1, X_2, \ldots, X_n$ are the water levels of a river on consecutive days, then the properties of the lower order statistics $X_{1:n}, X_{2:n}, \ldots, X_{k:n}$ are used for modeling droughts and the properties of the upper order statistics $X_{n-k:n}, X_{n-k+1:n}, \ldots, X_{n:n}$, or simply the maximum $X_{n:n}$, are used to study flood risks and the design of flood protection.   ■

Example 2.4. (Weakest link principle).   If the behaviour of a given system is determined by the weakest link principle (any series arrangement of elements has this property), the lifetime of the system coincides with the lifetime of the weakest element in the system, i.e. if $X_1, X_2, \ldots, X_n$ are the lifetimes of the components, then the lifetime of the system is given by

$$X_{(1)} = \text{Min}(X_1, X_2, \ldots, X_n)$$

Thus, the first order statistic (minimum) governs the life of the system.   ■

Example 2.5. (Earthquakes' intensities).   Let $X_1, X_2, \ldots, X_n$ be the sequence of earthquake intensities occurring at a given place during a period of time. Some codes define the design intensity, l_0, as the intensity such that the probability of the maximum value in the sequence to exceed that value, l_0, takes a given small value $(10^{-4}, 10^{-5}, 10^{-6}, \text{etc.})$. Thus, the maximum (last order statistic) is of interest in this case.   ■

Example 2.6. (r-out-of-n system).   Let us assume a system composed by n identical elements and such that the failure of the system occurs as soon as r of the elements fail (this is called an r-out-of-n system). Then, the r-th order statistic plays an important role, because the lifetime of the system is that of the r-th order statistic.   ■

2.3. Order Statistics from Independent and Identically Distributed Samples

In this section we assume that $X_1, X_2, \ldots, X_n$ are independent and identically distributed and that they come from a parent population with cumulative distribution function $F(x)$.

2.3.1. Distribution of one order statistic.

Let $m_n(x)$ be the number of elements in the sample with values $X_j \leq x$. This number is a random variable of binomial type, $B(n, F(x))$, because it coincides with the number of successes when n independent Bernouilli experiments are repeated. A Bernouilli experiment consisting here of drawing one value from the parent population at random, which can be smaller than or equal to x (success), with probability $p = F(x)$, or greater than x (failure), with probability $1\text{-}p$. Consequently, the cumulative distribution function of $m_n(x)$ coincides with the binomial, i.e. we have

$$(2.1) \qquad F_{m_n(x)}(r) = \text{Prob}[m_n(x) \leq r] = \sum_{k=0}^{r} \binom{n}{k} F^k(x)[1 - F(x)]^{n-k}$$

However, the event $\{X_{r:n} \leq x\}$, which represents that the r-th order statistic takes a value smaller than or equal to x, coincides with the event $\{m_n(x) \geq r\}$, which corresponds to r or more elements with values smaller than or equal to x in the sample. Thus, from (2.1) we get

$$F_{X_{r:n}}(x) = P[X_{r:n} \leq x] = 1 - F_{m_n(x)}(r-1) = \sum_{k=r}^{n} \binom{n}{k} F^k(x)[1 - F(x)]^{n-k}$$

$$(2.2) \qquad = r\binom{n}{r} \int_0^{F(x)} u^{r-1}(1-u)^{n-r}\, du = I_{F(x)}(r, n-r+1)$$

where $F_{X_{r:n}}(x)$ is the cumulative distribution function of $X_{r:n}$ and $I_p(a, b)$ is the incomplete beta function.

If the parent population is absolutely continuous, then $X_{r:n}$ has a probability density function given by the first derivative of (2.2) with respect to x

$$f_{X_{r:n}}(x) = r\binom{n}{r} F^{r-1}(x)[1 - F(x)]^{n-r} f(x)$$

$$(2.3) \qquad = F^{r-1}(x)[1 - F(x)]^{n-r} \frac{f(x)}{B(r, n-r+1)}$$

where $B(a, b)$ is the beta function.

Example 2.7. (Normal distribution). The probability density functions and their associated cumulative distribution functions of the order statistics of a sample of size 5 from a normal $N(0, 1)$ parent population, which can be obtained from (2.2) and (2.3) by substitution of $f(x)$ and $F(x)$ by the corresponding pdf and cdf, respectively, and setting $n = 5$, are shown in figure 2.1. Although the range of all of them is $(-\infty, \infty)$, it can be seen that, as could

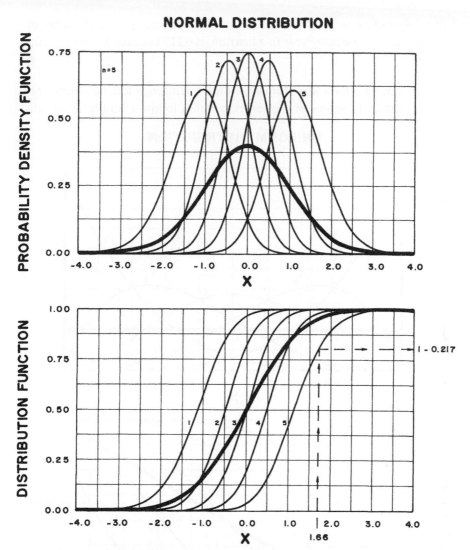

FIGURE 2.1. Pdf and cdf of the order statistics of a sample of size $n = 5$ from a normal $N(0, 1)$ parent population

be expected from the definition of order statistics, the interquartile range moves from left to right with increasing r.

If the total yearly rain, Z, in inches, in a given region is assumed to follow a normal distribution $N(10, 3^2)$, then the pdf and cdf of the amount of rain in the years with the highest, second highest, etc. amount of rain of a lustrum can be obtained from figure 2.1., because the random variable $X = (Z - 10)/3$ is

$N(0, 1)$. According to this, the probability of having more than 15 inches of rain during the most rainy year in a lustrum, is 0.217 because $X = (15 - 10)/3 = 1.66$.   ■

Example 2.8. (Uniform distribution).   One intelligence test has been normalized so that the scores of a given population follow a standard uniform

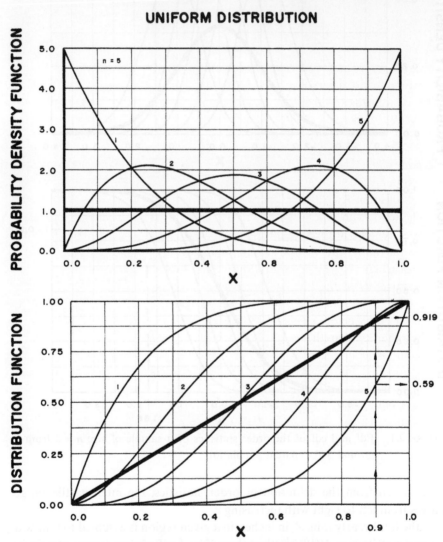

FIGURE 2.2.   Pdf and cdf of the order statistics of a sample of size $n = 5$ from a uniform $U(0, 1)$ parent population

distribution, $U(0, 1)$. If groups of 5 people are chosen at random, the scores of the person occupying the $(n - r + 1)$-th position when ranked by the test results have the following pdf and cdf

$$f_{X_{r:5}}(x) = r\binom{5}{r}x^{r-1}[1 - x]^{5-r}; \qquad 0 \le x \le 1$$

$$F_{X_{r:5}}(x) = \sum_{k=r}^{5}\binom{5}{k}x^{k}[1 - x]^{5-k}; \qquad 0 \le x \le 1$$

which for $r = 1, 2, \ldots, 5$ are represented in figure 2.2. According to this figure, the probability of a person ranked first in his group having a score greater than or equal to 0.9 is $1 - 0.59 = 0.41$, and the same probability for the second is $1 - 0.919 = 0.081$.   ∎

Example 2.9. (Exponential distribution).   The number of vehicles arriving at a given intersection of a main road and a secondary road follows a Poisson process of intensity 720 vehicles/hour. This implies that the time, T, between vehicles follows an exponential distribution $E[720]$ with mean $1/720$ hours. If a row of vehicles at the secondary road is waiting at the intersection in order to join the main road, and assuming they need a minimum time, t_0, between vehicles in order to join, and that no more than one vehicle can go between two consecutive cars from the main road (we really analyze conditional distributions given this event), the probability of passing at least r vehicles from the secondary road before n vehicles from the main road reach the intersection is the probability that the $(n - r + 1)$-th order statistic of the time between consecutive vehicles be greater than or equal to t_0. Note that r or more vehicles join the main road if the sequence of time intervals between vehicles has r or more elements larger than or equal to t_0, i.e. if the $(n - r + 1)$-th element in the ordered sequence $((n - r + 1)$-th order statistic) is larger than t_0. The probability density and the cumulative distribution functions of the last 5 order statistics of a sample of size 20 from an $E(1)$ parent population are shown in figure 2.3. According to this figure and (2.2), and taking into account that $720T$ is distributed as $E(1)$, the probability of passing 2 or more vehicles from the secondary road before 20 vehicles from the main road reach the intersection, assuming $t_0 = 10$ seconds, is

$$1 - \sum_{k=19}^{20}\binom{20}{k}(1 - e^{-2})^{k}e^{-2(20-k)} = 0.7747$$

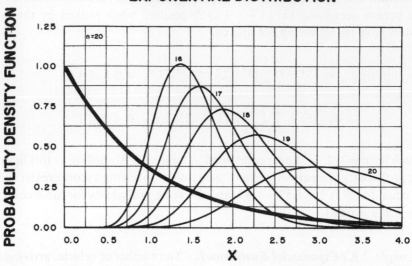

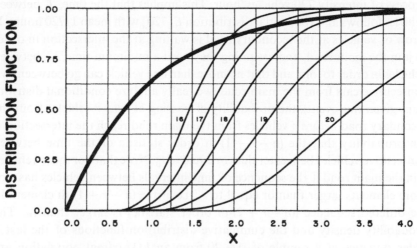

FIGURE 2.3.   Pdf and cdf of the last five order statistics of a sample of size $n = 20$
from an exponential $E(1)$ parent population ∎

Example 2.10. (Rayleigh distribution).   The height, a, of sea waves at a given
location is known to follow a Rayleigh distribution (Longuet-Higgins (1975)):

$$f(\beta) = \beta \exp\left[-\frac{\beta^2}{2}\right]; \qquad \beta \geq 0$$

where

$$\beta = \frac{a}{\mu_0^{1/2}}$$

and μ_0 is the zeroth moment of the energy spectrum (mean square wave amplitude).

Figure 2.4. shows the pdf and the cdf of the last 5 order statistics of the

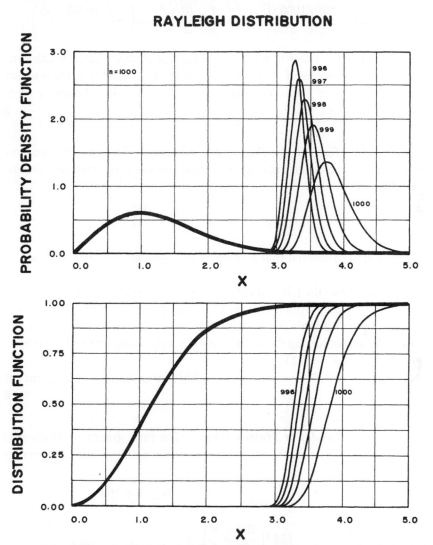

FIGURE 2.4.   Pdf and cdf of the last five order statistics of a sample of size $n = 1000$ from a Rayleigh parent population

heights of *a* series of 1000 waves:

$$f_{X_{r:1000}}(x) = r\binom{1000}{r}\left[1 - \exp\left(-\frac{\beta^2}{2}\right)\right]^{r-1}$$

$$\cdot \exp\left(-\frac{\beta^2(1000 - r)}{2}\right)\beta\exp\left(-\frac{\beta^2}{2}\right)$$

$$F_{X_{r:1000}}(x) = \sum_{k=r}^{1000}\binom{1000}{k}\left[1 - \exp\left(-\frac{\beta^2}{2}\right)\right]^k\exp\left(-\frac{\beta^2(1000 - k)}{2}\right)$$

If a rigid breakwater fails when $a/\mu_0^{1/2} = 4.8$, the probability of the breakwater surviving after a series of 1000 waves is 0.99. If a flexible rubble-mound breakwater fails after 5 waves higher than $a/\mu_0^{1/2} = 3.5$, then the above probability becomes 0.92.   ∎

Example 2.11. (Gumbel distribution for maxima).   The yearly maximum water level, X, at a given location of a river follows a Gumbel distribution with pdf.

$$F(x) = \exp\left\{-\exp\left[-\left(\frac{x - a}{b}\right)\right]\right\}$$

and it has been calculated that a value of $(x - a)/b$ greater than 4.0 leads to flooding.

Figure 2.5. shows the pdf and cdf of the last three order statistics of a sample of size 50 coming from a standard Gumbel population. Because the normalized random variable $Z = (x - a)/b$ follows a standard Gumbel model, the probabilities of having floods in 2 years or less during a period of 50 years is given by $1 - 0.77 = 0.23$.

Note that the cdf of the maximum ($r = 50$) is a translation of the parent cdf.
                                                                                  ∎

Example 2.12. (Grades in Mathematics).   The final grades in Math of a given population of students follow a distribution with pdf and cdf given by

(2.4) $$f(x) = \frac{12x^2(100 - x)}{10^8};\qquad 0 \le x \le 100$$

(2.5) $$F(x) = \frac{12x^3\left(\dfrac{100}{3} - \dfrac{x}{4}\right)}{10^8};\qquad 0 \le x \le 100$$

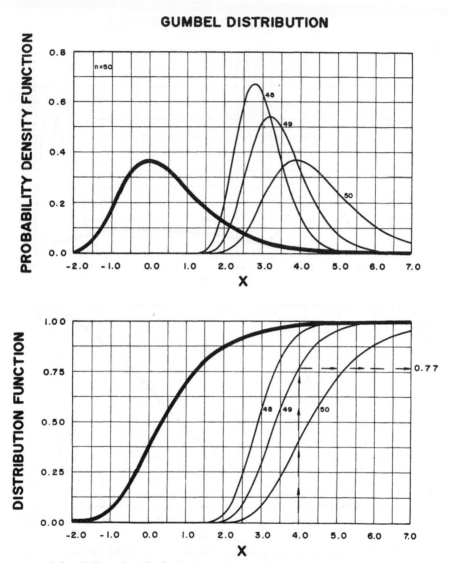

FIGURE 2.5.   Pdf and cdf of the last three order statistics of a sample of size
$n = 50$ from a standard Gumbel parent population

The pdf and cdf of the grades of the first, second and last students in a classroom of 40 students selected at random from the population can be obtained from (2.3) and (2.2) upon substitution of (2.4) and (2.5) and setting $n = 40$ and $r = 40$, 39 and 1 respectively. Figure 2.6. shows those functions.

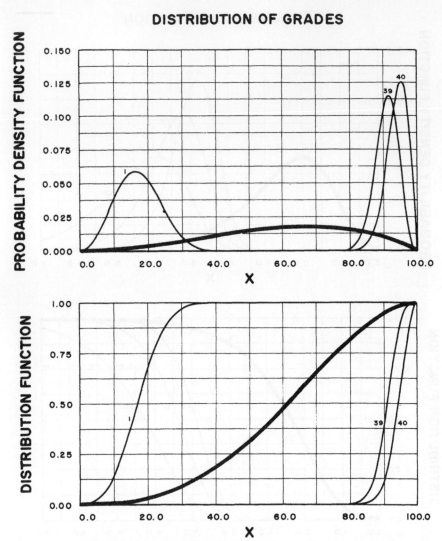

FIGURE 2.6.   Pdf and cdf of the grades of the first, second and last students in a
classroom with 40 students   ∎

2.3.1.1.   Distribution of maxima.   In this section we study the distribution
of the maximum of a random sample, which plays an important role in
the establishment of design operating conditions. Due to the fact that the
maximum is the last order statistic, by setting $r = n$ in (2.2) and (2.3), the
following cdf and pdf are obtained

$$(2.6) \qquad\qquad\qquad F_{X_{n:n}}(x) = F^n(x)$$

(2.7) $$f_{X_{n:n}}(x) = nF^{n-1}(x)f(x)$$

Example 2.13. (Normal distribution).   Figure 2.7 shows the pdf and cdf of
maxima of samples of sizes 1, 10, 50, 100 and 200 from a normal $N(0, 1)$ parent
population. If the yearly maximum of the load Y, of a certain structure is
assumed to be normal $N(30, 4^2)$, and the design load is forced by codes to be

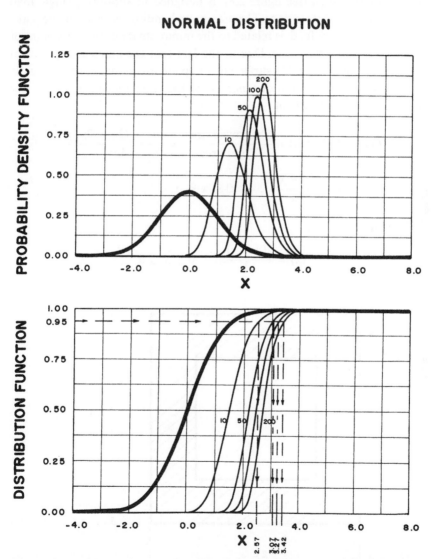

FIGURE 2.7.   Pdf and cdf of maxima of samples of sizes 1, 10, 50, 100 and 200 from
            a normal $N(0, 1)$ parent population

the value associated with a probability of being exceeded during its lifetime of 5×10^{-2}, then the design values for the load for 10, 50, 100 and 200 years are given by 2.57, 3.07, 3.27 and 3.42 for the normalized values and 40.28, 42.28, 43.08 and 43.68 for the unnormalized ones.   ∎

Example 2.14. (Bending moment of a plate).   A square plate resting on its four sides of length $2a$ (see figure 2.8), is designed to support a single load located at random on its surface. The maximum bending moment on the plate, produced by the point load, is related to the minimum distance of its point of application to the closest side. Due to the fact that the loads are applied at random and all points on the surface of the plate are assumed to be equally likely, the cdf of the random distance, X, to the closest side is given by (see Benjamin and Cornell (1970))

$$F(x) = P(X \leq x) = \frac{\text{Shadowed area}}{\text{Total area}} = \frac{4a^2 - (2a - 2x^2)}{4a^2}$$

$$= 1 - \left(\frac{a - x}{a}\right)^2; \qquad 0 \leq x \leq a$$

and the pdf, by

$$f(x) = \frac{2(a - x)}{a^2}; \qquad 0 \leq x \leq a$$

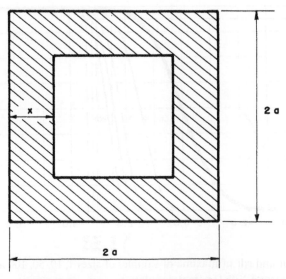

$2a$

$2a$

FIGURE 2.8.   Square plate resting on its four sides

Thus, the maximum distance of a series of n loads has a cdf given by

$$G_{max}(x) = F^n(x) = \left[1 - \left(\frac{a-x}{a}\right)^2\right]^n; \quad 0 \leq x \leq a$$

which can be used for design purposes.   ■

Example 2.15. (Exponential distribution).   A drought can be defined as a period of duration t_0 without a rain of intensity larger than I_0. If the rains

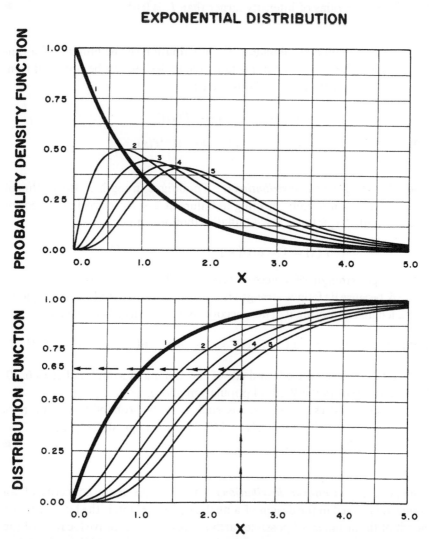

FIGURE 2.9.   Pdf and cdf of maxima of samples of sizes 1 to 5 from an exponential $E(1)$ parent population

of intensity larger than I_0 are assumed to behave as a poissonian process of mean μ rains/year, the probability of having no drought coincides with the probability of the maximum of a random sample of size n from an exponential $E(\mu)$ parent population to be smaller than t_0, because a drought does not occur if all the interval periods between consecutive rains (these periods are exponential $E(\mu)$ random variables) are smaller than t_0. Figure 2.9. shows the pdf and cdf of the maxima of samples of sizes 1 to 5 from an exponential $E(1)$ parent population. If $t_0 = 3$ months and $\mu = 10$, the probability of having no droughts after 5 rains of intensity larger than I_0 is 0.65.   ■

2.3.1.2.   Distribution of minima.   The distribution of the minima of a random sample, which plays an important role in design, can be obtained from (2.2) and (2.3) by setting $r = 1$. Thus, we get

(2.8) $$F_{X_{1:n}}(x) = 1 - [1 - F(x)]^n$$

(2.9) $$f_{X_{1:n}}(x) = n[1 - F(x)]^{n-1}f(x)$$

Example 2.16. (Weibull distribution).   The fatigue strength, X, (logarithm of the stress range, in N/mm^2, leading to failure at 2×10^6 cycles) of a piece of wire of length $2m$., follows a Weibull distribution with cdf.

$$F(x) = 1 - \exp\{-[3(x - 5.5) - 0.1]^5\}$$

The fatigue strength of a piece of $12m$. is the first order statistic of a sample of size 6, because its fatigue strength is the least of the strengths of the 6 pieces of $2m$. length composing it. Thus, according to (2.8), its cdf is given by

(2.10) $$F_{X_{1:6}}(x) = 1 - [1 - F(x)]^6 = 1 - \exp\{-6[3(x - 5.5) - 0.1]^5\}$$

Figure 2.10. shows the pdf and cdf of the minima of samples of sizes 6, 10, 20, 50 and 100 from the Weibull population above.

If the design stress range, X_0, is such that its probability of failure is 0.01, we have from (2.10) that $X_0 = 5.63$, i.e. the design stress range is $X_0 = \exp(5.63) = 277.6 \ N/mm^2$.   ■

Example 2.17. (Uniform distribution).   One of the main concerns of an engineer, involved in the design of a nuclear power plant, is the knowledge of the cdf of the distances of possible earthquakes to the tentative location of the plant. Due to the presence of a fault in the area surrounding this location, it has

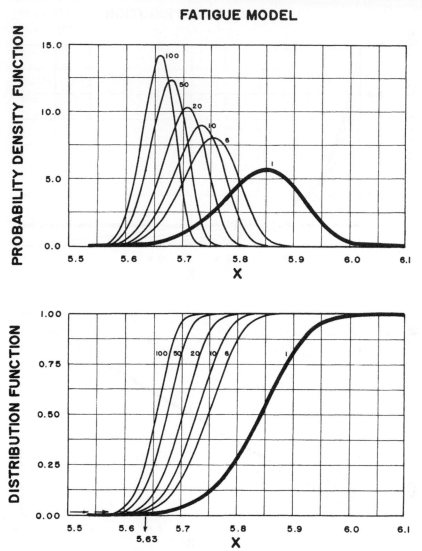

FIGURE 2.10.   Pdf and cdf of the minima of samples of sizes 6, 10, 20, 50 and 100 from
a Weibull parent population

UNIFORM DISTRIBUTION

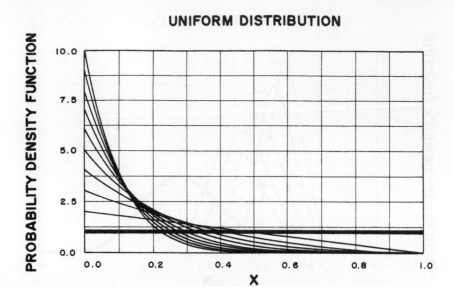

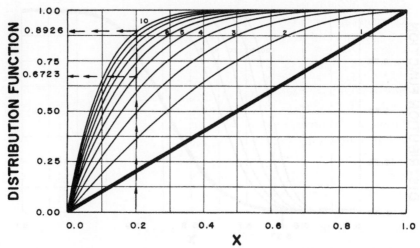

FIGURE 2.11.   Pdf and cdf of the minima of samples of sizes 1 to 10 from a uniform $U(0, 1)$ parent population

been established that the epicenter of an earthquake can occur, equally likely, along any point of the 50 Kms of the fault. If the tentative location is aligned with the fault and its closest extreme is 200 Kms away, then the distance of the epicenter to the plant can be assumed to be uniform $U(200, 250)$. Thus, the pdf and cdf of the closest earthquake to the plant in series of 5 and 10 earthquakes are given by

$$f_{X_{1:n}}(x) = \frac{n\left(1 - \dfrac{x - 200}{50}\right)^{n-1}}{50}; \quad n = 5, 10$$

$$F_{X_{1:n}}(x) = 1 - \left(1 - \frac{x - 200}{50}\right)^{n}; \quad n = 5, 10$$

The pdf and cdf of the minima of samples of sizes 1 to 10 from a standard $U(0, 1)$ parent population are shown in figure 2.11. From this figure we can see that the probability of having minimum distances less than 210 Kms which is equivalent to a standard value of $(210 - 200)/50 = 0.2$, are 0.6723 and 0.8926 for the series of 5 and 10 earthquakes, respectively. Note that due to the random character of the occurring earthquakes the above are conditional solutions given that $n = 5$ or 10. For an unconditional solution see section 2.3.3. ■

Example 2.18. (Normal distribution). Figure 2.12. shows the pdf and cdf of the minima of samples of sizes 1 to 10 from a standard normal parent population. If the times, in seconds, in which one athlete runs 100 m in a competition follow a normal distribution $N(10.4, 0.2^2)$, the pdf and cdf of the best time after 10 competitions can be obtained from the figure. As one example, the probability of this best time to be smaller than 10 seconds, which is equivalent to a normalized value of $(10 - 10.4)/0.2$, is 0.2059.

It is interesting to see that an athlete with larger mean value and more irregular performance (larger variance) can have a better chance to go below the 10 seconds. If the distribution of his times are $N(10.6, 0.4^2)$, the normalized value becomes $(10 - 10.6)/0.4 = -1.5$ and then, that probability becomes 0.4991. Consequently, the latter athlete, though more irregular and with smaller mean value should be chosen for the competition. This example illustrates how extreme value problems are different from central value problems.

NORMAL DISTRIBUTION

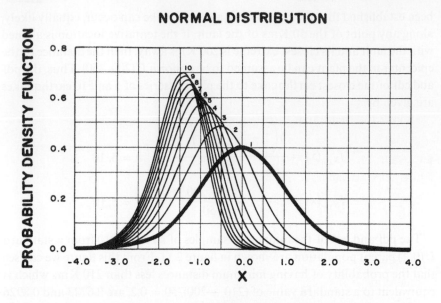

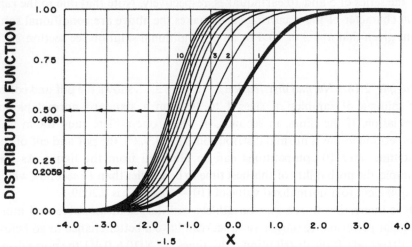

FIGURE 2.12.   Pdf and cdf of the minima of samples of sizes 1 to 10 from a normal $N(0, 1)$ parent population   ∎

2.3.2.   Joint distribution of several order statistics.

Let $X_{r_1:n}, X_{r_2:n}, \ldots, X_{r_k:n}$, with $r_1 < r_2 < \ldots < r_k$ be k order statistics from a random sample of size n coming from a population with pdf $f(x)$ and cdf $F(x)$. In order to derive the joint pdf of this set of order statistics, we consider

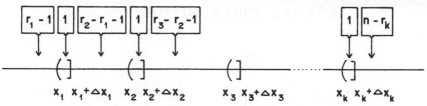

FIGURE 2.13.   Illustration of event leading to the joint pdf of two order statistics

the event $\{x_j \leq X_{r_j:n} < x_j + \Delta x_j; \; 1 \leq j \leq k\}$ for small $\Delta x_j, 1 \leq j \leq k$ (see figure 2.13.). That is, k of the elements in the sample lie in the intervals $(x_j, x_j + \Delta x_j)$ for $1 \leq j \leq k$ and the rest are distributed in such a manner that exactly $r_j - r_{j-1} - 1$ lie on the interval $(x_{j-1} + \Delta x_{j-1}, x_j)$ for $1 \leq j \leq k$, where $\Delta x_0 = 0, r_0 = 0, r_{k+1} = n + 1, x_0 = -\infty$ and $x_{k+1} = \infty$.

The following multinomial experiment with $2k + 1$ possible outcomes can be considered: Let us take n elements at random from the given population and determine to which of the $2k + 1$ intervals in figure 2.13. they belong. Because we assume independence and replacement, the number of elements in each interval is a multinomial random variable:

$$M\{n; f(x_1)\Delta x_1, f(x_2)\Delta x_2, \ldots, f(x_k)\Delta x_k, [F(x_1) - F(x_0)],$$

$$[F(x_2) - F(x_1)], \ldots, [F(x_{k+1}) - F(x_k)]\}$$

where the parameters are n (the sample size) and the probabilities associated with the $2k + 1$ intervals. In consequence, the properties of multinomial random variables can be used and the joint pdf of the set of statistics becomes

$$f_{r_1,r_2,\ldots,r_k:n}(x_1,x_2,\ldots,x_k) = n! \prod_{i=1}^{k} f(x_i) \prod_{j=1}^{k+1} \left(\frac{[F(x_j) - F(x_{j-1})]^{r_j - r_{j-1} - 1}}{(r_j - r_{j-1} - 1)!} \right)$$

(2.11) $x_1 \leq x_2 \leq \ldots \leq x_k$

Example 2.19. (Joint distribution of maxima and minima). By setting $k = 2$, $r_1 = 1$ and $r_2 = n$ in (2.11), the joint pdf of the maximum and minimum of a sample of size n becomes

(2.12) $f_{1,n:n}(x_1, x_2) = n(n - 1)f(x_1)f(x_2)[F(x_2) - F(x_1)]^{n-2};$ $x_1 \leq x_2$

The cases of a standard normal, uniform and Gumbel parents and $n = 5$ are shown in figures 2.14. to 2.16. respectively.   ∎

NORMAL DISTRIBUTION

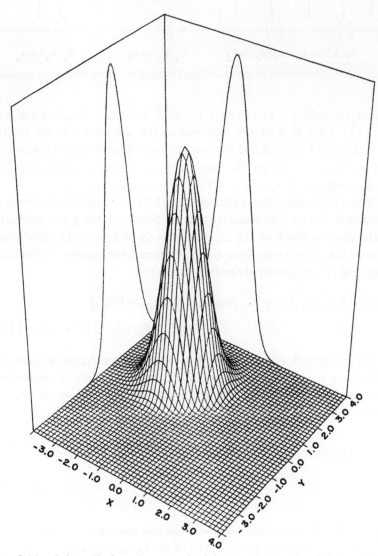

FIGURE 2.14.   Joint pdf of maximum and minimum of a sample of size $n = 5$ from a normal $N(0, 1)$ distribution

UNIFORM DISTRIBUTION

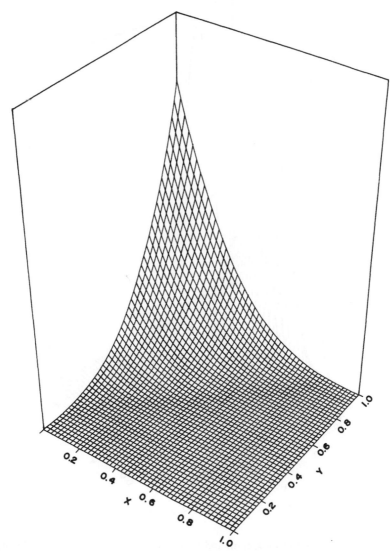

FIGURE 2.15.   Joint pdf of maximum and minimum of a sample of size $n = 5$ from a
uniform $U(0, 1)$ distribution

GUMBEL DISTRIBUTION

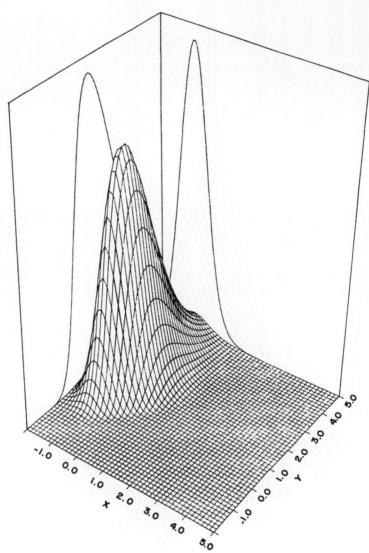

FIGURE 2.16.   Joint pdf of maximum and minimum of a sample of size $n = 5$ from a
Gumbel distribution

Example 2.20. (Joint distribution of two consecutive order statistics). By setting $k = 2$, $r_1 = i$ and $r_2 = i + 1$ in (2.11) the joint pdf of two consecutive order statistics becomes

$$(2.13) \quad f_{i,i+1:n}(x_1, x_2) = \frac{n! f(x_1) f(x_2) F^{i-1}(x_1)[1 - F(x_2)]^{n-i-1}}{(i-1)!(n-i-1)!} \qquad x_1 \leq x_2$$

■

Example 2.21. (Joint distribution of any two order statistics). The joint pdf of any two order statistics $X_{r:n}$ and $X_{s:n}$ $(r < s)$, according to (2.11) is

$$f_{r,s:n}(x_1, x_2) = n! f(x_1) f(x_2) \frac{F^{r-1}(x_1)[F(x_2) - F(x_1)]^{s-r-1}[1 - F(x_2)]^{n-s}}{(r-1)!(s-r-1)!(n-s)!}$$

$$(2.14) \qquad\qquad\qquad\qquad\qquad\qquad\qquad\qquad\qquad\qquad x_1 \leq x_2$$

which for $n = 5$, $r = 2$, $s = 4$ and the uniform distribution is shown in figure 2.17.

■

Example 2.22. (Joint distribution of all order statistics). The joint pdf of all order statistics becomes

$$(2.15) \quad f_{1,2,\ldots,n:n}(x_1, x_2, \ldots, x_n) = n! \prod_{i=1}^{n} f(x_i); \qquad x_1 \leq x_2 \leq \ldots \leq x_n$$

2.3.2.1. Conditional distribution of two order statistics.

One of the main interests of the conditional distributions of two order statistics is for simulation purposes. The pdf of $X_{r:n}$, given $X_{s:n} = x_2$, $(r < s)$, taking into account (2.14) and (2.3), is given by

$$f_{X_{r:n}/X_{s:n}}(x_1/x_2) = \frac{f_{r,s:n}(x_1, x_2)}{f_{s:n}(x_2)}$$

$$= \frac{(s-1)! f(x_1) F^{r-1}(x_1)[F(x_2) - F(x_1)]^{s-r-1} F^{1-s}(x_2)}{(r-1)!(s-r-1)!};$$

$$(2.16) \qquad\qquad\qquad\qquad\qquad\qquad\qquad\qquad x_2 \geq x_1 \qquad ■$$

This expression, when compared with (2.3), shows that the distribution of $X_{r:n}$, given $X_{s:n} = x_2$ is the distribution of the r-th order statistic in a sample of size $s - 1$ from the parent population, truncated on the right at x_2.

UNIFORM DISTRIBUTION

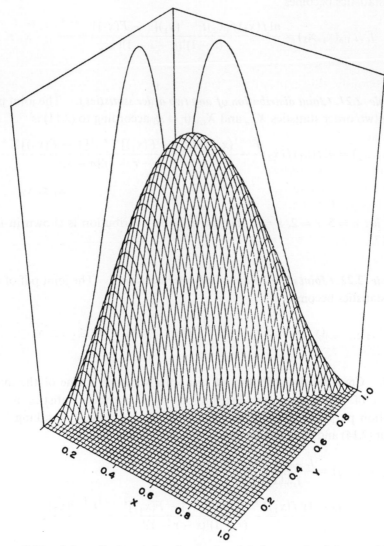

FIGURE 2.17.   Joint pdf of statistics of order 2 and 4 of a sample of size $n = 5$ from a uniform distribution

The pdf of $X_{s:n}$, given $X_{r:n} = x_1$, is

$$f_{X_{s:n}/X_{r:n}}(x_2/x_1) = \frac{f_{r,s:n}(x_1,x_2)}{f_{r:n}(x_1)}$$

(2.17)
$$= \frac{(n-r)!f(x_2)[1-F(x_2)]^{n-s}[F(x_2)-F(x_1)]^{s-r-1}[1-F(x_1)]^{r-n}}{(n-s)!(s-r-1)!}$$

Analogously, this expression when compared with (2.3) shows that the distribution of $X_{s:n}$, given $X_{r:n} = x_1$, is the distribution of the $(s-r)$-th order statistic in a sample of size $(n-r)$ from the parent population, truncated on the left at x_1.

Example 2.23. (Conditional distribution of two consecutive order statistics). For two consecutive order statistics, the pdf of $X_{i:n}$ conditiona! on $X_{i+1:n} = x_2$, taking into account (2.16), is given by

(2.18)
$$f_{X_{i:n}/X_{i+1:n}}(x_1/x_2) = \frac{f_{i,i+1:n}(x_1,x_2)}{f_{i+1:n}(x_2)} = \frac{if(x_1)F^{i-1}(x_1)}{F^i(x_2)}$$

and the cdf becomes

(2.19)
$$F_{X_{i:n}/X_{i+1:n}}(x_1/x_2) = \left(\frac{F(x_1)}{F(x_2)}\right)^i$$

The pdf and cdf of $X_{i+1:n}$ conditional on $X_{i:n} = x_1$ are

$$f_{X_{i+1:n}/X_{i:n}}(x_2/x_1) = \frac{f_{i,i+1:n}(x_1,x_2)}{f_{i:n}(x_1)}$$

(2.20)
$$= \frac{(n-i)f(x_2)[1-F(x_2)]^{n-i-1}}{[1-F(x_1)]^{n-i}}$$

and

$$F_{X_{i+1:n}/X_{i:n}}(x_2/x_1) = 1 - \left(\frac{1-F(x_2)}{1-F(x_1)}\right)^{n-1}; \qquad x_2 \geq x_1 \qquad \blacksquare$$

Example 2.24. (Uniform distribution). The cdfs of $X_{i:n}/X_{i+1:n} = x_2$ and $X_{i+1:n}/X_{i:n} = x_1$ for the standard $U(0,1)$ are respectively

(2.22)
$$F_{X_{i:n}/X_{i+1:n}}(x_1/x_2) = \left(\frac{x_1}{x_2}\right)^i; \qquad x_1 \leq x_2$$

(2.23)
$$F_{X_{i+1:n}/X_{i:n}}(x_2/x_1) = 1 - \left(\frac{1-x_2}{1-x_1}\right)^{n-i}; \qquad x_2 \geq x_1$$

These two cdfs play a very important role in simulation as will be seen in a subsequent section. ∎

2.3.2.2. Distribution of the difference of two order statistics.

In order to derive the distribution of the difference between two order statistics $X_{s:n}$ and $X_{r:n}$ ($r < s$), whose joint pdf is given by (2.14), we consider the transformation

(2.24)
$$U = X_{r:n} \atop V = X_{s:n} - X_{r:n} \Big\}$$

which has as inverse

(2.25)
$$X_{r:n} = U \atop X_{s:n} = U + V \Big\}$$

with jacobian

(2.26)
$$J = \begin{vmatrix} 1 & 0 \\ 1 & 1 \end{vmatrix} = 1$$

Thus, the joint pdf of (U, V) is given by

$$g(u, v) = \frac{n!}{(r - 1)!(s - r - 1)!(n - s)!} f(u) f(u + v) F^{r-1}(u)$$
$$\cdot [F(u + v) - F(u)]^{s-r-1} [1 - F(u + v)]^{n-s}$$

(2.27) $v \geq 0$

and then the V marginal has the following pdf

$$h(v) = \frac{n!}{(r - 1)!(s - r - 1)!(n - s)!} \int_{-\infty}^{\infty} f(u) f(u + v) F^{r-1}(u)$$
$$\cdot [F(u + v) - F(u)]^{s-r-1} [1 - F(u + v)]^{n-s} du$$

(2.28) $v \geq 0$

Example 2.25. (Parking lots).   A system of two parking areas A and B, with capacities of 100 and 50 vehicles, respectively, is organized in such a way that parking lot A is open at 7 am. in the morning and parking lot B is open as soon as parking lot A is full. It is also assumed that only 200 registered clients can use the parking areas on a first come first served basis (this means that every day 50 clients are not served). Assuming that the arrival of cars is poissonian with intensity 1000 vehicles/hour and that the number of cars leaving the parking areas during the early hours is negligible, the time in hours during which parking lot B is open is a random variable which is the difference between the 150-th and 100-th order statistics in a sample of size 200 from a exponential distribution with mean $1/1000$. Consequently, the pdf of that time can be obtained from (2.28), by setting $n = 200$, $r = 100$ and $s = 150$, $f(x) = 1000 \exp\{-1000u\}$ and $F(x) = 1 - \exp\{-1000u\}$. ∎

2.3.2.3. Distribution of the range. One particularly interesting case of the difference of two order statistics is the range, w, which is defined as the difference between the maximum and the minimum. Its pdf results from (2.28) by setting $r = 1$ and $s = n$, i.e.

$$(2.29) \qquad h(w) = n(n-1) \int_{-\infty}^{\infty} f(u)f(u+w)[F(u+w) - F(u)]^{n-2} \, du; \quad w \geq 0$$

and the cdf becomes

$$H(w) = \int_{-\infty}^{w} h(v) \, dv$$

$$(2.30) \qquad = n \int_{-\infty}^{\infty} f(u)[F(u+w) - F(u)]^{n-1} \, du; \qquad w \geq 0$$

Example 2.26. (Uniform distribution). The pdf of the range of a sample from a standard $U(0,1)$ parent population, according to (2.29) and taking into account that the region where $f(u)$ and $f(u+w)$ are simultaneously non-zero is

$$\left. \begin{matrix} 0 \leq u \leq 1 \\ 0 \leq u + w \leq 1 \end{matrix} \right\}$$

which is equivalent to

$$0 = \max(0, -w) \leq u \leq \min(1, 1 - w) = 1 - w$$

becomes

$$h(w) = n(n-1) \int_{0}^{1-w} w^{n-2} \, du$$

$$(2.31) \qquad = n(n-1)w^{n-2}(1-w); \qquad 0 \leq w \leq 1$$

2.3.2.4. Distribution of the midrange. The midrange is defined as the mean of the two extreme order statistics. Thus, taking into account of (2.12) and upon considering the transformation

$$U = X_{1:n}$$
$$V = \frac{X_{n:n} + X_{1:n}}{2}$$

which has an inverse

$$X_{1:n} = U$$
$$X_{n:n} = 2V - U$$

with jacobian $J = 2$, we get the following joint pdf for (U, V)

(2.32) $g(u, v) = 2n(n - 1)f(u)f(2v - u)[F(2v - u) - F(u)]^{n-2};$ $u \le v$

and then, the pdf of the midrange, V, becomes

$$h(v) = 2n(n - 1)\int_{-\infty}^{v} f(u)f(2v - u)[F(2v - u) - F(u)]^{n-2} \, du; \qquad u \le v$$

(2.33)

Example 2.27. (Uniform distribution).   The pdf of the midrange for the uniform $U(0, 1)$ is given by (2.33), taking into account that the region in which the subintegral part of (2.33) is non-zero is the intersection of the regions $\{0 \le u \le 1\}$ and $(2v - 1 \le u \le 2v)$, which can be written

$$\{\max(0, 2v - 1) \le u \le \min(1, 2v)\}.$$

Thus, the integration can be extended to the region $\{\max(0, 2v - 1) \le u \le \min(1, 2v, v) = v\}$, and we have

$$h(v) = \begin{cases} 2n(n-1)\displaystyle\int_0^v (2v - 2u)^{n-2} \, du = n(2v)^{n-1} & \text{if } 0 \le v \le \frac{1}{2} \\[2ex] 2n(n-1)\displaystyle\int_{2v-1}^v (2v - 2u)^{n-2} \, du = n[2(1 - v)]^{n-1} & \text{if } \frac{1}{2} \le v \le 1 \\[2ex] 0 & \text{otherwise} \end{cases}$$

(2.34)

and the cdf becomes

(2.35) $$H(v) = \begin{cases} 0 & \text{if } v < 0 \\[1.5ex] \dfrac{(2v)^n}{2} & \text{if } 0 \le v < \frac{1}{2} \\[2ex] 1 - \dfrac{[2(1 - v)]^n}{2} & \text{if } \frac{1}{2} \le v < 1 \\[2ex] 1 & \text{if } v \ge 1 \end{cases}$$

∎

2.3.3.   Order statistics from samples of random size.

In all the previous sections, the sample size was a fixed value n. However, there are many cases in practice where the sample size is random, and then, the given expressions are not valid any more.

If the distribution of the sample size n, is denoted by $p_N = P[n = N]$, and the statistic, X, under consideration (order statistic, difference of order statistics, midrange, etc.) has, for fixed sample size N, pdf $f_X(x; N)$ and cdf $F_X(x; N)$, then the total probability rule allows us to state that the pdf and the cdf of the statistic X are given by

$$(2.36) \qquad\qquad g_X(x) = \sum_N p_N f_X(x; N)$$

$$(2.37) \qquad\qquad G_X(x) = \sum_N p_N F_X(x; N)$$

where the summation sign is extended to all feasible values of N.

Example 2.28. (Earthquake intensities).   The occurrence of earthquakes in a given region is a Poisson process of intensity μ earthquakes/year, and their intensity, X ($X > 0$), is a random variable with cdf $H(x)$. Then, the intensity of the earthquake of maximum intensity in a period of duration t, according to (2.37) has cdf

$$F_{X_{max}}(x) = \begin{cases} \displaystyle\sum_{n=0}^{\infty} \exp\dfrac{\{-\mu t\}(\mu t)^n H^n(x)}{n!} = \exp\{-\mu t[1 - H(x)]\}; & x \geq 0 \\ 0 & \text{otherwise} \end{cases}$$

$$(2.38)$$

Note that it is a mixed type random variable. This is due to the fact that zero earthquakes can occur. ∎

Example 2.29. (Sea waves and breakwaters).   The number of sea waves occurring in a given location follows a Poisson process of intensity μ_1 waves/year. Assuming that the wave height, H, is an exponential $E(\mu_2)$ random variable, then the cdf of the maximum wave height in a period of t years is given by (see (2.38))

$$(2.39) \qquad F_{H_{max}}(x) = \begin{cases} \exp\{-\mu_1 t \exp[-\mu_2 x]\}; & x \geq 0 \\ 0 & \text{otherwise} \end{cases}$$

Thus, the probability of a rigid breakwater surviving that period, assuming it fails under 1 wave of height larger than h_0, is given by

$$p(h_0) = \exp\{-\mu_1 t \exp[-\mu_2 h_0]\}$$

Similarly, the cdf of the r-th order statistic in that period, according to (2.2) and (2.37) is

$$F_{X_{r:n}}(x) = \sum_{n=r}^{\infty} \exp\frac{[-\mu_1 t](\mu_1 t)^n}{n!}(n-r+1)\binom{n}{r}\int_0^{F(x)} u^{n-r}(1-u)^{r-1}\, du$$

$$= \exp\frac{[-\mu_1 t](\mu_1 t)^r}{(r-1)!}\int_0^{F(x)}(1-u)^{r-1}\sum_{n=r}^{\infty}\frac{(\mu_1 tu)^{n-r}}{(n-r)!}\, du$$

$$= \frac{(\mu_1 t)^r}{(r-1)!}\int_0^{F(x)}(1-u)^{r-1}\exp\{-\mu_1 t(1-u)\}\, du$$

$$= \frac{(\mu_1 t)^r}{(r-1)!}\int_{1-F(x)}^1 v^{r-1}\exp\{-\mu_1 tv\}\, dv$$

$$= H_{G(r,\mu_1 t)}(1) - H_{G(r,\mu_1 t)}(1-F(x))$$

$$= H_{G(r,\mu_1 t)}(1) - H_{G(r,\mu_1 t)}(\exp(-\mu_2 x))$$

where $H_{G(r,\mu_1 t)}(x)$ is the cdf of the gamma distribution $G(r,\mu_1 t)$.

Consequently, the probability of a flexible breakwater surviving that period of t years, assuming it fails after N waves of height larger than h_0, is

$$p_f = H_{G(n-N+1,\mu_1 t)}(1) - H_{G(n-N+1,\mu_1 t)}(\exp(-\mu_2 h_0)) \qquad \blacksquare$$

2.3.4.   Order statistics of one sample from a uniform parent.

In this section the order statistics of a sample of size n, coming from a uniform $U(0,1)$ parent population are studied. The importance of this population is based on the fact that any other continuous population can be easily transformed to the uniform and some properties of these order statistics can be directly translated to properties of the order statistics of a sample drawn from the initial population.

The r-th order statistic has the following pdf (see (2.3)):

(2.40) $$f_{X_{r:n}}(x) = \frac{x^{r-1}(1-x)^{n-r}}{B(r,n-r+1)}; \qquad 0 \le x \le 1$$

which is a Beta $B(r, n - r + 1)$ and has the following mean and variance

(2.41)
$$\mu_{r:n} = \int_0^1 \frac{x^r(1 - x)^{n-r} dx}{B(r, n - r + 1)} = \frac{r}{n + 1}$$

$$\sigma_{r:n}^2 = \int_0^1 \frac{x^{r+1}(1 - x)^{n-r} dr}{B(r, n - r + 1)} - \left(\frac{r}{n + 1}\right)^2$$

(2.42)
$$= \frac{r(n - r + 1)}{(n + 1)^2(n + 2)}$$

Equation (2.42) implies that for a given sample size, n, the variance increases as r approaches the central values (the maximum is for $r = (n + 1)/2$ or one of the closest integers) and for given r, the variance decreases with increasing n.

The difference of the statistics $X_{s:n}$ and $X_{r:n}$ $(r < s)$ has pdf (see (2.28)):

$$h(v) = n! \int_0^{1-v} \frac{u^{r-1} v^{s-r-1}(1 - u - v)^{n-s} du}{(r - 1)!(s - r - 1)!(n - s)!}$$

(2.43)
$$= \frac{v^{s-r-1}(1 - v)^{n-s+r}}{B(s - r, n - s + r + 1)}; \qquad 0 \le v \le 1$$

which is also a Beta $B(s - r, n - s + r + 1)$ and has mean and variance

(2.44)
$$\mu_{vr,s} = \frac{s - r}{n + 1}$$

(2.45)
$$\sigma_{vr,s}^2 = \frac{(s - r)(n - s + r + 1)}{(n + 1)^2(n + 2)}$$

Expression (2.43) shows that the pdf depends only on the difference $(s - r)$ and not on the particular s and r values.

In the following section we demonstrate that the quotients of two consecutive order statistics are independent random variables.

Let us consider the following transformation

(2.46)
$$\left. \begin{array}{l} Y_j = \dfrac{X_{(j)}}{X_{(j+1)}}; \qquad j = 1, 2, \ldots, n - 1 \\[2mm] Y_n = X_{(n)} \end{array} \right\}$$

which has as inverse

(2.47)
$$X_{(j)} = \prod_{s=j}^n Y_s; \qquad j = 1, 2, \ldots, n$$

with jacobian

$$
J = \begin{vmatrix}
Y_2 Y_3 \dots Y_n & \# & \# & \dots & \# & \# \\
0 & Y_3 Y_4 \dots Y_n & \# & \dots & \# & \# \\
0 & 0 & Y_4 Y_5 \dots Y_n & \dots & \# & \# \\
\dots & \dots & \dots & \dots & \dots & \dots \\
0 & 0 & 0 & \dots & Y_n & Y_{n-1} \\
0 & 0 & 0 & \dots & 0 & Y_n
\end{vmatrix}
$$

$$= Y_n^{n-1} Y_{n-1}^{n-2} \dots Y_2$$

(2.48)

where the $\#$ are values not needed in evaluating the determinant.

Then, the joint pdf of the new variables is

(2.49) $$g(y_1, y_2, \dots, y_n) = n! y_2 y_3^2 \dots y_{n-1}^{n-2} y_n^{n-1}$$

which is defined in the region

$$0 \le Y_1 Y_2 \dots Y_n \le Y_2 Y_3 \dots Y_n \le \dots \le Y_{n-1} Y_n \le Y_n \le 1$$

or, equivalently, in the region

$$0 \le Y_j \le 1; \qquad j = 1, 2, \dots, n$$

This shows that the Y_i are independent.

The marginal distribution of Y_j is

(2.50) $$h_j(y) = n! y^{j-1} \prod_{\substack{i=2 \\ i \ne j}}^{n} \int_0^1 y_i^{i-1} \, dy_i = j y^{j-1}; \qquad 0 \le y \le 1$$

and its cdf

(2.51) $$H_j(y) = y^j; \qquad 0 \le y \le 1$$

The following property is important for simulation purposes.

Let $Z_1, Z_2, \dots, Z_{n+1}$ be an independent random sample of size $n + 1$ drawn from a unit exponential variable. Then, the joint pdf of $(Z_1, Z_2, \dots, Z_{n+1})$ is

(2.52) $$f(z_1, z_2, \dots, z_{n+1}) = \exp\left\{ -\sum_{i=1}^{n+1} z_i \right\}; \quad 0 \le z_i \le \infty; \quad i = 1, 2, \dots, n + 1$$

Now, consider the transformation

(2.53) $$
\left.
\begin{aligned}
U_j &= \frac{Z_1 + Z_2 + \dots + Z_j}{Z_1 + Z_2 + \dots + Z_{n+1}}; \quad j = 1, 2, \dots, n \\
U_{n+1} &= Z_1 + Z_2 + \dots + Z_{n+1}
\end{aligned}
\right\}
$$

which has as inverse

$$(2.54) \qquad \left.\begin{array}{l} Z_1 = U_1 U_{n+1} \\ Z_j = (U_j - U_{j-1})U_{n+1}; \qquad j = 1, 2, \ldots, n \\ Z_{n+1} = (1 - U_n)U_{n+1} \end{array}\right\}$$

with jacobian

$$J = \begin{vmatrix} U_{n+1} & 0 & 0 & 0 & \cdots & 0 & U_1 \\ -U_{n+1} & U_{n+1} & 0 & 0 & \cdots & \cdots & (U_2 - U_1) \\ \cdots & \cdots & \cdots & \cdots & \cdots & \cdots & \cdots \\ 0 & 0 & 0 & 0 & \cdots & -U_{n+1} & (1 - U_n) \end{vmatrix}$$

$$(2.55) \qquad = U_{n+1}^n$$

Then, the pdf of the new variables becomes

$$(2.56) \qquad g(u_1, u_2, \ldots, u_{n+1}) = \exp\{-u_{n+1}\}u_{n+1}^n$$

in the domain

$$\left.\begin{array}{l} 0 \le U_1 U_{n+1} < \infty \\ 0 \le (U_j - U_{j-1})U_{n+1} < \infty; \qquad j = 2, \ldots, n \\ 0 \le (1 - U_n)U_{n+1} < \infty \end{array}\right\}$$

which, due to the fact that $U_{n+1} > 0$, is equivalent to

$$\left.\begin{array}{l} 0 \le U_1 \le U_2 \le \ldots \le U_n \le 1 \\ 0 \le U_{n+1} < \infty \end{array}\right\}$$

Finally, the joint pdf of $(U_1, U_2, \ldots, U_n)$ becomes

$$h(u_1, u_2, \ldots, u_n) = \int_0^\infty g(u_1, u_2, \ldots, u_{n+1})\, du_{n+1}$$

$$= \int_0^\infty \exp\{-u_{n+1}\}u_{n+1}^n\, du_{n+1}$$

$$(2.57) \qquad = \Gamma(n+1) = n!; \qquad u_1 \le u_2 \le \ldots \le u_n$$

We thus established the following theorem:

Theorem 2.1. *The joint distribution of $(U_1, U_2, \ldots, U_n)$, where $U_1, U_2, \ldots, U_n$ are defined by (2.53), coincides with the joint distribution of the order statistics $(X_{1:n}, X_{2:n}, \ldots, X_{n:n})$ of a sample of size n drawn from a uniform $U(0,1)$ parent population.* ∎

2.3.5.   Simulation of order statistics.

In many applications it is necessary to generate order statistics of samples from some specified populations. In this section we describe four different methods for this generation.

Because there is a general concern about the efficiencies of several simulation techniques we want to point out initially that an analysis and comparison of the efficiencies of these methods is not an easy matter. The main reason is that these efficiencies depend on the relative time that a given computer requires for different operations (summation, multiplication, raising to a power, comparison, etc.) and on the different algorithms used for some elemental operations, as sorting, for example. In other words, hardware and software conditions are involved. These factors actually have a large influence on the scores of the different methods and new advances on either or both components can change them. Thus, any ordering of a set of different simulation techniques must clearly state hardware and software conditions. This is the reason why some comparisons can be contradictory.

At this point, it can be said that, although some general rules can be given, the final decision about optimality of a given method for a given system must be carefully established.

On the other hand, in addition to the time consuming criteria, other facts must be considered. It is not the same to design a program to be used for many users and many times as to design a program to be used only once; in the latter, the time spent in finding the least time consuming method is usually not compensated by the saving, and the use of a ready-made subroutine can be considered the optimal solution, even though it could be the worst designed given some other criteria. On the other hand, time requirements, which were very important in the past when central processing units were slow, are becoming unimportant with the exponential increase in the power of new computers.

Finally, we want to point out here that the following methods assume that the pdf, the cdf or both, of the population under study are known.

Because most of the simulation techniques for arbitrary distributions are based on uniform distribution, $U(0, 1)$, we shall study this particular case first.

2.3.5.1.   Case of a uniform parent.   In this section we give several methods for simulating samples from a uniform parent. From the very many available possibilities we have selected four that we consider to be the most representative.

THE DIRECT METHOD. Conceptually the simplest and the most widely used method is the so-called direct method. It is based on generating an independent sample of size n from an uniform parent, followed by sorting of the sample.

The first step of generation of the uniform sample is straightforward and very fast, but the sorting is usually very time consuming for large sample sizes. However, new sorting techniques, as those based on the random access method, are favoring this method.

A flow chart for this method is shown in figure 2.18.

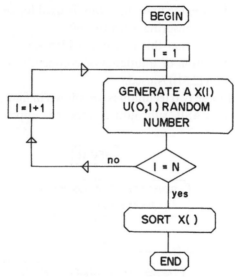

FIGURE 2.18. Flow chart for the direct method

THE SEQUENTIAL METHOD. This method, initially proposed by Schucany (1972), Lurie and Hartley (1972) and Reeder (1972), is based on the distribution of one order statistic, given the previous, or the posterior, order statistic (see section 2.3.2.1, expressions (2.22) and (2.23)).

One of the versions (Lurie and Hartley (1972)) starts with the simulation of the minimum of the sample, which has cdf (see (2.40))

$$F_{1:n}(x) = 1 - [1 - x]^n$$

and then, the following order statistics are generated, one at a time, by means of the conditional distribution of consecutive order statistics (expression (2.23)), until the last is obtained.

The other version, which is simpler and less time consuming, starts with the maximum of the sample, which has cdf (see (2.40))

$$F_{n:n}(x) = x^n$$

and then the other order statistics in descending order are generated by means of the conditional distribution given in expression (2.22).

The main advantage of these two methods is that no sorting is needed, because the generation technique gives them in the desired order. However, real powers, which are time consuming to obtain, must be calculated.

One of the clearest applications of these techniques is to the case of censored samples. In this case, the first algorithm is used for censoring from the right, and the second, for censoring from the left.

It is interesting to note that this method is identical to the method that results when the distribution of quotients of consecutive order statistics, in section 2.3.4 (expression (2.51)) is used.

The flow charts of these two techniques are shown in figures 2.19. and 2.20., and the corresponding BASIC subroutines appear in appendix B.

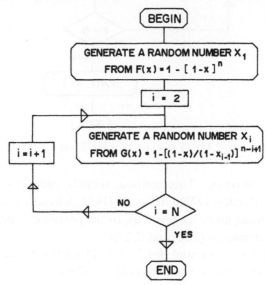

FIGURE 2.19.   Flow chart for the sequential method (version 1)

THE EXPONENTIAL SPACING METHOD.   This method is based on theorem 2.1, and consists of generating an independent unit exponential sample $Z_1, Z_2, \ldots, Z_{n+1}$ of size $n+1$, and then obtaining the order statistics $U_1, U_2, \ldots, U_n$ from (2.53).

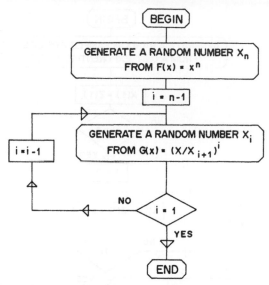

FIGURE 2.20.   Flow chart for the sequential method (version 2)

The main advantage of this method lies in the fact that no sorting is necessary, because the U_i $(i = 1, 2, \ldots, n)$ are already sorted. The only short-coming is related to the generation of the initial exponential sample, which usually requires either logarithmic calculations or the rejection technique, which are both time consuming, though the last is faster.

The flow chart for this method is shown in figure 2.21.

THE GROUPING METHOD.   This method, proposed by Gerontidis and Smith (1982), is the result of a careful study in order to avoid some computations, as powers and logarithms, and to reduce the sorting problem. The first step of the method consists of generating a multinomial vector $(M_1, M_2, \ldots, M_k)$ such that $\Sigma M_j = n$ and with equally likely cells. The second step is the generation of uniform samples $U[(j - 1)/k, j/k]$ of sizes M_j for $j = 1, 2, \ldots, k$ by the direct method and the final step consists of obtaining the ordered sample by simply putting together the k previous ordered samples.

Gerontidis and Smith recommend values of k satisfying the following condition

$$\frac{n}{7} \le k \le \frac{n}{2}$$

However, some actual sorting techniques use very similar tricks as the one mentioned here, and then, the direct method can be superior.

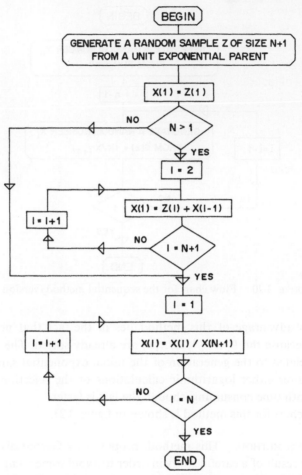

FIGURE 2.21.   Flow chart for the exponential spacing method

2.3.5.2.   Case of an arbitrary parent.

Suppose that a set of order statistics corresponding to a sample of size n drawn from a population with cdf $F(x)$ is required. The most widely known method for this generation is the inversion method, which consists of generating first a set of order statistics, $X_{(i)}$, for the uniform $U(0, 1)$ population and then transforming these by $F^{-1}(X_{(i)})$.

This method is very convenient when an analytic expression is available for the inverse of the distribution function. When this is not the case, the method becomes less attractive and some alternate methods can be superior. Among them, an extension of the grouping method to arbitrary parents is a good option. This method starts with a partition, of k equally likely intervals, of the range of the distribution (this step can be done once and for all simulations) and, similarly to the case of the uniform parent, generation of the multinomial vector $(M_1, M_2, \ldots, M_k)$. Then, for each interval j $(j = 1, 2, \ldots, k)$, a sample of size M_j, with associated pdf, that is, with the truncated pdf of the initial distribution corresponding to that interval is drawn by the rejection method, and sorted. Finally, by putting together all subsamples, the desired sample is obtained.

2.3.6. Moments of order statistics.

This section deals with means, variances and covariances of order statistics. In a few practical situations, when the parent population is known or assumed to be known, and the analytical expressions for moments become simple, exact values for the moments can be found. However, in most of the cases, where analytical expressions become complicated or the parent population is not known, these exact values are not available and some approximate formulas or bounds are required.

On the other hand, there are parent populations such that moments of order larger than some value, r say, do not exist. In this case, some moments of the order statistics are not finite either.

Moments play a very important role in estimation through linear functions of order statistics.

2.3.6.1. Finite moments. In the sequel, following David (1981), we shall use the notation $\mu_{r:n}^{(k)}$, $\mu_{rs:n}$ and $\sigma_{rs:n}$ for the k-th moment of the r-th order statistic, the product moment of $X_{r:n}$ and $X_{s:n}$ and the covariance of $X_{r:n}$ and $X_{s:n}$, respectively. That is,

$$(2.58) \qquad \mu_{r:n}^{(k)} = E[X_{r:n}^k] = \int_{-\infty}^{\infty} x^k f_{X_{r:n}}(x)\, dx$$

$$(2.59) \qquad \mu_{rs:n} = E[X_{r:n} X_{s:n}]$$

$$(2.60) \qquad \sigma_{rs:n} = E[(X_{r:n} - \mu_{r:n})(X_{s:n} - \mu_{s:n})]$$

For $k = 1$, the superscript is omitted for simplicity.

According to (2.3), expression (2.58) becomes

$$(2.61) \qquad \begin{aligned} \mu_{r:n}^{(k)} &= \int_{-\infty}^{\infty} \frac{x^k F^{r-1}(x)[1 - F(x)]^{n-r} f(x)\, dx}{B(r, n-r+1)} \\ &\leq \int_{-\infty}^{\infty} \frac{|x|^k f(x)\, dx}{B(r, n-r+1)} = \frac{v_k'}{B(r, n-r+1)} \end{aligned}$$

where v_k' is the absolute moment of order k.

Sen (1959) has demonstrated that if v'_s $(s > 0)$ is finite, then $\mu^{(k)}_{r:n}$ also is finite for

(2.62)
$$\frac{k}{s} \leq r \leq n + 1 - \frac{k}{s}$$

Setting $k = s$ we have that the moment $\mu^{(k)}_{r:n}$ is finite if v'_k is finite. Consequently, if the parent population has finite moment of order k, then the same is true for any order statistic of a finite sample.

Example 2.30. (Pareto distribution).   Let us assume the parent population is Pareto, that is, its pdf

$$f(x) = \begin{cases} \dfrac{2}{x^3} & \text{if } 1 \leq x < \infty \\ 0 & \text{otherwise} \end{cases}$$

and, thus, its cdf

$$F(x) = \begin{cases} 0 & \text{if } x < 1 \\ 1 - \dfrac{1}{x^2} & \text{if } 1 \leq x \leq \infty \end{cases}$$

This has a finite mean, and finite moments v_k for $k < 2$, but its variance is not finite.

According to (2.62) the r-th order statistic of a sample of size n has finite mean $(k = 1)$ if

$$\frac{1}{2} < r < n - \frac{1}{2} + 1 = n + \frac{1}{2}$$

i.e. all order statistics have a finite mean. However, for the variance $(k = 2)$ from (2.62) we get

$$1 < r < n$$

Thus, (2.62) does not guarantee the finiteness of the variances for the first and last order statistics. In fact,

$$\sigma_{rr:n} = \mu^{(2)}_{r:n} - \mu^2_{r:n} = \int_1^\infty \frac{2x^2 \left[1 - \dfrac{1}{x^2} \right]^{r-1} \left[\dfrac{1}{x^2} \right]^{n-r}}{x^3}\, dx - \mu^2_{r:n}$$

$$= 2 \int_1^\infty \frac{x^2 [x^2 - 1]^{r-1}}{x^{2n+1}}\, dx - \mu^2_{r:n}$$

which for $r = 1$ and $n > 1$ is finite, but for $r = n$ is not. ∎

The mean and variance of the difference of two order statistics $X_{s:n}$ and $X_{r:n}$ (the range is a particular case) are given by

(2.63)
$$E[X_{s:n} - X_{r:n}] = \mu_{s:n} - \mu_{r:n}$$

and

(2.64)
$$\text{Var}[X_{s:n} - X_{r:n}] = \sigma_{ss:n} - 2\sigma_{sr:n} + \sigma_{rr:n}$$

Their finiteness depends on the finiteness of the terms on the right hand sides of the equations.

2.3.6.2.   Moments of a uniform parent.

In the case of a uniform $U(0,1)$ parent, because its range is bounded, all moments exist for all order statistics. It can be demonstrated (see David and Johnson (1954)) that

(2.65)
$$E\left[\prod_{i=1}^{k} X_{r_i:n}^{\alpha_i}\right] = n! \prod_{i=1}^{k} \frac{\left(r_i - 1 + \sum_{j=1}^{i} \alpha_j\right)!}{\left(r_i - 1 + \sum_{j=1}^{i-1} \alpha_j\right)!} \bigg/ \left[n + \sum_{i=1}^{k} \alpha_i\right]!$$

In particular, the mean values are

(2.66)
$$\mu_{r:n} = \int_0^1 \frac{x x^{r-1}(1-x)^{n-r}\,dx}{B(r, n-r+1)} = \frac{r}{n+1}$$

and the covariances

(2.67)
$$\sigma_{rs:n} = \frac{r(n-s+1)}{(n+1)^2(n+2)}; \qquad r \le s$$

The mean and variance of the range in this case become (see (2.63) and (2.64))

(2.68)
$$\mu_W = \mu_{n:n} - \mu_{1:n} = \frac{n-1}{n+1}$$

and

(2.69)
$$\text{Var}[W] = \sigma_{nn:n} - 2\sigma_{n1:n} + \sigma_{11:n} = \frac{2(n-1)}{(n+1)^2(n+2)}$$

2.3.6.3.   Recurrence relations.

In order to avoid long calculations for the moments it is helpful to know some recurrence relations among moments of order statistics. We include here two of them without proof (the interested reader is referred to David (1981) for the proofs of these relations).

Recurrence relation I:

(2.70) $(n - r)\mu_{r:n}^{(k)} + r\mu_{r+1:n}^{(k)} = n\mu_{r:n-1}^{(k)}$; $r = 1, 2, \ldots, n - 1; k = 1, 2, \ldots$

Recurrence relation II:

(2.71) $$\mu_{r:n}^{(k)} = \sum_{i=r}^{n} \binom{i-1}{r-1}\binom{n}{i}(-1)^{i-r}\mu_{i:i}^{(k)}$$

Recurrence relation I allows us to obtain the k-th moment of all other statistics if the k-th moments of minima are known.

Recurrence relation II allows us to obtain the k-th moment of the r-th order statistics if the k-th moments of all maxima for samples sizes between r and n are known.

Example 2.31. (Uniform distribution).   For the uniform $U(0, 1)$ distribution, the moment of order k of the maximum of one sample of size n is

(2.72) $$\mu_{n:n}^{(k)} = n \int_0^1 x^k x^{n-1} \, dx = \frac{n}{n+k}$$

and then, according to (2.71)

(2.73) $$\mu_{r:n}^{(k)} = \frac{\sum_{i=r}^{n} \binom{i-1}{r-1}\binom{n}{i} i(-1)^{i-r}}{i+k};   k = 1, 2, \ldots$$

2.3.6.4.   Bounds and approximations for moments.   In some cases, because the parent distribution is not known, the formulas of the preceding sections are not applicable. In other cases, when n is large, the recurrence expressions (2.70) and (2.71) can lead to large rounding errors, which make them useless. In such situations the only remaining alternatives are the use of approximation formulas or bounds for the moments.

When the parent distribution is unknown we shall give some bounds which are called distribution-free bounds because they are valid for any distribution. On the contrary, if that distribution is known, the formulas in the next section will allow an approximation as exactly as desired.

The orthogonal inverse expansion method.   The name of this method comes from the fact that it is based on the orthogonal expansion of the powers of the inverse function, $U(u)$, of the cdf parent. In other words, an orthonormal system of functions

$$\{\psi_k(u)\}   (k = 0, 1, \ldots)$$

is considered, i.e. a system satisfying the following conditions

$$(2.74) \qquad \psi_0(u) = 1; \qquad \int_0^1 \psi_k(u)\, du = 0; \qquad \int_0^1 \psi_k^2(u)\, du = 1; \qquad k = 1, 2, \dots$$

$$(2.75) \qquad \int_0^1 \psi_k(u)\psi_{k'}(u)\, du = 0; \qquad k \neq k'$$

The main advantage of the method consist of the possibility of getting bounds as approximate, to the exact value, as desired.

Two different groups of formulas are given in the following sections. One is for general parents, and the other for symmetric parents. The derivation of the formulas by the method of Sugiura (1962,1964) can be followed in appendix A.

CASE OF A GENERAL PARENT.   Under this denomination we refer to any parent not necessarily symmetric. In this case the approximating formula for the α-moment with respect to the origin of $X_{r:n}$ is

$$\left| \mu_{r:n}^{(\alpha)} - \sum_{k=0}^{m} a_k b_k \right| \leq \left(\mu^{(2\alpha)} - \sum_{k=0}^{m} a_k^2 \right)^{1/2} \left(\frac{B(2r-1,\, 2n-2r+1)}{B^2(r,\, n-r+1)} - \sum_{k=0}^{m} b_k^2 \right)^{1/2}$$

where $\mu^{(2\alpha)}$ is the moment of order 2α with respect to the origin of the parent distribution, $B(p, q)$ is, as usual, the Beta function and

$$(2.77) \qquad a_k = \int_0^1 f(u)\psi_k(u)\, du; \qquad b_k = \int_0^1 g(u)\psi_k(u)\, du; \qquad k = 0, 1, \dots$$

with

$$(2.78) \qquad f(u) = U^\alpha(u)$$

$$(2.79) \qquad g(u) = \frac{u^{r-1}(1-u)^{n-r}}{B(r,\, n-r+1)}$$

This expression, if $U(u)$ is known, allows us to obtain approximations for $\mu_{r:n}^{(\alpha)}$ to a desired precision by taking m large enough, because if $\{\psi_k(u)\}$ is a complete system, the right hand side of (2.76) can be made zero. Note that the two factors on the right hand side of (2.76) decrease with increasing m.

If $m = 0$, (2.76) gives

$$(2.80) \qquad |\mu_{r:n}^{(\alpha)} - \mu^{(\alpha)}| \leq [\mu^{(2\alpha)} - (\mu^{(\alpha)})^2]^{1/2} \left(\frac{B(2r-1,\, 2n-2r+1)}{B^2(r,\, n-r+1)} - 1 \right)^{1/2}$$

which is a distribution free bound, i.e. valid for any distribution.

These last two expressions are especially interesting because they give bounds for the mean and second moments for $\alpha = 1$ and $\alpha = 2$ respectively.

Similarly, for the difference of expectations of powers of order statistics we get the following approximating formula (see appendix A)

$$\left| E[X_{s:n}^{(\alpha)} - X_{r:n}^{(\alpha)}] - \sum_{k=0}^{m} a_k b_k \right| \le \left(\mu^{(2\alpha)} - [\mu^{(\alpha)}]^2 - \sum_{k=1}^{m} a_k^2 \right)^{1/2}.$$

$$\left\{ \frac{B(2s-1, 2n-2s+1)}{B^2(s, n-s+1)} + \frac{B(2r-1, 2n-2r+1)}{B^2(r, n-r+1)} \right.$$

(2.81)
$$\left. - 2\frac{B(r+s-1, 2n-s-r+1)}{B(s, n-s+1)B(r, n-r+1)} - \sum_{k=1}^{m} b_k^2 \right\}^{1/2}$$

where now a_k and b_k are given by (2.77) with

(2.82) $$f(u) = U^{\alpha}(u)$$

(2.83) $$g(u) = \frac{u^{s-1}(1-u)^{n-s}}{B(s, n-s+1)} - \frac{u^{r-1}(1-u)^{n-r}}{B(r, n-r+1)}$$

which for $m = 0$ gives the following distribution free bound

$$|E[X_{s:n}^{(\alpha)} - X_{r:n}^{(\alpha)}]| \le \{\mu^{(2\alpha)} - [\mu^{(\alpha)}]^2\}^{1/2}$$

$$\cdot \left(\frac{B(2s-1, 2n-2s+1)}{B^2(s, n-s+1)} + \frac{B(2r-1, 2n-2r+1)}{B^2(r, n-r+1)} \right.$$

(2.84)
$$\left. - 2\frac{B(r+s-1, 2n-s-r+1)}{B(s, n-s+1)B(r, n-r+1)} \right)^{1/2}$$

Finally, for product moments we have the following approximating formula

$$\left| E\left(X_{r:n}X_{s:n} - \frac{1}{2} \sum_{\lambda,v=0}^{k} a_\lambda a_v(b_{\lambda,v} + b_{v,\lambda}) \right) \right|$$

$$\le \left(\sigma^4 - \sum_{\lambda,v=1}^{k} a_\lambda^2 a_v^2 \right)^{1/2} \cdot \left(\frac{B(2r-1, 2s-2r-1, 2n-2s+1)}{2B^2(r, s-r, n-s+1)} \right.$$

$$- \frac{B(2r-1, 2n-2r+1)}{2B^2(r, n-r+1)} - \frac{B(r+s-1, 2n-r-s+1)}{B(r, n-r+1)B(s, n-s+1)}$$

$$\left. - \frac{B(2s-1, 2n-2s+1)}{2B^2(s, n-s+1)} + 1 - \frac{1}{4} \sum_{\lambda,v=1}^{k}(b_{\lambda,v} + b_{v,\lambda}) \right)^{1/2}$$

(2.85) $$1 \le r < s \le n$$

where

$$(2.86) \qquad a_\lambda = \int_0^1 U(u)\psi_k(u)\,du$$

$$(2.87) \qquad B(p,q,r) = \frac{\Gamma(p)\Gamma(q)\Gamma(r)}{\Gamma(p+q+r)}$$

$$(2.88) \qquad b_{\lambda,v} = \frac{1}{B(r, s-r, n-s+1)} \iint\limits_{0<u<v<1}$$

$$\cdot u^{r-1}(v-u)^{s-r-1}(1-v)^{n-s}\psi_\lambda(u)\psi_v(v)\,du\,dv$$

CASE OF A SYMMETRIC PARENT. We now turn to the case of symmetric parent populations for which $U(u) = -U(1-u)$. In this case the above bounds can be improved (see appendix A). For the α-moment with respect to the origin of $X_{r:n}$ we get

$$\left| \mu_{r:n}^{(\alpha)} - \sum_{k=0}^m a_k b_k \right| \le \left(\mu^{(2\alpha)} - \sum_{k=0}^m a_k^2 \right)^{1/2}$$

$$(2.89) \qquad \cdot \left(\frac{B(2r-1, 2n-2r+1) + (-1)^\alpha B(n,n)}{2B^2(r, n-r+1)} - \sum_{k=0}^m b_k^2 \right)^{1/2}$$

where a_k and b_k $(k = 0, 1, 2, \ldots)$ are given by (2.81) with

$$(2.90) \qquad f(u) = U^\alpha(u)$$

$$(2.91) \qquad g(u) = \frac{h_r(u) + (-1)^\alpha h_r^*(u)}{2}$$

$$(2.92) \qquad h_r(u) = \frac{u^{r-1}(1-u)^{n-r}}{B(r, n-r+1)}$$

$$(2.93) \qquad h_r^*(u) = \frac{u^{n-r}(1-u)^{r-1}}{B(r, n-r+1)} = h_{n-r+1}(u)$$

Upon making $m = 0$ we get

$$(2.94) \quad |\mu_{r:n}^{(\alpha)}| \le \{\mu^{(2\alpha)} - [\mu^{(\alpha)}]^2\}^{1/2} \left(\frac{B(2r-1, 2n-2r+1) + (-1)^\alpha B(n,n)}{2B^2(r, n-r+1)} \right)^{1/2}$$

which is a bound valid for any symmetric distribution.

Finally, for the differences of expectations of order statistics we get

$$\left| E[X_{s:n}^{(\alpha)} - X_{r:n}^{(\alpha)}] - \sum_{k=1}^{m} a_k b_k \right|$$

$$\le \left(\mu^{(2\alpha)} - [\mu^{(\alpha)}]^2 - \sum_{k=1}^{m} a_k^2 \right)^{1/2} \left(\frac{B(2s-1, 2n-2s+1) - B(n,n)}{2B^2(s, n-s+1)} \right.$$

$$+ \frac{B(2r-1, 2n-2r+1) - B(n,n)}{2B^2(r, n-r+1)}$$

(2.95)
$$\left. - \frac{B(s+r-1, 2n-r-s+1) - B(n+s-r, n-s+r)}{B(r, n-r+1)B(s, n-s+1)} - \sum_{k=1}^{m} b_k^2 \right)^{1/2}$$

where now

(2.96)
$$f(u) = U^{\alpha}(u)$$

(2.97)
$$g(u) = \frac{h_s(u) - h_s^*(u) - h_r(u) + h_r^*(u)}{2}$$

which for $m = 0$ gives the following distribution free bound for symmetric distributions

$$|E[X_{s:n}^{(\alpha)} - X_{r:n}^{(\alpha)}]|$$

$$\le \{\mu^{(2\alpha)} - [\mu^{(\alpha)}]^2\}^{1/2} \left(\frac{B(2s-1, 2n-2s+1) - B(n,n)}{2B^2(s, n-s+1)} \right.$$

$$+ \frac{B(2r-1, 2n-2r+1) - B(n,n)}{2B^2(r, n-r+1)}$$

(2.98)
$$\left. - \frac{B(s+r-1, 2n-r-s+1) - B(n+s-r, n-s+r)}{B(r, n-r+1)B(s, n-s+1)} \right)^{1/2}$$

NUMERICAL SOLUTIONS.   In previous sections we have given several formulas to get bounds and approximations of moments of order statistics. These formulas were derived in appendix A. However, in order to get some practical results, some numerical treatment is needed.

The first step consists of selecting some complete orthogonal system in the interval $(0, 1)$. Some complete systems are:

1. The Legendre polynomials

(2.99)
$$\psi_k(u) = \frac{\sqrt{2k+1}}{k!} \frac{d^k}{du^k} [u^k (1-u)^k]; \qquad k = 0, 1, \ldots$$

2. The trigonometric functions

$$\psi_0(u) = 1$$

(2.100)
$$\psi_k(u) = \frac{\cos(k\pi u)}{\sqrt{2}}; \qquad k = 2, 4, \ldots$$

$$\psi_k(u) = \frac{\sin[(k+1)\pi u]}{\sqrt{2}}; \qquad k = 1, 3, \ldots$$

For the numerical application we only need the coefficients a_k, b_k ($k = 1, 2, \ldots, m$). But these coefficients in the case of Legendre polynomials (see (2.77) become

(2.101)
$$a_k = \int_0^1 f(u)\psi_k(u)\,du = \sum_{i=0}^{k} C_{k,i} \int_0^1 u^i f(u)\,du$$

(2.102)
$$b_k = \int_0^1 g(u)\psi_k(u)\,du$$

where $C_{k,i}$ are the coefficients of u^i in ψ_k, i.e.

(2.103)
$$\psi_k(u) = \sum_{i=0}^{k} C_{k,i} u^i$$

In order to facilitate the calculus of b_k we note, as indicated by Sugiura (1962), that

$$\int_0^1 h_r(u)\psi_k(u)\,du = \int_0^1 \sum_{i=0}^{k} \frac{C_{k,i} u^i u^{r-1}(1-u)^{n-r}\,du}{B(r, n-r+1)}$$

(2.104)
$$= \sum_{i=0}^{k} C_{k,i} \frac{B(r+i, n-r+1)}{B(r, n-r+1)} = \sum_{i=0}^{k} C_{k,i} D_{r,i,n}$$

where

(2.105)
$$D_{r,i,n} = \frac{B(r+i, n-r+1)}{B(r, n-r+1)} = \frac{(r-1)r(r+1)\ldots(r+i-1)}{n(n+1)(n+2)\ldots(n+i)} \frac{n}{r-1}$$

Note that the term $n/(r-1)$ in (2.105) can be simplified for $i \geq 1$. For the Legendre polynomials we have

(2.106)
$$b_{\lambda,v} = \sum_{\substack{i=0,\lambda \\ j=0,v}} (-1)^v C_{\lambda,i} C_{v,j} \frac{\Gamma(r+i)\Gamma(n-s+1+j)\Gamma(n+1)}{\Gamma(r)\Gamma(n-s+1)\Gamma(n+i+j+1)}$$

Example 2.32. (Normal distribution).   The expectations of the order statistics of a sample of size 8 coming from a standard normal population have
been approximated by means of expression (2.89) for different values of m. The
results are shown in table 2.1. Due to the symmetry of the parent population
and to the properties of Legendre polynomials, the odd terms in (2.89) are zero.

Parent	r	$m = 2$	$m = 4$	$m = 6$
N	1	-1.3160 ± 0.120	-1.4192 ± 0.0058	-1.4235 ± 0.00004
O	2	-0.9400 ± 0.093	-0.8670 ± 0.0190	-0.8524 ± 0.00029
R	3	-0.5640 ± 0.122	-0.4610 ± 0.0140	-0.4722 ± 0.00088
M	4	-0.1880 ± 0.057	-0.1440 ± 0.0127	-0.1535 ± 0.00148
A	5	0.1880 ± 0.057	0.1440 ± 0.0127	0.1535 ± 0.00148
L	6	0.5640 ± 0.122	0.4610 ± 0.0140	0.4722 ± 0.00088
	7	0.9400 ± 0.093	0.8670 ± 0.0190	0.8524 ± 0.00029
	8	1.3160 ± 0.120	1.4192 ± 0.0058	1.4235 ± 0.00004

TABLE 2.1   Different approximations to the expectations of order statistics from
samples of size 8 from a normal parent

Similarly, $E[X_{j:r}^2]$ $(i = 1, 2, \ldots, n)$ have also been approximated by expression (2.89) with $\alpha = 2$ and the resulting values are shown in table 2.2. From
these the variances of the order statistics can be calculated.

Finally, the covariance of $X_{i:n}$ and $X_{j:n}(i, j = 1, 2, \ldots, n)$ with $i \neq j$ have been
approximated by expression (2.85) with the help of (2.106), and the resulting

Parent	r	$m = 2$	$m = 4$	$m = 6$
N	1	2.2860 ± 0.147	2.3969 ± 0.0038	$2.3995 \pm 0.6 \times 10^{-7}$
O	2	1.1830 ± 0.270	0.9780 ± 0.019	$0.9657 \pm 0.4 \times 10^{-7}$
R	3	0.4480 ± 0.082	0.4010 ± 0.034	$0.4243 \pm 0.2 \times 10^{-7}$
M	4	0.0810 ± 0.019	0.2230 ± 0.019	$0.2104 \pm 0.2 \times 10^{-7}$
A	5	0.0810 ± 0.019	0.2230 ± 0.019	$0.2104 \pm 0.2 \times 10^{-7}$
L	6	0.4480 ± 0.082	0.4010 ± 0.034	$0.4243 \pm 0.2 \times 10^{-7}$
	7	1.1830 ± 0.270	0.9780 ± 0.019	$0.9657 \pm 0.4 \times 10^{-7}$
	8	2.2860 ± 0.147	2.3969 ± 0.0038	$2.3995 \pm 0.6 \times 10^{-7}$

TABLE 2.2.   Different approximations to $E[X_{r:n}^2]$ from samples of size 8 from a
normal parent

covariance matrix is

$$
\begin{pmatrix}
0.372 \\
0.186 & 0.239 \\
0.126 & 0.163 & 0.201 \\
0.095 & 0.123 & 0.152 & 0.187 \\
0.075 & 0.098 & 0.121 & 0.149 & 0.187 \\
0.060 & 0.079 & 0.098 & 0.121 & 0.152 & 0.201 \\
0.048 & 0.063 & 0.079 & 0.098 & 0.123 & 0.163 & 0.239 \\
0.037 & 0.048 & 0.060 & 0.075 & 0.095 & 0.126 & 0.186 & 0.373
\end{pmatrix}
$$

■

Example 2.33. (Gumbel distribution).   The same approximations as those in example 2.32 for a Gumbel parent with cdf

$$F(x) = 1 - \exp\{-\exp(x)\}$$

were performed. The resulting values for univariate moments are shown in tables 2.3. and 2.4. and the covariance matrix is

$$
\begin{pmatrix}
1.645 \\
0.613 & 0.647 \\
0.347 & 0.373 & 0.398 \\
0.231 & 0.246 & 0.266 & 0.290 \\
0.163 & 0.173 & 0.187 & 0.207 & 0.232 \\
0.117 & 0.125 & 0.136 & 0.149 & 0.169 & 0.198 \\
0.083 & 0.089 & 0.097 & 0.107 & 0.120 & 0.144 & 0.184 \\
0.054 & 0.058 & 0.063 & 0.070 & 0.079 & 0.094 & 0.123 & 0.200
\end{pmatrix}
$$

Parent	r	$m = 2$	$m = 4$	$m = 6$
	1	-2.469 ± 0.22	-2.6488 ± 0.012	-2.6566 ± 0.00009
G	2	-1.772 ± 0.22	-1.615 ± 0.038	-1.5887 ± 0.00062
U	3	-1.153 ± 0.22	-0.9872 ± 0.032	-1.0100 ± 0.0018
M	4	-0.612 ± 0.14	-0.5773 ± 0.026	-0.5899 ± 0.0031
B	5	-0.1499 ± 0.14	-0.2479 ± 0.026	-0.2295 ± 0.0031
E	6	0.2337 ± 0.22	0.0895 ± 0.032	0.1018 ± 0.0018
L	7	0.5387 ± 0.22	0.4740 ± 0.038	0.4531 ± 0.00062
	8	0.7653 ± 0.22	0.8954 ± 0.012	0.90207 ± 0.000089

TABLE 2.3.   Different approximations to the expectations of order statistics from samples of size 8 from a Gumbel parent

Parent	r	m = 2	m = 4	m = 6
	1	7.55 ± 1.76	8.644 ± 0.12	8.7023 ± 0.001
G	2	4.48 ± 1.73	3.375 ± 0.40	3.172 ± 0.007
U	3	2.15 ± 1.72	1.22 ± 0.33	1.412 ± 0.022
M	4	0.55 ± 1.12	0.579 ± 0.27	0.650 ± 0.037
B	5	−0.29 ± 1.12	0.416 ± 0.27	0.270 ± 0.037
E	6	−0.41 ± 1.72	0.272 ± 0.33	0.2174 ± 0.022
L	7	0.21 ± 1.73	0.2577 ± 0.40	0.386 ± 0.007
	8	1.57 ± 1.76	1.056 ± 0.12	1.013 ± 0.001

TABLE 2.4.   Different approximations to $E[X_{r:n}^2]$ from samples of size 8 from a Gumbel parent

Bounds for the mean of order statistics based on parent percentiles.   The bounds given by the orthogonal inverse expansion method range between two extremes: complete knowledge of the parent population or complete ignorance of it. In the first case the method allows us to obtain the moments of the order statistics as accurately as desired; in the second case, the bounds are distribution free and in many cases, the symmetric case included, they become useless. In this paragraph we give some intermediate bounds which are valid for sets of families of distributions much more restricted than the symmetric case. This allows better bounds to be obtained.

Table 2.5, where $U(p) = F^{-1}(p)$, gives bounds and the sets of distributions to which they are applicable. We do not include here the proofs (the interested reader can find them in Van Zwet (1964) or David (1981)). However, we include some definitions which are needed to understand Table 2.5.

Definition 2.2. (Convex function).   A function $g(x)$ is said to be convex on the interval (a, b) if for any $0 \le \lambda \le 1$ and any $x_1, x_2 \in (a, b)$ it satisfies

$$(2.107) \qquad g(\lambda x_1 + (1 - \lambda)x_2) \le \lambda g(x_1) + (1 - \lambda)g(x_2) \qquad \blacksquare$$

If $g'(x)$ and $g''(x)$ exist, this is equivalent to saying that $g''(x) \ge 0$ or $g'(x)$ is nondecreasing.

Definition 2.3. (Hazard function).   The hazard function, $h(x)$, associated with an absolutely continuous cdf, $F(x)$, with pdf $f(x)$ is defined by

$$(2.108) \qquad\qquad h(x) = \frac{f(x)}{1 - F(x)} \qquad\qquad \blacksquare$$

Condition	Upper bond	Lower bound
$F(x)$ convex $(f'(x) \geq 0)$	$\mu_{r:n} \leq U\left(\dfrac{r}{n+1}\right)$	—
$F(x)$ concave $(f'(x) \leq 0)$	—	$\mu_{r:n} \geq U\left(\dfrac{r}{n+1}\right)$
$\dfrac{1}{F(x)}$ concave $(r > 1)$ $(2f^2(x) - f'(x)F(x) \leq 0)$	$\mu_{r:n} \leq U\left(\dfrac{r-1}{n}\right)$	—
$\dfrac{1}{F(x)}$ convex $(r > 1)$ $(2f^2(x) - f'(x)F(x) \geq 0)$	—	$\mu_{r:n} \geq U\left(\dfrac{r-1}{n}\right)$
$\dfrac{1}{1-F(x)}$ convex $(r < n)$ $(f'(x)(1 - F(x)) + 2f^2(x) \geq 0)$	$\mu_{r:n} \leq U\left(\dfrac{r}{n}\right)$	—
$\dfrac{1}{1-F(x)}$ concave $(r < n)$ $(f'(x)(1 - F(x)) + 2f^2(x) \leq 0)$	—	$\mu_{r:n} \geq U\left(\dfrac{r}{n}\right)$
$F(x)$ with increasing hazard function $(f'(x)(1 - F(x)) + f^2(x) \geq 0)$	$\mu_{r:n} \leq$ $\leq U\left[1 - \exp\left(-\Sigma\dfrac{1}{n-i}\right)\right]$ $< U\left(\dfrac{r}{n + \frac{1}{2}}\right)$	—
$f(x)$ symmetric U-shaped $r \geq \dfrac{n+1}{2}$	$\mu_{r:n} \leq U\left(\dfrac{r}{n+1}\right)$	—
$f(x)$ symmetric and unimodal $r \geq \dfrac{n+1}{2}$	—	$\mu_{r:n} \geq U\left(\dfrac{r}{n+1}\right)$

TABLE 2.5.   Bounds based on parent percentiles

Example 2.34. (Parabolic distribution). Let us assume that the parent population has the following cdf

$$F(x) = \begin{cases} 0 & \text{if } x < 0 \\ x^2 & \text{if } 0 \le x < 1 \\ 1 & \text{if } x \ge 1 \end{cases}$$

Then, on the interval $(0, 1)$ we have

$$f'(x) = 2 \ge 0$$

$$2f^2(x) - f'(x)F(x) = 2(2x)^2 - 2x^2 = 6x^2 \ge 0$$

$$f'(x)[1 - F(x)] + 2f^2(x) = 2(1 - x^2) + 2(2x)^2 = 2 - 2x^2 + 8x^2 = 2 + 6x^2 > 0$$

$$f'(x)[1 - F(x)] + f^2(x) = 2(1 - x^2) + 4x^2 = 2 + 2x^2 > 0$$

Consequently, $F(x)$, $1/F(x)$ and $1/[1 - F(x)]$ are convex and the hazard function is increasing. Thus, the following bounds result

$$[(r - 1)/n]^{1/2} = U\left(\frac{r-1}{n}\right) \le \mu_{r:n} \le U\left\{\min\left[\frac{r}{n+1}, \frac{r}{n}, 1 - \exp\left(-\sum_{i=0}^{r-1}\frac{1}{n-i}\right)\right]\right\}$$

$$= \left\{\min\left[\frac{r}{n+1}, 1 - \exp\left(-\sum_{i=0}^{r-1}\frac{1}{n-i}\right)\right]\right\} \qquad ■$$

The Taylor inverse expansion method.   Credit for the original idea seems to be due to K. Pearson (1931) and consists of using the Taylor expansion of the inverse of the cdf, i.e. of the function $U(u)$.

We know that

(2.109) $$\qquad\qquad X_{r:n} = F^{-1}(U_{r:n}) = U(U_{r:n})$$

where $U(u)$ stands for $F^{-1}(u)$, as before, and $U_{r:n}$ is the r-th order statistic of the uniform $U(0, 1)$ parent. We also know that asympotically

(2.110) $$\qquad E[X_{r:n}] \quad \rightarrow \quad U(E(U_{r:n})) = U\left(\frac{r}{n+1}\right) = U(p_r)$$

where

(2.111) $$\qquad\qquad\qquad\qquad p_r = \frac{r}{n+1}$$

If we now expand $X_{r:n}$ by Taylor's series in a neighbourhood of p_r

(2.112) $$\qquad\qquad X_{r:n} = U(U_{r:n}) = \sum_{i=0}^{\infty} \frac{U^{(i)}(p_r)(U_{r:n} - p_r)^i}{i!}$$

and take expectation in (2.112) we get

(2.113) $$E[X_{r:n}] = \sum_{i=0}^{\infty} \frac{U^{(i)}(p_r)E[(U_{r:n} - p_r)^i]}{i!}$$

which when truncated at $U^{(4)}(p_r)$, and taking into account of (2.65) − (2.67) gives the approximate formula

$$\mu_{r:n} = E[X_{r:n}] = U(p_r) + \frac{p_r q_r}{2(n + 2)} U''(p_r)$$

(2.114) $$+ \frac{p_r q_r}{(n + 2)^2} \left(\frac{1}{3}(q_r - p_r)U'''(p_r) + \frac{1}{8} p_r q_r U^{(iv)}(p_r) \right)$$

where

(2.115) $$q_r = 1 - p_r$$

Analogously, taking covariances in (2.113) we get

$$\text{Cov}[X_{r:n}, X_{s:n}] = \frac{p_r q_s}{n + 2} U'(p_r)U'(p_s)$$

$$+ \frac{p_r q_s}{(n + 2)^2} \Big((q_r - p_r)U''(p_r)U'(p_s) + (q_s - p_s)U''(p_s)U'(p_r)$$

$$+ \frac{1}{2} p_r q_r U'''(p_r)U'(p_s)$$

(2.116) $$+ \frac{1}{2} p_s q_s U'(p_r)U''(p_s) + \frac{1}{2} p_r q_s U''(p_r)U''(p_s) \Big); \qquad r \leq s$$

Example 2.35. (Normal distribution). The approximations to the expectations and to the covariance matrix of the order statistics of a sample of size 8 coming from a standard normal population, obtained by the Taylor expansion method (expressions (2.114) and (2.116) are given by

$$(-1.4231 \quad -0.8511 \quad -0.4722 \quad -0.1523 \quad 0.1523 \quad 0.4722 \quad 0.8511 \quad 1.4231)$$

0.3576							
0.1819	0.2328						
0.1238	0.1597	0.1962					
0.0934	0.1210	0.1493	0.1832				
0.0739	0.0959	0.1187	0.1462	0.1832			
0.0596	0.0775	0.0961	0.1187	0.1493	0.1962		
0.0479	0.0624	0.0775	0.0959	0.1210	0.1597	0.2328	
0.0368	0.0479	0.0596	0.0739	0.0934	0.1238	0.1819	0.3576

It is interesting to compare these values with those in example 2.32.   ■

Example 2.36. (Gumbel distribution).   The Taylor approximations to the
expectations and the covariance matrix of the order statistics of a sample of
size 8 from a Gumbel parent are

$$(-2.643 \quad -1.583 \quad -1.009 \quad -0.587 \quad -0.231 \quad 0.103 \quad 0.453 \quad 0.903)$$

$$\begin{pmatrix}
1.395 \\
0.564 & 0.598 \\
0.331 & 0.352 & 0.379 \\
0.221 & 0.236 & 0.255 & 0.279 \\
0.157 & 0.168 & 0.182 & 0.200 & 0.225 \\
0.114 & 0.122 & 0.132 & 0.146 & 0.164 & 0.194 \\
0.082 & 0.087 & 0.095 & 0.105 & 0.119 & 0.140 & 0.180 \\
0.054 & 0.058 & 0.063 & 0.069 & 0.079 & 0.093 & 0.121 & 0.198
\end{pmatrix}$$

It is interesting to compare these values with those in example 2.33.   ∎

2.3.7.   Estimation based on order statistics.

In this section we analyse the problem of parameter estimation by linear
combinations of order statistics. These linear estimators provide a very useful
alternative to other methods, as maximum likelihood for example, because
their efficiencies are very high and they are much simpler to use. Even in the
case of calculating the coefficients, the required computer programs are much
simpler and some convergence or slow convergence problems are avoided.
Only for large samples would we face some precision problems.

2.3.7.1.   Best linear unbiased estimators (BLUES).   The best linear unbiased
estimators (BLUES) are those linear estimators which are unbiased and
minimize the variance. They coincide with the so-called least-squares estimators and are mainly used for families of distributions which depend on
two parameters: scale, σ, and location, μ.

In addition to the normal family, one example of special relevance is the
Gumbel family. If X is assumed to belong to one of these families, with cdf
$F((x - \mu)/\sigma)$, the random variable $Y = (x - \mu)/\sigma$ has cdf $F(y)$, which does not
depend on μ and σ.

The mean values and the variance-covariance matrix of the order statistics
$Y_{r:n}$ ($k = 1, 2, \ldots, n$) do not depend on μ and σ

(2.117)
$$\left. \begin{array}{r} E[Y_{r:n}] = \alpha_r \\ \text{Cov}[Y_{r:n}, Y_{s:n}] = \beta_{rs} \end{array} \right\}$$

and the same values for $X_{r:n}$ become

$$
(2.118) \qquad \left.\begin{array}{l} \mu_{r:n} = E[X_{r:n}] = \mu + \sigma\alpha_r \\ \mathrm{Cov}[X_{r:n}, X_{s:n}] = \sigma^2\beta_{rs} \end{array}\right\}
$$

which in matrix form can be written as

$$
(2.119) \qquad E[X] = A\theta
$$

where

$$
(2.120) \qquad X = \begin{pmatrix} X_{1:n} \\ X_{2:n} \\ \vdots \\ X_{n:n} \end{pmatrix}; \quad E[X] = \begin{pmatrix} \mu_{1:n} \\ \mu_{2:n} \\ \vdots \\ \mu_{n:n} \end{pmatrix}; \quad A = \begin{pmatrix} 1 & \alpha_1 \\ 1 & \alpha_2 \\ \vdots & \\ 1 & \alpha_n \end{pmatrix}; \quad \theta = \begin{pmatrix} \mu \\ \sigma \end{pmatrix}
$$

Thus, the weighted least-squares method, taking into account that the covariance matrix of X is $\sigma^2\beta$, leads to the minimization of

$$
(2.121) \qquad (X - A\theta)\beta^{-1}(X - A\theta)'
$$

from which the estimator, θ^*, of θ becomes

$$
(2.122) \qquad \theta^* = \begin{pmatrix} \mu^* \\ \sigma^* \end{pmatrix} = (A'\beta^{-1}A)^{-1}A\beta^{-1}X
$$

and the covariance matrix of θ^* equals

$$
(2.123) \qquad \sigma^2(A'\beta^{-1}A)^{-1}
$$

where for practical purposes σ can be substituted by σ^*.

Example 2.37. (Normal distribution). The coefficients of the BLUES for one sample of size 8 from a normal parent, obtained by the method above, are given by

$$(\quad 0.125 \quad\quad 0.125 \quad\quad 0.125 \quad\quad 0.125 \quad 0.125 \quad 0.125 \quad 0.125 \quad 0.125)$$

$$(-0.248 \quad -0.129 \quad -0.072 \quad -0.023 \quad 0.023 \quad 0.072 \quad 0.129 \quad 0.248)$$

for μ and σ, respectively, and the covariance matrix of the estimators is

$$
\sigma^2 \begin{pmatrix} 0.1250 & 0.0000 \\ 0.0000 & 0.0746 \end{pmatrix} \qquad \blacksquare
$$

Example 2.38. (Gumbel distribution). The same coefficients for a Gumbel parent are

(0.037   0.055   0.077   0.096   0.122   0.150   0.189   0.274)

(−0.102   −0.107   −0.105   −0.086   −0.060   −0.0093   0.074   0.395)

$$\sigma^2 \begin{pmatrix} 0.1420 & -0.026 \\ -0.026 & 0.093 \end{pmatrix}$$

2.3.7.2.   Best linear invariant estimators (BLIES).   The name of best linear invariant estimators comes from the fact that these estimators have minimum mean square error and that they do not depend on the location parameter μ. They are related to the best linear unbiased estimators by the following relation

(2.124) $$\mu^{**} = \mu^* - \sigma^*\left(\frac{C}{1 + B}\right)$$

(2.125) $$\sigma^{**} = \frac{\sigma^*}{1 + B}$$

where μ^{**} and σ^{**} are the BLIES, μ^* and σ^* are the BLUES and C and B are given by

(2.126) $$\begin{pmatrix} \text{Var}(\mu^*) & \text{Cov}(\mu^*, \sigma^*) \\ \text{Cov}(\mu^*, \sigma^*) & \text{Var}(\sigma^*) \end{pmatrix} = \sigma^2 \begin{pmatrix} A & C \\ C & B \end{pmatrix}$$

The mean square errors of these estimators are

(2.127) $$E(\mu^{**} - \mu)^2 = \sigma^2[A - C^2(1 + B)^{-1}]$$

(2.128) $$E(\sigma^{**} - \sigma)^2 = \frac{\sigma^2 B}{1 + B}$$

(2.129) $$E[(\mu^{**} - \mu)(\sigma^{**} - \sigma)] = \frac{\sigma^2 C}{1 + B}$$

where A, B and C are defined by (2.126).

2.3.8.   Characterization of distributions.

The distributions of the order statistics have been analyzed for a number of population distributions in the previous sections. When the figures for these distributions are compared, one immediately recognizes the sensitivity of the order statistics for the choice of the population distribution. One, there-fore, has to be careful with building the model for a particular problem. An even more evident example is provided by formula (2.6) for pointing out the danger of arbitrarily choosing the population distribution when the maximum (flood, parallel systems, and others) determines the underlying distribution for a problem. For example, if $G(x)$ is mistakenly chosen as population distribution instead of the true distribution $F(x)$, then $G^n(x)$ is used for the distribution of the maximum in place of $F^n(x)$. Now, if $F(x) = 0.99$ and $G(x) = 0.98$ for some x, then a seemingly small error is committed by the wrong choice, but, with $n = 100$, the error for the maximum is in fact (see (2.6))

$$G^{100}(x) = 0.98^{100} = 0.133 \text{ versus } F^{100}(x) = 0.99^{100} = 0.366$$

This numerical example clearly demonstrates that even small errors in the choice of the population distribution cannot be permitted in extreme value theory. The branch of probability theory when simple assumptions lead to a unique population distribution is called characterization of probability distributions. In the present section, a few characterization results are listed in the form of theorems and examples. The interested reader can find more involved results and a systematic theory in the books Kagan, Linnik and Rao (1973) and Galambos and Kotz (1978). Scattered, but significant results, are also given in the conference proceedings in Calgary (1975) and Trieste (1981), which are published as Volumes 3 and 4 of the series, Statistical Distributions in Scientific Work.

We start with two basic characterizations.

Theorem 2.2. *Let $x \geq 0$ represent random lifetime. Assume the distribution of X satisfies the no-aging property*

(2.130) $\qquad P(X \geq s + t / X \geq t) = P(X \geq s) \qquad$ *for all $s, t \geq 0$*

Then the distribution of X is exponential, i.e.

$$F(x) = P(X \leq x) = 1 - \exp(-\lambda x), \qquad x \geq 0, \lambda > 0$$

Proof. From the concept of conditional probabilities, (2.130) has the equivalent form

$$1 - F(s + t) = [1 - F(s)][1 - F(t)] \qquad \text{for all } s, t \geq 0$$

which, as is well known in elementary mathematics (see Aczél (1966)), has a unique solution (up to a parameter). Clearly, the unique solution must be exponential, which, for distribution functions, takes the claimed form. ∎

Example 2.39. (Insurance policy).   Assume that, from prior estimates, the expected time to a particular accident is 20 years. What is the probability that a person, owning an insurance policy for such an accident, will have no claim in 25 years.

If X is the random time (lifetime) up to an accident, then clearly, the distribution of X satisfies the no-aging property (this property can, in fact, be used as the definition of an accident). Therefore, X has the distribution function $1 - \exp(-\lambda x)$, and, consequently, $E(x) = 1/\lambda$. We assumed $E(x) = 20$, so $\lambda = 1/20$ (years). We thus have

$$P(X \geq 25) = 1 - F(25) = 1 - \exp\left(\frac{-25}{20}\right) = 0.713.$$

Just as in the example, Theorem 2.2 is the basis of accident insurance calculations. ∎

Theorem 2.3. (Measurement errors).   *Measurement errors ε due to the inaccuracy of instruments are normally distributed.*

Proof.   Measurement errors have the characteristic that when measurements are made in segments, the errors add up, they are independent, and the aggregate error is of the same type as the error in a simple measurement. Think of measuring a length. First, measure the length in one step, then measure this same length from one end to the middle, and then from the middle to the other end. We got the first time ε, and in the latter case ε_1 and ε_2. If ε has the same type of distribution as $\varepsilon_1 + \varepsilon_2$, and if ε_1 and ε_2 are independent, we can write $a\varepsilon + b = \varepsilon_1 + \varepsilon_2$ in distribution. But then, if we have (perhaps hypothetically only) 4 independent copies of ε, ε_j, $1 \leq j \leq 4$, $\varepsilon_1 + \varepsilon_2 = a\varepsilon + b$ and $\varepsilon_3 + \varepsilon_4 = a\varepsilon^* + b$, i.e.

$$\varepsilon_1 + \varepsilon_2 + \varepsilon_3 + \varepsilon_4 = a(\varepsilon + \varepsilon^*) + b = a(a\varepsilon^{**} + b) + b = a_1\varepsilon^{**} + b_1,$$

say, in distribution. Arguing sequentially, we get

$$\varepsilon_1 + \varepsilon_2 + \cdots + \varepsilon_N = a_{n-1}\varepsilon + b_{n-1}, \qquad N = 2^n, \text{ in distribution,}$$

or, equivalently,

$$(\varepsilon_1 + \varepsilon_2 + \cdots + \varepsilon_N - b_{n-1})/a_{n-1} = \varepsilon$$

Now, the right hand side does not depend on n (or N), while the left hand side is asymptotically normal under the evident assumptions that $E(\varepsilon_j)$ and $V(\varepsilon_j)$ are finite (the classical central limit theorem). Thus, by letting $n \to \infty$, we obtain ε to be normal. ∎

Example 2.40. (Wire length).   From a long wire, 150 pieces of 50 cm. in length are to be cut. These are subject to a measurement error with expectation 0 and variance 0.25 cm^2. Find the probability that the largest error exceeds 1 cm.

Because the errors are normally distributed, by (2.6),

$$P(\varepsilon_{max} > 1) = 1 - P(\varepsilon_{max} \leq 1) = 1 - \Phi^{150}\left(\frac{1}{0.5}\right)$$

$$= 1 - \Phi^{150}(2) = 1 - 0.9722^{150} = 0.9486$$

where $\Phi(x)$ signifies the standard normal distribution function. Clearly, the length 50 cm plays no role in the computation.   ∎

The next characterization result is in terms of failure rate, or hazard rate. Recall the definition of hazard rate at (2.108). Let X be a positive random variable, so it can represent lifetime. By definition, $h(x)$ is the failure rate (Hazard rate) of X, or of its distribution function, if

$$h(x) = \lim_{\Delta x \to 0} \frac{P(x + \Delta x \geq X/X > x)}{\Delta x}$$

When the density, $f(x)$, of X exists, the above relation becomes

(2.131) $$h(x) = \frac{f(x)}{1 - F(x)} = \frac{d}{dx}\{-\log[1 - F(x)]\}$$

Integration thus yields

(2.132) $$F(x) = 1 - \exp\left(-\int_A^x h(t)\,dt\right), \qquad x > 0$$

assuming that $F(A) = 0$. An immediate consequence of (2.132) is that the failure rate function uniquely determines the distribution function. Thus, if $h(x)$ is constant, $F(x)$ is exponential.

Example 2.41. (Machine fail).   Assume that a machine "fails" at the decreasing rate of $r(x) = 3/x$ during the running-in period of $(1, 3)$ time units Find the probability that the machine fails in less than 2 time units.

By (2.132), during the running-in period,

$$F(x) = 1 - \frac{1}{x^3}$$

and thus, $F(2) = 1 - (\frac{1}{2})^3 = \frac{7}{8}$.

Failure here might means the need for simple repairs. In fact, during running-in periods, $r(x)$ usually decreases, then for a period of normal operations, $r(x)$ is constant (thus exponential, i.e. during this period, aging is not affecting operation), and finally, the machine "becomes old", reflected in an increasing failure rate. For each period of operation, (2.132) is applicable.

■

We have referred to the figures of the previous sections, showing sensitivity of the distributions of order statistics on the population distribution. In fact, it is not an accident that all of these figures are different. Namely, the distribution of a single order statistic uniquely determines the population distribution. Formulas (2.6) and (2.8) clearly imply this claim for the maximum and the minimum. For a general r, $1 < r < n$, formula (2.2) proves our claim.

Example 2.42. (Lifetime of electric bulbs).   In a test-room, 150 electric bulbs are to be tested. It turns out that the lifetime of the first bulb to burn out is exponentially distributed with expected life of 2 hours. What is the expected lifetime of the average bulb?

Because $X_{1:150}$ has the distribution function $1 - \exp(-x/2)$, formula (2.8) becomes

$$1 - \exp\left(\frac{-x}{2}\right) = 1 - [1 - F(x)]^{150}$$

i.e.

$$1 - F(x) = \exp\left(\frac{-x}{300}\right)$$

which again is exponential. Its expected life is 300 hours.   ■

Out of the many known characterizations we now formulate two without proof. Their proof can be found in the references listed in the first paragraph of the present section.

The set $E(X_{r:n})$, $1 \leq r \leq n$, $n \geq 1$, of expectations of all order statistics characterizes the population distribution.

Example 2.43. (Uniform distribution).   If $E(X_{r:n}) = r/(n + 1)$ for all r and n, then the population is uniform on $(0, 1)$ (see (2.66)). In other words, if the expected values of a division of $(0, 1)$ by the order statistics divide $(0, 1)$ into identical segments for all n, then the population is uniform.

While the above conclusion appears quite natural by intuition, its implication is quite surprising. Namely, by the recurrence relation (2.70), if

$E(X_{1:n})$ is given for all n, then $E(X_{r:n})$ is given for all r and all n. Thus, if $E(X_{1:n}) = 1/n$ for $n \geq 1$, then the population must be unit exponential, while if $E(X_{1:n}) = 1/(n + 1)$, $n \geq 1$, then the population is uniform on $(0, 1)$.   ■

Another characterization of the exponential distribution is in terms of its remarkable property, as demonstrated earlier, that the differences of order statistics are independent. No other population has this property.

2.4.   Order Statistics from Dependent Samples

In this section we turn to the case of dependence and derive some formulas which are used to obtain the exact or approximate cdfs of order statistics in the general case.

2.4.1.   Inclusion-exclusion formula.

If A and B are given events of a probability space, it is well known from probability theory that

$$(2.133) \qquad P(A \cup B) = P(A) + P(B) - P(A \cap B)$$

This very simple formula, that can be found in any elementary probability book, can be easily generalized to the case of n events in the following way: Let $C_1, C_2, \ldots, C_n$ be n arbitrary events. Then, the following relation (inclusion-exclusion formula) holds:

$$(2.134) \quad P\left(\bigcup_{i=1}^{n} C_i\right) = \sum_{i=1}^{n} P(C_i) - \sum_{1 \leq i_1 < i_2 \leq n} P(C_{i_1} \cap C_{i_2})$$

$$+ \sum_{1 \leq i_1 < i_2 < i_3 \leq n} P(C_{i_1} \cap C_{i_2} \cap C_{i_3}) - \cdots + (-1)^{n+1}$$

$$\cdot \sum_{1 \leq i_1 < i_2 < \ldots < i_n \leq n} P(C_{i_1} \cap C_{i_2} \cap \ldots \cap C_{i_n})$$

and calling

$$(2.135) \qquad S_{k,n} = \sum_{1 \leq i_1 < i_2 < \ldots < i_k \leq n} P(C_{i_1} \cap C_{i_2} \cap \ldots \cap C_{i_k})$$

and remembering that

$$(2.136) \qquad P\left(\bigcup_{i=1}^{n} C_i\right) = 1 - P\left(\bigcap_{i=1}^{n} \bar{C}_i\right)$$

where $\bar{C}_i$ is the complement of C_i, expression (2.134) can be written

(2.137) $$P\left(\bigcap_{i=1}^{n} \bar{C}_i\right) = 1 - P\left(\bigcup_{i=1}^{n} C_i\right) = \sum_{i=0}^{n} (-1)^i S_{i,n}$$

where

(2.138) $S_{0,n} = 1$

Because of the alternating plus and minus sign in (2.137) and the definition of $S_{i,n}$, its truncated sums give upper and lower bounds to the left hand side values. More precisely,

$$\sum_{i=0}^{2s+1} (-1)^i S_{i,n} \leq P(m_n = 0) = P\left(\bigcap_{i=1}^{n} \bar{C}_i\right) \leq \sum_{i=0}^{2s} (-1)^i S_{i,n};$$

(2.139) $0 \leq s \leq \text{int}\left(\dfrac{n-1}{2}\right)$

where m_n is the number of C_i which occur and $\text{int}[x]$ means the integer part of x.

The probability of the event $\{m_n = t\}$ can be written in terms of $S_{i,n}$ ($i = t$, $t + 1, \ldots, n$) by means of (see Galambos (1978) p. 19)

(2.140) $$P(m_n = t) = \sum_{i=0}^{n-t} (-1)^i \binom{i+t}{t} S_{i+t,n}$$

2.4.2.   Distribution of one order statistic.

The previous formulas are completely general, i.e. for any set of events $\{C_1, C_2, \ldots, C_n\}$. Their application to lower and upper order statistics is possible by an adequate selection of the above sets. In fact, if

(2.141) $C_i(x) = \{X_i \leq x\}$

(2.142) $\bar{C}_i(x) = \{X_i > x\}$

we have

(2.143) $F_{X_{r:n}}(x) = P[X_{r:n} \leq x] = P[m_n(x) \geq r] = P[\underline{m}_n(x) \leq n - r]$

(2.144) $F_{X_{n-r+1:n}}(x) = P[X_{n-r+1:n} \leq x] = P[\underline{m}_n(x) < r]$

where $m_n(x)$ and $\underline{m}_n(x)$ are the number of C_i and $\bar{C}_i$ ($i = 1, 2, \ldots, n$) which occur in the sample, respectively. They satisfy the relation

(2.145) $m_n(x) + \underline{m}_n(x) = n$

Because now $C_i(x)$ and $\bar{C}_i(x)$ are dependent on x, the associated $S_{i,n}$ and $\bar{S}_{i,n}$ will be denoted by $S_{i,n}(x)$ and $\bar{S}_{i,n}(x)$, respectively.

Example 2.44. (Distribution of maxima).   For $r = 1$, expression (2.144) gives

$$F_{X_{n:n}}(x) = P[X_{n:n} \le x] = P[\underline{m}_n(x) < 1] = P[\underline{m}_n(x) = 0]$$

and substitution in (2.139) leads to

$$(2.146) \qquad \sum_{i=0}^{2s+1} (-1)^i \bar{S}_{i,n}(x) \le F_{X_{n:n}}(x) \le \sum_{i=0}^{2s} (-1)^i \bar{S}_{i,n}(x)$$

where $\bar{S}_{i,n}(x)$ are the values in (2.135) when the C_i are substituted by $\bar{C}_i$. Both sides in the above expression become an equality for $n = 2s + 1$ and $n = 2s$, respective and give the cdf of the maximum in the general case (dependence case included).

For the i.i.d. case we have

$$\bar{S}_{i,n}(x) = \binom{n}{i}[1 - F(x)]^i$$

and for $2s = n$, it degenerates to the simple formula

$$F_{X_{n:n}}(x) = \sum_{i=0}^{n}(-1)^i \binom{n}{i}[1 - F(x)]^i = \{1 - [1 - F(x)]\}^n = F^n(x)$$

which is the well known expression (2.6).   ■

Example 2.45. (Distribution of minima).   For $r = 1$, expression (2.143) leads to

$$F_{X_{1:n}}(x) = P[X_{1:n} \le x] = P[m_n(x) \ge 1] = 1 - P[m_n(x) = 0]$$

and substitution into (2.139) gives

$$(2.147) \qquad \sum_{i=0}^{2s+1} (-1)^i S_{i,n}(x) \le 1 - F_{X_{1:n}}(x) \le \sum_{i=0}^{2s} (-1)^i S_{i,n}(x)$$

which gives bounds for the cdf of the minimum of the sample and exact values for $n = 2s + 1$ or $n = 2s$.   ■

Example 2.46. (Independence case).   In the case of independence

$$(2.148) \quad S_{k,n}(x) = \sum_{1 \le i_1 < i_2 < \ldots < i_k \le n} P(X_{i_1} \le x, X_{i_2} \le x, \ldots, X_{i_k} \le x) = \binom{n}{k} F^k(x)$$

Thus, for the r-th order statistic (2.140) and (2.143) lead to

$$F_{X_{r:n}}(x) = P[m_n(x) \geq r] = \sum_{t=r}^{n} \sum_{i=0}^{n-t} (-1)^i \binom{i+t}{t} S_{i+t,n}$$

$$= \sum_{t=r}^{n} \sum_{i=0}^{n-t} (-1)^i \binom{i+t}{t} \binom{n}{i+t} F^{i+t}(x)$$

$$= \sum_{t=r}^{n} \binom{n}{t} F^t(x) \sum_{i=0}^{n-t} (-1)^i \binom{n-t}{i} F^i(x)$$

$$= \sum_{t=r}^{n} \binom{n}{t} F^t(x)[1 - F(x)]^{n-t}$$

which is the well known expression (2.2).   ∎

Example 2.47. (Multivariate Mardia's distribution). Assume a system made of n equal components such that their multivariate lifetimes follow the multivariate Mardia's distribution, i.e. a hazard function

$$G(x_1, x_2, \ldots, x_n) = \left[\sum_{j=1}^{n} \exp\left(\frac{x_j}{a}\right) - n + 1 \right]^{-a}$$

The hazard function of the minimum of the lifetimes of the n elements is given by

$$G_{\min}(x) = 1 - F_{\min}(x) = P[\min(X_1, X_2, \ldots, X_n) > x]$$

$$= P[X_i > x; i = 1, 2, \ldots, n] = G(x, x, \ldots, x)$$

$$= \left[n \exp\left(\frac{x}{a}\right) - n + 1 \right]^{-a}$$

Hence, the cdf of the minimum is

$$F_{\min}(x) = 1 - \left[n \exp\left(\frac{x}{a}\right) - n + 1 \right]^{-a}.$$ ∎

Example 2.48. (Marshall-Olkin model). Assume that the components of a five-component system fail after receiving a fatal shock. Independent Poisson processes affecting one, two, three, four or all five components govern the occurrence of shocks. Assume that processes affecting any given r components have an intensity λ^r. Then, the joint distribution of the lifetimes of the com-

K	X			
	0.25	0.50	1.00	2.00
0	1	1	1	1
1	2.65547	1.41031	0.39779	0.0316485
2	3.48301	1.21313	0.14717	0.0021659
3	2.62911	0.69122	0.04778	0.00022828
4	1.08980	0.23753	0.01128	0.00002547
5	0.19235	0.03700	0.00137	0.00000187

TABLE 2.6.   Values of $\bar{S}_{k,n}(x)$ for different values of k and x

ponents is given by the Marshall-Olkin model (1967a) with hazard function

$$G(x_1, x_2, \ldots, x_5) = P[X_1 > x_1, X_2 > x_2, \ldots, X_5 > x_5]$$

$$= \exp\left(-\sum_{i=1}^{5} \lambda x_i - \sum_{i<j} \lambda^2 \max(x_i, x_j) - \sum_{i<j<k} \lambda^3 \max(x_i, x_j, x_k) \right.$$

$$\left. - \sum_{i<j<k<l} \lambda^4 \max(x_i, x_j, x_k, x_l) - \lambda^5 \max(x_1, x_2, x_3, x_4, x_5) \right)$$

and the one-dimensional marginal hazard functions are given by

$$G(x) = \exp\{-(\lambda + 4\lambda^2 + 6\lambda^3 + 4\lambda^4 + \lambda^5)x\}$$

We also have, from the k-dimensional hazard marginals, that

$$\alpha_k(x) = G(x, x, \ldots, x) = \exp\left\{ \left[-\sum_{r=1}^{n} \binom{n}{r} \lambda^r - \sum_{r=1}^{n} \binom{n-k}{r} \lambda^r \right] x \right\}$$

from which we can write

$$\bar{S}_{k,n}(x) = \binom{n}{k} \alpha_k(x)$$

Table 2.6. shows the values of $\bar{S}_{k,n}(x)$ for several values of x.
From (2.140) and (2.143) we get

$$F_{X_{r:n}}(x) = \sum_{t=0}^{n-r} \sum_{i=0}^{n-t} (-1)^i \binom{i+t}{t} \binom{n}{i+t} \alpha_{i+t}(x); \qquad r = 1, 2, \ldots, 5$$

which gives the cdfs of the order statistics for this case.

Assuming $\lambda = 0.5$ and an unit period of time, the cdfs shown in Figure 2.22. were obtained.

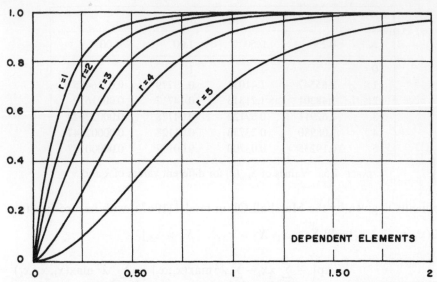

FIGURE 2.22.   Cdf of the order statistics of a sample of size $n = 5$ (Marshall-Olkin model)

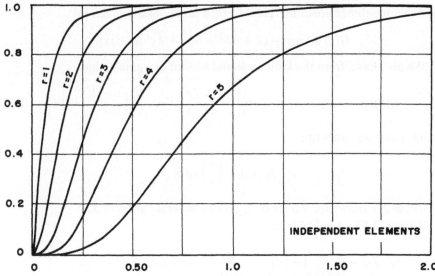

FIGURE 2.23.   Cdf of the order statistics of a sample of size $n = 5$ (Assuming independence and the same marginals as in Marshall-Olkin model)

If the elements had been treated as independent, the cdfs would have been (see expression (2.2))

$$F_{X_{r:n}}(x) = \sum_{k=r}^{n} \binom{n}{k} F^k(x)[1 - F(x)]^{n-k}$$

where $F(x)$ is the marginal distribution of the lifetime of a single element, i.e.

$$F(x) = 1 - G(x) = 1 - \exp\{(0.5 + 4 \cdot 0.5^2 + 6 \cdot 0.5^3 + 4 \cdot 0.5^4 + 0.5^5)x]$$

$$= 1 - \exp(-2.53125x)$$

Finally, figure 2.23. shows the cdfs of the order statistics in this case. Comparison of the last two figures allows studying the influence of dependence on the cdfs. ∎

For the events $\{m_n(x) \geq r\}$ Galambos (1977) proved the inequalities

$$\sum_{k=r}^{r+2u-1} (-1)^{k-r} \binom{k-1}{r-1} S_{k,n}(x) + 2u/(n-r) \binom{r+2u-1}{r-1} S_{r+2u,n}(x)$$

$$\leq P[m_n(x) \geq r] \leq \sum_{k=r}^{r+2u} (-1)^{k-r} \binom{k-1}{r-1} S_{k,n}(x)$$

(2.149)
$$- \frac{(2u+1)\binom{r+2u}{r-1} S_{r+2u+1,n}(x)}{(n-r)}$$

where u is any non-negative number such that $2 \leq 2u \leq n - r - 1$, and $m_n(x)$ is the number of C_j ($j = 1, 2, \ldots, n$) which occur in the sample.
Expression (2.149) when $r = 1$ reduces to

$$\sum_{k=1}^{2u} (-1)^{k-1} S_{k,n}(x) + \frac{2uS_{2u+1,n}(x)}{(n-1)} \leq P[m_n(x) \geq 1]$$

(2.150)
$$\leq \sum_{k=1}^{2u+1} (-1)^{k-1} S_{k,n}(x) - \frac{(2u+1)S_{2u+2,n}(x)}{(n-1)}$$

and when only $S_{1,n}(x)$ and $S_{2,n}(x)$ are known we get

(2.151)
$$S_{1,n}(x) - S_{2,n}(x) \leq P[m_n(x) \geq 1] \leq S_{1,n}(x) - \frac{S_{2,n}(x)}{n-1}$$

which is useful when only two-dimensional marginals are known.
Kwerel (1975a,b,c) and Galambos (1977b) improved one of the bounds in

(2.151) giving

(2.152) $$\frac{2S_{1,n}(x)}{k_0 + 1} - \frac{2S_{2,n}(x)}{k_0(k_0 + 1)} \le P[m_n(x) \ge 1]$$

where $k_0 = \text{int}[2S_{2,n}(x)/S_{1,n}(x)] + 1$. Sathe et al. (1980) extended to $r > 1$ this bound when

(2.153) $$2S_{2,n}(x) < (n + r - 2)S_{1,n}(x) - n(r - 1)$$

to

(2.154) $$P[m_n(x) \ge r] \ge 2\frac{(k_r - 1)[S_{1,n}(x) - r + 1] - S_{2,n}(x) + \dfrac{(r - 1)(r - 2)}{2}}{(k_r - r)(k_r - r + 1)}$$

where

(2.155) $$k_r = \text{int}\left(\frac{2S_{2,n}(x) - (r - 1)(r - 2)}{S_{1,n}(x) - r + 1}\right) + 3 - r$$

In many (practically all) occasions, the n-dimensional cumulative distribution or the hazard functions are unknown. However, one and two-dimensional marginals can be, either exactly or approximately, known. In such cases, the above bounds become very useful in order to estimate probabilities. This is specially useful when one is in need of estimating the probabilities of order statistics in the tails, because in this case some of the bounds give very good approximations. Note, however, that a partial knowledge of these marginals is needed, because the only terms involved in the bounds are the $S_{k,n}(x)$, for $k = 1, 2$.

Another bound, due to Galambos (1969), is

(2.156) $$P[m_n(x) \ge r] \ge \frac{[S_{1,n}(x) - r + 1]S_{r,n}(x)}{(r + 1)S_{r+1,n}(x) + rS_{r,n}(x)}$$

Use of expressions (2.143) and (2.144) together with the above bounds gives the bounds for the cdf of upper and lower order statistics.

Example 2.49. (Marshall and Olkin model). Assume the Marshall-Olkin model in example 2.48. Table 2.7. shows exact and upper and lower bounds for the probabilities $P[\underline{m}_n(x) \ge r]$ where $\underline{m}_n(x)$ is the number of elements in the system with a lifetime larger than x and $r = 1$ to 2. These bounds have been obtained by expressions (2.149), (2.151), (2.152), (2.154) and (2.156).

TABLE 2.7. Values of $P[\underline{m}_n(x) \geq r]$

X	r	Lower Bounds					Exact Value	Upper Bounds	
		(2.149) (u = 1)	(2.151)	(2.152)	(2.154)	(2.156)		(2.151)	(2.149) (u = 1)
0.25	1	0.48702	−0.82753	0.74723	0.74723	0.73289	0.90412	1.78472	0.98422
0.25	2	0.40439			0.52315	0.38819	0.72479		0.72480
0.50	1	0.54278	0.19717	0.53583	0.53583	0.51842	0.68786	1.1070	0.71025
0.50	2	0.30575				0.11061	0.39530		0.39530
1.00	1	0.27451	0.25062	0.25062	0.25062	0.22862	0.28849	0.36100	0.28994
1.00	2	0.07418				−0.2024	0.0800		0.08000
2.00	1	0.02959	0.02948	0.02948	0.02948	0.02783	0.02968	0.03110	0.02969
2.00	2	0.00176				−0.41807	0.00177		0.00177

Note that in some cases upper or lower bounds can lie outside the range $[0, 1]$, and they become useless. Note also that the bounds improve for decreasing values of $P[\underline{m}_n(x) \geq r]$. In particular, it is interesting to note the good bounds for large values of x obtained by means of expressions (2.151), (2.152) and (2.154), which are only based on bivariate marginals. This is a very common stituation in applications.

III   Asymptotic Distribution of Sequences of Independent Random Variables

Chapter 3

Asymptotic Distributions of Maxima and Minima (I.I.D. Case)

3.1. Introduction and Motivation

In chapter 2, the pdf and cdf of maxima and minima for the independent case was derived. A first look at the problem of extremes can lead to the conclusion that this chapter gives such powerful tools that practically any problem can be solved. In fact, the parent distribution and the sample size, if they are known, provide sufficient tools for solving most practical problems, because substitution of $F(x)$ and n in expressions (2.6), (2.7), (2.8) or (2.9) gives the desired cdf or pdf for maxima and minima, respectively. However, there are still some important cases where these expressions become insufficient, as when one of the following conditions holds:

(i) The sample size goes to infinity.
(ii) The cdf, $F(x)$, of the parent distribution is unknown.
(iii) The sample size is unknown.

The first case is important in order to derive exact and approximate models. The key question in relation to this case is: Is there any limit distribution or family of distributions for maxima and minima? If the answer is "yes", many other questions arise such as: How many limit distributions or distribution families exist? Which are they? Can two different parent cdfs lead to the same

97

limit distribution? If so, how can we determine which limit distribution is associated with a given parent distribution?

Case (ii) is very frequently encountered in practice, where some sample is known, but the parent distribution is unknown. Note that in this case expressions (2.6) to (2.9) become useless and an alternative method of solution to the problem under consideration must be applied. Some typical questions to be raised in this situations are: What methods are applicable when $F(x)$ is unknown? What is the reliability of these methods?, etc.

Finally, in case (iii) where the sample size n, is unknown, the value of the contents of chapter 2 is considerably limited. It is worthwhile mentioning that the frequency of this situation is greater that may appear at first glance because n may be unknown due to direct or indirect causes. For example, a random variable may be registered only when some threshold value, X_0, is exceeded. When this happens, the size of the original sample is unknown and if X_0 is also unknown, the censored cdf

$$(3.1) \qquad\qquad G(x) = \frac{F(x) - F(x_0)}{1 - F(x_0)}$$

cannot be determined.

The questions associated with this case are similar to these: What is to be done when n is unknown? How important is the ignorance of n?, etc.

These and other questions of practical relevance will be answered in this and the following chapters.

Finally, in this book we refer to continuous limit distributions. Thus, the weak convergence (convergence in law) will be substituted by pointwise convergence.

3.2. Statement of the Problem

Let us begin this paragraph by defining the end points of a cdf $F(x)$.

Definition 3.1. (*Lower and upper end points of a distribution*).   The lower end point, $\alpha(F)$, of the cdf $F(x)$ is defined by

$$(3.2) \qquad\qquad \alpha(F) = \inf\{x: F(x) > 0\}$$

Similarly, the upper end point, $\omega(F)$, is defined by

$$(3.3) \qquad\qquad \omega(F) = \sup\{x: F(x) < 1\} \qquad\qquad\blacksquare$$

From a practical point of view, the end points of a cdf are the minimum and maximum values that the associated random variable can have. If the random variable is unbounded on one or both tails they become $+\infty$ or $-\infty$.

As shown in chapter 2, the cdf of the maximum, Z_n, of a sample of size n from a parent distribution with cdf $F(x)$ is

(3.4) $$H_n(x) = \text{Prob}[Z_n \le x] = F^n(x)$$

Similarly, the minimum, W_n, has cdf

(3.5) $$L_n(x) = \text{Prob}[W_n \le x] = 1 - [1 - F(x)]^n$$

The structure of these two functions shows that the percentiles of the maximum and minimum move to the right and left, respectively, with increasing n, approaching the upper and lower end points, or going to $+\infty$ or $-\infty$ if they become unbounded.

When n goes to infinity we have

(3.6) $$\lim_{n \to \infty} F^n(x) = \begin{cases} 1 & \text{if } F(x) = 1 \\ 0 & \text{if } F(x) < 1 \end{cases}$$

(3.7) $$\lim_{n \to \infty} 1 - [1 - F(x)]^n = \begin{cases} 0 & \text{if } F(x) = 0 \\ 1 & \text{if } F(x) \le 1 \end{cases}$$

This means that the limit distribution takes only values 0 and 1, i.e. it degenerates.

In order to avoid this degeneration we look for linear transformations

$$Y = a_n + b_n x$$

where a_n and b_n are constants depending on n and such that the limit distributions

(3.8) $$\lim_{n \to \infty} H_n(a_n + b_n x) = \lim_{n \to \infty} F^n(a_n + b_n x) = H(x); \qquad \text{for all } x$$

(3.9) $$\lim_{n \to \infty} L_n(c_n + d_n x) = \lim_{n \to \infty} 1 - [1 - F(c_n + d_n x)]^n = L(x); \qquad \text{for all } x$$

become non-degenerated. In other words, we try to follow $H_n(x)$ and $L_n(x)$ in such a manner that degeneration does not occur.

Note that we assume that $H(x)$ and $L(x)$ are continuous and thus, (3.8) and (3.9) are equivalent to weak convergence (convergence in law).

Definition 3.2. (*Domain of attraction of a distribution*). A given cdf, $F(x)$, is said to belong to, or lie in, the domain of attraction for maxima of a given cdf, $H(x)$, when it satisfies (3.8) for given sequences $\{a_n\}$ and $\{b_n > 0\}$. Analogously, when $F(x)$ satisfies (3.9) we say that it belongs to the domain of attraction for minima of $L(x)$.   ∎

At this point, the problems of limit distributions can be stated as:

(a) Find conditions under which (3.8) and (3.9) hold.
(b) Give rules to construct the sequences a_n and b_n.
(c) Find which cdfs can occur as $H(x)$ and $L(x)$.

3.3.   Limit Distributions and Domains of Attraction

The answer to the problems stated above is given by the following theorems, that we give without proof (the interested reader is referred to Fisher and Tippett (1928) or Galambos (1978)).

Theorem 3.1. (Feasible limit distributions for maxima).   *The only three types of non-degenerated distributions, $H(x)$, satisfying (3.8) are*

(3.10)   FRECHET: $H_{1,\gamma}(x) = \begin{cases} \exp(-x^{-\gamma}) & \text{if } x > 0 \\ 0 & \text{otherwise} \end{cases}$

(3.11)   WEIBULL: $H_{2,\gamma}(x) = \begin{cases} 1 & \text{if } x \geq 0 \\ \exp(-(-x)^{\gamma}) & \text{otherwise} \end{cases}$

(3.12)   GUMBEL: $H_{3,0}(x) = \exp[-\exp(-x)]$ $-\infty < x < \infty$

Theorem 3.2. (Feasible limit distributions for minima).   *The only three types of non-degenerated distributions, $L(x)$, satisfying (3.9) are*

(3.13)   FRECHET: $L_{1,\gamma}(x) = \begin{cases} 1 - \exp[-(-x)^{-\gamma}] & \text{if } x < 0 \\ 1 & \text{otherwise} \end{cases}$

(3.14)   WEIBULL: $L_{2,\gamma}(x) = \begin{cases} 1 - \exp(-x^{\gamma}) & \text{if } x > 0 \\ 0 & \text{otherwise} \end{cases}$

(3.15)   GUMBEL: $L_{3,0}(x) = 1 - \exp[-\exp(x)]$ $-\infty < x < \infty$

Theorem 3.3. (Domain of attraction for maxima of a given distribution).   *The distribution $F(x)$ lies in the domain of attraction for maxima of*

(i) $H_{1,\gamma}(x)$ if, and only if, $\omega(F) = \infty$ and

(3.16) $$\lim_{t \to \infty} \frac{1 - F(tx)}{1 - F(t)} = x^{-\gamma}; \qquad \gamma > 0$$

(ii) $H_{2,\gamma}(x)$ *if, and only if, $\omega(F) < \infty$ and the function*

$$(3.17)\qquad F^*(x) = F\left(\omega(F) - \frac{1}{x}\right); \qquad x > 0$$

satisfies (3.16).

(iii) $H_{3,0}(x)$ *if, and only if,*

$$(3.18)\qquad \lim_{n\to\infty} n\{1 - F[X_{1-1/n} + x(X_{1-1/(ne)} - X_{1-1/n})]\} = \exp(-x)$$

where X_α is the 100α percentile of $F(x)$.

   The normalizing constants a_n and b_n can be chosen as

(i) (3.19) $\qquad\qquad a_n = 0; \qquad b_n = F^{-1}\left(1 - \frac{1}{n}\right)$

(ii) (3.20) $\qquad a_n = \omega(F); \qquad b_n = \omega(F) - F^{-1}\left(1 - \frac{1}{n}\right)$

(iii) $\qquad\qquad\qquad a_n = F^{-1}\left(1 - \frac{1}{n}\right)$

(3.21)
$$b_n = [1 - F(a_n)]^{-1} \int_{a_n}^{\omega(F)} [1 - F(y)]\, dy$$

or

(3.21a) $\qquad\qquad\qquad b_n = F^{-1}\left(1 - \frac{1}{ne}\right) - a_n$

Note that one way of obtaining normalizing constants is by equating two percentiles of $F^n(a_n + b_n x)$ and $H(x)$.

Theorem 3.4. (Domain of attraction for minima of a given distribution). *The distribution $F(x)$ lies in the domain of attraction for minima of*

(i) $L_{1,\gamma}(x)$ *if, and only if, $\alpha(F) = -\infty$ and*

$$(3.22)\qquad\qquad \lim_{t\to-\infty} \frac{F(tx)}{F(t)} = x^{-\gamma}; \qquad \gamma > 0$$

(ii) $L_{2,\gamma}(x)$ *if, and only if, $\alpha(F) > -\infty$ and the function*

$$(3.23)\qquad\qquad F^*(x) = F\left(\alpha(F) - \frac{1}{x}\right); \qquad x < 0$$

satisfies (3.22).

(iii) $L_{3,0}(x)$ *if, and only if,*

(3.24) $$\lim_{n \to \infty} n\{F[X_{1/n} + x(X_{1/(ne)} - X_{1/n})]\} = \exp(-x)$$

The normalizing constants c_n and d_n can be chosen as

(i) (3.25) $$c_n = 0; \qquad d_n = \left| F^{-1}\left(\frac{1}{n}\right) \right|$$

(ii) (3.26) $$c_n = \alpha(F); \qquad d_n = F^{-1}\left(\frac{1}{n}\right) - \alpha(F)$$

(iii) (3.27) $$c_n = F^{-1}\left(\frac{1}{n}\right); \qquad d_n = F(c_n)^{-1} \int_{\alpha(F)}^{c_n} F(y)\, dy$$

or

(3.27a) $$d_n = c_n - F^{-1}\left(\frac{1}{ne}\right)$$

Some very important practical implications of the above theorems are:

(a) Only three distributions (Frechet, Weibull and Gumbel) can occur as limit distributions for maxima or minima of independent trials.
(b) Rules for determining if any given $F(x)$ lies in the domain of attraction of any of those distributions are given. Although there exist parent distributions such that the limit does not exist, for most continuous distributions in practice one of them actually occurs.
(c) Rules for determining the sequences a_n and b_n or c_n and d_n ($i = 1, 2, \ldots$) are given.
(d) A parent distribution with non-finite end-point in the tail of interest cannot lie in a Weibull type domain of attraction.
(e) A parent distribution with finite end-point in the tail of interest cannot lie in a Frechet type domain of attraction.

Note that the fact that the parent distribution has a finite end in the tail of interest is not enough to guarantee that it lies in the domain of attraction of the Weibull distribution. In fact the lognormal distribution which is limited in the left tail lies in the domain of attraction for minima of the Gumbel distribution in that tail. A similar conclusion can be obtained for the cdf

$$F(x) = \begin{cases} 0 & \text{if } x < 0 \\ \exp\left(\dfrac{-1}{x^2}\right) & \text{if } x \geq 0 \end{cases}$$

whose lower end is $\alpha(F) = 0$. The reader can check this statement by means of theorems 3.4, 3.7 or 3.9.

Expressions (3.8) and (3.9) together with the above theorems allow, for large enough n, substitution of $F''(a_n + b_n x)$ by $H(x)$ of $F''(x)$ by $H((x - a_n)/b_n)$ or, what is equivalent (see section 3.10), for large enough x, substitution of $F(x)$ by $H^{1/n}[(x - a_n)/b_n]$. The practical importance of this substitution is that for any continuous distribution with cdf, $F(x)$, only three families are possible. Thus, for extremes, the infinitely many degrees of freedom we had with the parent distribution have been reduced now to three parametric families with parameters a and b, associated with a_n and b_n. Similar arguments are valid for $L(x)$.

Example 3.1. (Cauchy distribution).   For the Cauchy distribution with cdf.

$$F(x) = \frac{1}{2} + \frac{\arctan(x)}{\pi}; \qquad -\infty < x < \infty$$

expression (3.22) gives

$$\lim_{t \to -\infty} \frac{F(tx)}{f(t)} = \lim_{t \to -\infty} \frac{\dfrac{1}{2} + \dfrac{\arctan(tx)}{\pi}}{\dfrac{1}{2} + \dfrac{\arctan(t)}{\pi}}$$

$$= \lim_{t \to -\infty} \frac{\dfrac{x}{1 + (tx)^2}}{\dfrac{1}{1 + t^2}}$$

$$= \lim_{t \to -\infty} \frac{x(1 + t^2)}{1 + t^2 x^2} = x^{-1}$$

Consequently, $\gamma = 1$ and we can say that it lies in the domain of attraction for minima of the Frechet distribution.

The normalizing constants, according (3.25) are

$$c_n = 0$$

$$d_n = \tan\left[\pi \left(\frac{1}{2} - \frac{1}{n} \right) \right]$$

By symmetry, we find that it also belongs to the domain of attraction for maxima of the Frechet distribution and the normalizing constants are (see (3.19))

$$a_n = 0$$

$$b_n = \tan\left[\pi\left(\frac{1}{2} - \frac{1}{n}\right)\right]$$ ∎

Example 3.2. (Uniform distribution). For the uniform distribution $\omega(F) = 1$ and expression (3.17) gives

$$F^*(x) = 1 - \frac{1}{x}$$

thus, (3.16) becomes

$$\lim_{t \to \infty} \frac{1 - F^*(tx)}{1 - F^*(t)} = \lim_{t \to \infty} \frac{\frac{1}{tx}}{\frac{1}{t}} = x^{-1}$$

Thus, the uniform distribution belongs to the domain of attraction for maxima of the Weibull distribution $H_{2,1}(x)$.

According to (3.20), the normalizing constants can be chosen as

$$a_n = \omega(F) = 1$$

$$b_n = \omega(F) - F^{-1}\left(1 - \frac{1}{n}\right)$$

$$= 1 - 1 + \frac{1}{n} = \frac{1}{n}$$

By symmetry, it also belongs to the domain of attraction for minima of the Weibull distribution and

$$c_n = 0$$

$$d_n = \frac{1}{n}$$ ∎

Example 3.3. (Exponential distribution). The exponential distribution has a lower end-point $\alpha(F) = 0$. Thus, function $F^*(x)$ becomes

$$F^*(x) = 1 - \exp\left(\frac{1}{x}\right)$$

and (3.22) gives

$$\lim_{t \to -\infty} \frac{F^*(tx)}{F^*(t)} = \lim_{t \to -\infty} \frac{1 - \exp\left(\dfrac{1}{tx}\right)}{1 - \exp\left(\dfrac{1}{t}\right)} = \lim_{t \to -\infty} \frac{\exp\left(\dfrac{1}{tx}\right)\left(\dfrac{1}{t^2}\right)\left(\dfrac{1}{x}\right)}{\exp\left(\dfrac{1}{t}\right)\left(\dfrac{1}{t^2}\right)} = x^{-1}$$

Consequently, the exponential distribution belongs to the domain of attraction for minima of the Weibull distribution.

The normalizing constants are (see (3.26))

$$c_n = \alpha(F) = 0$$

$$d_n = F^{-1}\left(\frac{1}{n}\right) - \alpha(F) = -\log\left(1 - \frac{1}{n}\right) \approx \frac{1}{n} \qquad \blacksquare$$

Example 3.4. (Rayleigh distribution).   The cdf of the Rayleigh distribution, frequently used for wave heights, is

$$F(x) = 1 - \exp\left(-\frac{x^2}{a^2}\right)$$

Thus, the percentiles in expression (3.18) become

$$X_{1 - 1/n} = \left[-a^2 \log\left(\frac{1}{n}\right)\right]^{1/2} = [a^2 \log(n)]^{1/2}$$

$$X_{1 - 1/(ne)} = [a^2 \log(ne)]^{1/2} = [a^2 \log(n) + a^2]^{1/2}$$

and expression (3.18) can be written

$$\lim_{n \to \infty} n\left[\exp\left(-\frac{[(a^2 \log n)^{1/2} + x[(a^2 \log n + a^2)^{1/2} - (a^2 \log n)^{1/2}]]^2}{a^2}\right)\right]$$

$$= \lim_{n \to \infty} n\{\exp\{-[(\log n)^{1/2} + x[(\log n + 1)^{1/2} - (\log n)^{1/2}]]^2\}\}$$

$$= \lim_{n \to \infty} n\left\{\exp\left\{-\log n\left[1 - x + x\left(\frac{\log n + 1}{\log n}\right)^{1/2}\right]^2\right\}\right\}$$

$$= \lim_{n \to \infty} n\left\{\exp\left[-\log n\left(1 + \frac{x}{2\log n}\right)^2\right]\right\}$$

$$= \lim_{n \to \infty} n\left\{\exp\left[-\log n\left(1 + \frac{x^2}{4\log^2 n} + \frac{x}{\log n}\right)\right]\right\} \approx \exp(-x)$$

This proves that the Rayleigh distribution belongs to the domain of attraction for maxima of the Gumbel distribution.

The normalizing constants, according to (3.21), can be chosen as

$$a_n = F^{-1}\left(1 - \frac{1}{n}\right) = \left[-a^2 \log\left(\frac{1}{n}\right)\right]^{1/2} = a\sqrt{\log n}$$

$$b_n = \left[1 - 1 + \exp\left(\frac{-a^2 \log n}{a^2}\right)\right]^{-1} \int_{a\sqrt{\log n}}^{\infty} \exp\left(\frac{-y^2}{a^2}\right) dy$$

$$= \frac{na}{\sqrt{2}} \int_{\sqrt{2 \log n}}^{\infty} \exp\left(\frac{-u^2}{2}\right) du$$

$$= na\sqrt{\pi}[1 - \Phi(\sqrt{2 \log n})]$$

$$= \frac{na\sqrt{\pi} \exp(-\log n)}{\sqrt{2 \log n} \sqrt{2\pi}} = \frac{a}{2\sqrt{\log n}}$$

where $\Phi(x)$ is the cdf of the standard normal distribution, which has been approximated using the relation

$$1 - \Phi(x) \approx \frac{\exp\left(\frac{-x^2}{2}\right)}{x\sqrt{2\pi}}$$

b_n could also be obtained by means of (3.21a)

$$b_n = F^{-1}\left(1 - \frac{1}{ne}\right) - a_n = \left[-a^2 \log\left(\frac{1}{ne}\right)\right]^{1/2} - a_n$$

$$= a\sqrt{\log(ne)} - a\sqrt{\log n}$$

$$= a\sqrt{\log(n)}[\sqrt{(1 + 1/\log n)} - 1]$$

$$= \frac{a\sqrt{\log n}}{2 \log n} = \frac{a}{2\sqrt{\log n}}$$

Another way of obtaining some normalizing constants is by means of expression (3.8). For x large we have

$$F(x) = 1 - \exp\left(\frac{-x^2}{a^2}\right)$$

$$\approx \exp\left[-\exp\left(\frac{-x^2}{a^2}\right)\right]$$

and taking into account (3.8) we get

$$F^n(a_n + b_n x) = \exp\left[-n \exp\left(\frac{-(a_n + b_n x)^2}{a^2}\right)\right]$$

$$= \exp\left[-\exp\left(\log n - \frac{a_n^2 + b_n^2 x^2 + 2a_n b_n x}{a^2}\right)\right]$$

$$\approx \exp[-\exp(-x)]$$

Now, by identifying coefficients of x^2, x and the constant term we get

$$a^2 \log n - a_n^2 = 0$$

$$\lim_{n \to \infty} b_n^2 x^2 = 0$$

$$\frac{2a_n b_n}{a^2} = 1$$

and then

$$a_n = a\sqrt{\log n}$$

$$b_n = \frac{a}{2\sqrt{\log n}} \qquad\qquad \blacksquare$$

Example 3.5. (Exponential distribution).   For the exponential distribution with cdf

$$F(x) = 1 - \exp\left\{\frac{-x}{a}\right\}; \qquad x > 0$$

the percentiles in (3.18) are

$$X_{1-1/n} = -a \log\left(\frac{1}{n}\right) = a \log(n)$$

$$X_{1-1/(ne)} = -a \log\left(\frac{1}{ne}\right) = a \log(ne)$$

and expression (3.18) gives

$$\lim_{n \to \infty} n\left[\exp\left(-\frac{[a \log(n) + x(a \log(ne) - a \log(n))]}{a}\right)\right]$$

$$= \lim_{n \to \infty} n\{\exp[-(\log n - x)]\} = \lim_{n \to \infty} \frac{n \exp(-x)}{n} = \exp(-x)$$

Thus, it belongs to the domain of attraction for maxima of the Gumbel distribution.

A possible choice of the normalizing constants is given by (3.21)

$$a_n = F^{-1}\left(1 - \frac{1}{n}\right) = a \log n$$

$$b_n = [1 - F(a_n)]^{-1} \int_{a_n}^{\omega(F)} [1 - F(y)] \, dy = \left[\exp\left(\frac{-a \log n}{a}\right)\right]^{-1}$$

$$\cdot \int_{a \log n}^{\infty} \exp\left(\frac{-y}{a}\right) dy = n\left[-a \exp\left(\frac{-y}{a}\right)\right]\Big|_{a \log n}^{\infty} = a$$

or from (3.21a)

$$b_n = F^{-1}\left(1 - \frac{1}{ne}\right) - F^{-1}\left(1 - \frac{1}{n}\right) = X_{1 - 1/(ne)} - X_{1 - 1/n}$$

$$= a \log n - a \log(ne) = a \qquad\qquad \blacksquare$$

3.4.   Von-Mises Forms

The three limit distributions (3.10)–(3.12), when the above parameters, a_n and b_n, are passed to the right hand side of (3.8) can be included in the single analytical expression

$$(3.28) \quad H_c(x; \lambda, \delta) = \exp\left\{-\left[1 + c\left(\frac{x - \lambda}{\delta}\right)\right]^{-1/c}\right\}; \quad 1 + c\left(\frac{x - \lambda}{\delta}\right) \geq 0$$

which is called the Von-Mises form.

For $c > 0$, $c < 0$ or $c = 0$ we get the Frechet, Weibull and Gumbel families, respectively.

Note that for $c = 0$ (3.28) must be interpreted in a limit sense, i.e. for $c = 0$

$$H_0(x; \lambda, \delta) = \exp\left[-\exp\left(\frac{-(x - \lambda)}{\delta}\right)\right]$$

Similarly, the three limit distributions (3.13)–(3.15) can be included in the single Von-Mises form

$$(3.29) \quad L_c(x; \lambda, \delta) = 1 - \exp\left\{-\left[1 + c\left(\frac{\lambda - x}{\delta}\right)\right]^{-1/c}\right\}; \quad 1 + c\left(\frac{\lambda - x}{\delta}\right) \geq 0$$

where for $c > 0$, $c < 0$ and $c = 0$ we get the Frechet, Weibull and Gumbel families, respectively.

For $c = 0$ we get

$$L_0(x; \lambda, \delta) = 1 - \exp\left[-\exp\left(\frac{-(\lambda - x)}{\delta}\right)\right]$$

In the following section $H_c(x)$ and $L_c(x)$ will be used instead of $H_c(x; \lambda, \delta)$ and $L_c(x; \lambda, \delta)$, for simplicity.

3.5. Normalizing Constants

In section 3.3 we gave specific choices for the normalizing sequences of constants a_n and b_n. However, these sequences are not unique and many different choices can be made. In order to demonstrate the degree of freedom we have in this choice we give, without proof the following theorem from Galambos (1978)

Theorem 3.5. (Feasible sequences of normalizing constants). *If a_n and b_n are sequences that satisfy (3.8) and a_n^* and b_n^* are sequences such that*

$$(3.30) \qquad \lim_{n \to \infty} \frac{a_n - a_n^*}{b_n} = 0$$

$$(3.31) \qquad \lim_{n \to \infty} \frac{b_n^*}{b_n} = 1$$

then, (3.8) holds when a_n^* and b_n^* are substituted by a_n and b_n, respectively. ∎

An identical theorem can be stated for c_n and d_n. Note that in example 3.3, $d_n = -\log(1 - 1/n)$ was substituted by $d_n^* = 1/n$.

This freedom in the choice of the normalizing constants allows some discussion about which choice is the best according to different criteria, as speed of convergence for example; however, we do not discuss this topic in the present book.

3.6. Domain of Attraction of a Given Distribution

Theorems 3.3 and 3.4 in section 3.3 are for determining the domain of attraction of maxima and minima, respectively, of a given cdf, $F(x)$.

In this section we give two additional theorems which give different criteria for identifying the domain of attraction of a given distribution. The main advantage of these theorems is that a single rule, instead of the three different rules appearing in theorems 3.3 and 3.4, is valid for Gumbel, Weibull or Frechet type domains of attraction.

Theorem 3.6. (Domain of attraction for maxima of a given distribution).   *Let* p_1, p_2, p_3 *and* p_4 *be four real numbers in the interval* $(0, 1)$. *A necessary and sufficient condition for a continuous cdf,* $F(x)$, *to belong to the domain of attraction for maxima of* $H_c(x)$ *(see* (3.28)) *is given by*

$$(3.32) \qquad \lim_{n \to \infty} \frac{F^{-1}(p_1^{1/n}) - F^{-1}(p_2^{1/n})}{F^{-1}(p_3^{1/n}) - F^{-1}(p_4^{1/n})} = \frac{(-\log p_1)^{-c} - (-\log p_2)^{-c}}{(-\log p_3)^{-c} - (-\log p_4)^{-c}}$$

Note that (3.32) includes all three limit distributions together because of the Von-Mises form, i.e. if $c = 0$, the domain of attraction is the Gumbel type, if $c > 0$ is the Frechet type and if $c < 0$ it is the Weibull type.

Note also that when n goes to infinity, $p_i^{1/n}(i = 1, 2, 3, 4)$ goes to 1 and this means that all terms in expression (3.32) belong to the right tail of $F(x)$. Thus, the limit in (3.32) is only dependent on the right tail properties.

Finally, we point out that when $c \to 0$, the right hand side of (3.32) becomes

$$\frac{\log(-\log p_1) - \log(-\log p_2)}{\log(-\log p_3) - \log(-\log p_4)}$$

Because (3.32) is true for any set of values p_1, p_2, p_3 and p_4 they can be selected and the function in the right hand side of (3.32) becomes a function of c, $Q(c; p_1, p_2, p_3, p_4)$ say. Thus, for a given distribution $F(x)$, the limit, a, in the left hand side can be calculated and c can be obtained by means of

$$(3.33) \qquad c = Q^{-1}(a; p_1, p_2, p_3, p_4)$$

The corresponding theorem for minima is the following

Theorem 3.7. (Domain of attraction for minima of a given distribution).   *Let* p_1, p_2, p_3 *and* p_4 *be four real numbers in the interval* $(0, 1)$. *A necessary and sufficient condition for* $F(x)$ *to belong to the domain of attraction for minima of* $L_c(x)$ *(see* (3.29)) *is given by*

$$(3.34) \qquad \lim_{n \to \infty} \frac{F^{-1}(1 - p_1^{1/n}) - F^{-1}(1 - p_2^{1/n})}{F^{-1}(1 - p_3^{1/n}) - F^{-1}(1 - p_4^{1/n})} = \frac{(-\log p_1)^{-c} - (-\log p_2)^{-c}}{(-\log p_3)^{-c} - (-\log p_4)^{-c}}$$

Note that now, when n goes to infinity, $(1 - p_i^{1/n}) (i = 1, 2, 3, 4)$ goes to zero and all terms in expression (3.34) belong to the left tail of $F(x)$. Thus, (3.34) depends only on the left tail properties of $F(x)$.

Example 3.6. (Uniform distribution). If we apply theorem 3.6 to the uniform distribution we get

$$\lim_{n \to \infty} \frac{F^{-1}(p_1^{1/n}) - F^{-1}(p_2^{1/n})}{F^{-1}(p_3^{1/n}) - F^{-1}(p_4^{1/n})} = \lim_{n \to \infty} \frac{p_1^{1/n} - p_2^{1/n}}{p_3^{1/n} - p_4^{1/n}}$$

$$= \lim_{n \to \infty} \frac{p_1^{1/n} \log p_1 - p_2^{1/n} \log p_2}{p_3^{1/n} \log p_3 - p_4^{1/n} \log p_4}$$

$$= \frac{\log p_1 - \log p_2}{\log p_3 - \log p_4}$$

which is the right hand side of expression (3.32) for $c = -1$. Consequently, the uniform distribution belongs to the domain of attraction for maxima of $H_{2,1}(x)$. ∎

Example 3.7. (Exponential distribution). For the exponential distribution

$$F^{-1}(u) = -\log(1 - u)$$

and expression (3.32) gives

$$\lim_{n \to \infty} = \frac{-\log(1 - p_1^{1/n}) + \log(1 - p_2^{1/n})}{-\log(1 - p_3^{1/n}) + \log(1 - p_4^{1/n})}$$

$$= \frac{\log\left(\dfrac{1 - p_2^{1/n}}{1 - p_1^{1/n}}\right)}{\log\left(\dfrac{1 - p_4^{1/n}}{1 - p_3^{1/n}}\right)}$$

$$= \frac{\log(-\log p_1) - \log(-\log p_2)}{\log(-\log p_3) - \log(-\log p_4)}$$

which corresponds to $c = 0$. Then, it belongs to the domain of attraction for maxima of the Gumbel distribution. ∎

Example 3.8. (Uniform distribution). If we apply theorem 3.7 to the uniform distribution we get

$$\lim_{n \to \infty} \frac{p_2^{1/n} - p_1^{1/n}}{p_4^{1/n} - p_3^{1/n}} = \frac{\log p_2 - \log p_1}{\log p_4 - \log p_3}$$

which is the right hand side of (3.34) for $c = -1$. Thus, the uniform distribution belongs to the domain of attraction for minima of the Weibull distribution. ∎

Example 3.9. (Rayleigh distribution).   For the Rayleigh distribution

$$F(x) = 1 - \exp\left(-\frac{x^2}{a^2}\right)$$

we have

$$F^{-1}(p) = a[-\log(1 - p)]^{1/2}$$

and substitution in (3.34) gives

$$\lim_{n \to \infty} \frac{a[-\log p_1^{1/n}]^{1/2} - a[-\log p_2^{1/n}]^{1/2}}{a[-\log p_3^{1/n}]^{1/2} - a[-\log p_4^{1/n}]^{1/2}} = \frac{[-\log p_1]^{1/2} - [-\log p_2]^{1/2}}{[-\log p_3]^{1/2} - [-\log p_4]^{1/2}}$$

from which $c = -1/2$, and then the Rayleigh distribution belongs to the domain of attraction for minima of the Weibull distribution.   ∎

Example 3.10. (Pareto distribution).   For the Pareto distribution with cdf

$$F(x) = 1 - x^{-\beta}; \qquad x \geq 1; \qquad \beta > 0$$

we have

$$F^{-1}(u) = (1 - u)^{-1/\beta}$$

and substitution in (3.34) leads to

$$\lim_{n \to \infty} \frac{p_1^{-1/(\beta n)} - p_2^{-1/(\beta n)}}{p_3^{-1/(\beta n)} - p_4^{-1/(\beta n)}} = \frac{\log p_1 - \log p_2}{\log p_3 - \log p_4}$$

$$= \frac{(-\log p_1) - (-\log p_2)}{(-\log p_3) - (-\log p_4)}$$

from which $c = -1$. Consequently, the Pareto distribution lies in the domain of attraction for minima of $L_{2,1}(x)$.   ∎

    The following theorem is inspired by Pickands (1975) and gives an alternative to theorem 3.6. In general, the practical application of this theorem becomes simpler that the associated with theorem 3.6, as a comparison of examples 3.6 to 3.10 and 3.11 to 3.18 shows.

Theorem 3.8. (Domain of attraction for maxima of a given distribution).   *A necessary and sufficient condition for a continuous cdf, $F(x)$, to belong to the* *domain of attraction for maxima of* $H_c(x)$ *is that*

$$(3.35) \qquad \lim_{\varepsilon \to 0} \frac{F^{-1}(1 - \varepsilon) - F^{-1}(1 - 2\varepsilon)}{F^{-1}(1 - 2\varepsilon) - F^{-1}(1 - 4\varepsilon)} = 2^c$$

This implies that

if $c < 0$ *then* $F(x)$ *belongs to a Weibull type domain of attraction,*
if $c = 0$ *then* $F(x)$ *belongs to a Gumbel type domain of attraction,*
if $c > 0$ *then* $F(x)$ *belongs to a Frechet type domain of attraction.*

Example 3.11. (Uniform distribution).   For the uniform distribution we have

$$F^{-1}(\alpha) = \alpha$$

and then (3.35) gives

$$\lim_{\varepsilon \to 0} \frac{1 - \varepsilon - 1 + 2\varepsilon}{1 - 2\varepsilon - 1 + 4\varepsilon} = 2^{-1}$$

Then, $c = -1$ and it belongs to the domain of attraction for maxima of the Weibull distribution.   ■

Example 3.12. (Exponential distribution).   For the exponential distribution

$$F^{-1}(\alpha) = -\log(1 - \alpha)$$

and then

$$\lim_{\varepsilon \to 0} \frac{-\log(1 - (1 - \varepsilon)) + \log(1 - (1 - 2\varepsilon))}{-\log(1 - (1 - 2\varepsilon)) + \log(1 - (1 - 4\varepsilon))}$$

$$= \lim_{\varepsilon \to 0} \frac{-\log(\varepsilon) + \log 2 + \log(\varepsilon)}{-\log 2 - \log(\varepsilon) + \log 4 + \log(\varepsilon)}$$

$$= 1 = 2^0$$

Then $c = 0$ and the domain of attraction for maxima is Gumbel type.   ■

Example 3.13. (Cauchy distribution).   For the Cauchy distribution

$$F^{-1}(p) = \text{tg}\left[\left(p - \frac{1}{2}\right)\pi\right]$$

and then

$$\lim_{\varepsilon \to 0} \frac{\text{tg}\left[\left(1 - \varepsilon - \frac{1}{2}\right)\pi\right] - \text{tg}\left[\left(1 - 2\varepsilon - \frac{1}{2}\right)\pi\right]}{\text{tg}\left[\left(1 - 2\varepsilon - \frac{1}{2}\right)\pi\right] - \text{tg}\left[\left(1 - 4\varepsilon - \frac{1}{2}\right)\pi\right]}$$

$$= \lim_{\varepsilon \to 0} \frac{\text{tg}\left[\left(\frac{1}{2} - \varepsilon\right)\pi\right] - \text{tg}\left[\left(\frac{1}{2} - 2\varepsilon\right)\pi\right]}{\text{tg}\left[\left(\frac{1}{2} - 2\varepsilon\right)\pi\right] - \text{tg}\left[\left(\frac{1}{2} - 4\varepsilon\right)\pi\right]} = 2^1$$

Then, $c = 1$ and the domain of attraction for maxima is the Frechet type.

∎

The theorem corresponding to 3.8 for the left tail is the following.

Theorem 3.9. (Domain of attraction for minima of a given distribution). *A necessary and sufficient condition for a continuous cdf, $F(x)$, to lie in the domain of attraction for minima of $L_c(x)$ is that*

$$(3.36) \qquad \lim_{\varepsilon \to 0} \frac{F^{-1}(\varepsilon) - F^{-1}(2\varepsilon)}{F^{-1}(2\varepsilon) - F^{-1}(4\varepsilon)} = 2^c$$

This means that

if $c < 0$ then $F(x)$ belongs to a Weibull type domain of attraction,
if $c = 0$ then $F(x)$ belongs to a Gumbel type domain of attraction,
if $c > 0$ then $F(x)$ belongs to a Frechet type domain of attraction.

Example 3.14. (Exponential distribution). For the exponential distribution expression (3.36) becomes

$$\lim_{\varepsilon \to 0} \frac{-\log(1 - \varepsilon) + \log(1 - 2\varepsilon)}{-\log(1 - 2\varepsilon) + \log(1 - 4\varepsilon)}$$

$$= \lim_{\varepsilon \to 0} \frac{\varepsilon - 2\varepsilon}{2\varepsilon - 4\varepsilon} = 2^{-1}$$

Then, $c = -1$ and the domain of attraction for minima is the Weibull type.

∎

Example 3.15. (Cauchy distribution). For the Cauchy distribution, expression (3.36) gives

$$\lim_{\varepsilon \to 0} \frac{\text{tg}\left[\left(\varepsilon - \frac{1}{2}\right)\pi\right] - \text{tg}\left[\left(2\varepsilon - \frac{1}{2}\right)\pi\right]}{\text{tg}\left[\left(2\varepsilon - \frac{1}{2}\right)\pi\right] - \text{tg}\left[\left(4\varepsilon - \frac{1}{2}\right)\pi\right]} = \lim_{\varepsilon \to 0} \frac{-\dfrac{1}{2\varepsilon}}{-\dfrac{1}{4\varepsilon}} = 2^1$$

Consequently, $c = 1$ and the domain of attraction for minima is the Frechet type. ∎

Example 3.16. (Gumbel distribution for maxima). For the Gumbel distribution for maxima we have

$$F(x) = \exp[-\exp(-x)]$$

from which

$$F^{-1}(u) = -\log[-\log u]$$

Substitution in (3.36) gives

$$\lim_{\varepsilon \to 0} \frac{-\log(-\log \varepsilon) + \log(-\log(2\varepsilon))}{-\log(-\log(2\varepsilon)) + \log(-\log(4\varepsilon))}$$

$$= \lim_{\varepsilon \to 0} \frac{\log\left[1 + \dfrac{\log 2}{\log \varepsilon}\right]}{\log\left[1 + \dfrac{\log 2}{\log(2\varepsilon)}\right]}$$

$$= \lim_{\varepsilon \to 0} \frac{\log 2 + \log \varepsilon}{\log \varepsilon} = 2^0 = 1$$

from which $c = 0$ and then it belongs to the domain of attraction for minima of the Gumbel distribution. ∎

Example 3.17. (Weibull distribution for maxima). For the Weibull distribution for maxima we have

$$F(x) = \exp[-(-x)^\beta]; \qquad x \le 0$$

and then

$$F^{-1}(u) = -(-\log u)^{1/\beta}$$

Substitution now in (3.36) leads to

$$\lim_{\varepsilon \to 0} \frac{-[-\log(\varepsilon)]^{1/\beta} + [-\log(2\varepsilon)]^{1/\beta}}{-[-\log(2\varepsilon)]^{1/\beta} + [-\log(4\varepsilon)]^{1/\beta}}$$

$$= \lim_{\varepsilon \to 0} \frac{1 - [\log(\varepsilon)/\log(2\varepsilon)]^{1/\beta}}{[\log(4\varepsilon)/\log(2\varepsilon)]^{1/\beta} - 1}$$

$$= \lim_{\varepsilon \to 0} \frac{1 - \left[1 - \dfrac{\log 2}{\log(2\varepsilon)}\right]^{1/\beta}}{\left[1 + \dfrac{\log 2}{\log(2\varepsilon)}\right]^{1/\beta} - 1} = 1 = 2^0$$

Consequently, $c = 0$ and belongs to the domain of attraction for minima of the Gumbel distribution. ∎

Example 3.18. (Frechet distribution for maxima). The cdf of the Frechet distribution for maxima is

$$F(x) = \exp[-x^{-\beta}]; \qquad x \geq 0$$

and its inverse

$$F^{-1}(u) = (-\log u)^{-1/\beta}$$

Thus, (3.36) becomes

$$\lim_{\varepsilon \to 0} \frac{[-\log \varepsilon]^{-1/\beta} - [-\log(2\varepsilon)]^{-1/\beta}}{[-\log(2\varepsilon)]^{-1/\beta} - [-\log(4\varepsilon)]^{-1/\beta}}$$

$$= \lim_{\varepsilon \to 0} \frac{\left[1 - \dfrac{\log 2}{\log(2\varepsilon)}\right]^{1/\beta} - 1}{1 - \left[1 + \dfrac{\log 2}{\log(2\varepsilon)}\right]^{-1/\beta}} = 1 = 2^0$$

which shows that the Frechet distribution for maxima lies in the domain of attraction for minima of the Gumbel distribution. ∎

Theorem 3.10. *Let $F(x)$ be a cdf with continuous and finite right derivatives at $\alpha(F)$ until the order $k + 1$ included, such that the first non-zero right derivative at $\alpha(F)$ has order k. If $\alpha(F)$ is finite, then $F(x)$ belongs to the domain of attraction for minima of the Weibull distribution, and the γ value in (3.22) is k.*

Proof.   If $\alpha(F)$ is finite, in a neighbourhood of $\alpha(F)$ we can write

(3.37)
$$F(x) \approx \frac{[x - \alpha(F)]^k F^k(\alpha(F))}{k!}$$

and

$$F^*(x) = F\left(\alpha(F) - \frac{1}{x}\right)$$

(3.38)
$$= \frac{\left(-\dfrac{1}{x}\right)^k F^{(k)}(\alpha(F))}{k!}$$

and then, (3.22) gives

(3.39)
$$\lim_{t \to \infty} \frac{F^*(tx)}{F^*(t)} = \frac{\left(-\dfrac{1}{tx}\right)^k}{\left(-\dfrac{1}{t}\right)^k} = x^{-k}$$

which proves the theorem for the Weibull case.   ∎

Example 3.15.   Let $F(x)$ be defined by

$$F(x) = \begin{cases} 0 & \text{if } x < 0 \\ x^m & \text{if } 0 \le x \le 1 \\ 1 & \text{if } x > 1 \end{cases}$$

where m is an integer. All derivatives of $F(x)$ are finite and the first non-zero derivative at $\alpha(F) = 0$ is the m-th derivative. Then, it belongs to the domain of attraction for minima of the Weibull distribution.   ∎

Theorem 3.11.   *Let $F_X(x)$ be the cdf of the random variable X. Let Y be the random variable defined by*

(3.40)
$$Y = h(X) \quad \Leftrightarrow \quad X = g(Y)$$

where $g(y)$ is invertible, non-decreasing and with finite right derivatives at $h(\alpha(F_X))$ and such that the first non-null derivative at $h(\alpha(F_X))$ is finite and of order k. Then, if X belongs to the domain of attraction for minima of the Weibull distribution, Y also belongs to the same domain of attraction, and if c_n and d_n are normalizing constants for X, then some normalizing constants for

Y are given by

(3.41) $$c_n^* = h(c_n)$$

(3.42) $$d_n^* = h(c_n + d_n) - h(c_n)$$

Proof.   By assumption, X belongs to the domain of attraction for minima of the Weibull distribution, then

(3.43) $$\lim_{t \to -\infty} \frac{F_X\left(\alpha(F_X) - \dfrac{1}{tx}\right)}{F_X\left(\alpha(F_X) - \dfrac{1}{t}\right)} = x^{-\gamma}$$

In addition, we have

(3.44) $$F_Y(y) = P[Y \le y] = P[h(X) \le y] = P[X \le g(y)] = F_X[g(y)]$$

But for Y we can write

$$\lim_{t \to -\infty} \frac{F_Y^*(ty)}{F_Y^*(t)} = \lim_{t \to -\infty} \frac{F_Y\left(\alpha(F_Y) - \dfrac{1}{ty}\right)}{F_Y\left(\alpha(F_Y) - \dfrac{1}{t}\right)}$$

$$= \lim_{t \to -\infty} \frac{F_X\left[g\left(\alpha(F_Y) - \dfrac{1}{ty}\right)\right]}{F_X\left[g\left(\alpha(F_Y) - \dfrac{1}{t}\right)\right]}$$

$$= \lim_{t \to -\infty} \frac{F_X\left[g(\alpha(F_Y)) - \left(\dfrac{1}{ty}\right)^k g^{(k)}(\alpha(F_Y))\right]}{F_X\left[g(\alpha(F_Y)) - \left(\dfrac{1}{t}\right)^k g^{(k)}(\alpha(F_Y))\right]}$$

(3.45) $$= \lim_{t \to -\infty} \frac{F_X\left[\alpha(F_X) - \left(\dfrac{1}{ty}\right)^k g^{(k)}(\alpha(F_Y))\right]}{F_X\left[\alpha(F_X) - \left(\dfrac{1}{t}\right)^k g^{(k)}(\alpha(F_Y))\right]} = y^{-\gamma}$$

which demonstrates the proof.

According to (3.26), the normalizing constants c_n^* and d_n^* are given by

(3.46) $$c_n^* = \alpha(F_Y) = h(\alpha(F_X)) = h(c_n)$$

(3.47)
$$d_n^* = F_Y^{-1}\left(\frac{1}{n}\right) - \alpha(F_Y)$$

but we have

(3.48)
$$\frac{1}{n} = F_Y(y) = F_X(g(y)) \quad \Rightarrow \quad y = h\left[F_X^{-1}\left(\frac{1}{n}\right)\right]$$

and

(3.49)
$$d_n = F_X^{-1}\left(\frac{1}{n}\right) - c_n$$

and then

(3.50)
$$d_n^* = h(c_n + d_n) - c_n^* \qquad \blacksquare$$

Example 3.20. (Square of exponential variables). Let Y be a random variable defined as the square of an exponential variable X. Then

$$Y = h(X) = X^2$$

and due to the fact that $h(X)$ is invertible, non-decreasing and with finite derivatives at the lower-end $\alpha(F_X) = 0$, then, the cdf $F_Y(y)$ belongs to the domain of attraction for minima of the Weibull distribution and the sequences of constants c_n^* and d_n^* are given by (see example 3.2))

$$c_n^* = h(c_n) = 0$$

$$d_n^* = h(c_n + d_n) - h(c_n) = \frac{1}{n^2} \qquad \blacksquare$$

Finally, in table 3.1. we summarize the domains of attraction of some known distributions.

3.7. Asymptotic Joint Distribution of Maxima and Minima

In this section we study the asymptotic joint distribution of maxima and minima, which is given by the following theorem (see Galambos (1978)).

Theorem 3.12. (Asymptotic joint distribution of maxima and minima). *Let us assume that there are sequences a_n, b_n, c_n and d_n such that*

(3.51)
$$\lim_{n \to \infty} F^n(a_n + b_n x) = H_{c_1}(x)$$

$$(3.52) \qquad \lim_{n \to \infty} 1 - [1 - F(c_n + d_n x)]^n = L_{c_2}(x)$$

for some given c_1 and c_2. Then

$$(3.53) \qquad \lim_{n \to \infty} P(Z_n \leq a_n + b_n x, W_n \leq c_n + d_n x) = H_{c_1}(x)L_{c_2}(x)$$

This theorem states that the maximum and minimum, suitably normalized, are asymptotically independent and that the normalization sequences appearing in the bivariate distribution coincide with those in the univariate cases.

Domain of Attraction Type		
Distribution	For maxima	For minima
Normal	Gumbel	Gumbel
Exponential	Gumbel	Weibull
Log-normal	Gumbel	Gumbel
Gamma	Gumbel	Weibull
$Gumbel_M$	Gumbel	Gumbel
$Gumbel_m$	Gumbel	Gumbel
Rayleigh	Gumbel	Weibull
Uniform	Weibull	Weibull
$Weibull_M$	Weibull	Gumbel
$Weibull_m$	Gumbel	Weibull
Cauchy	Frechet	Frechet
Pareto	Frechet	Weibull
$Frechet_M$	Frechet	Gumbel
$Frechet_m$	Gumbel	Frechet

M = for maxima
m = for minima

TABLE 3.1. Domains of attraction of some known distributions

3.8. Asymptotic Distributions of Range and Midrange

The asymptotic distribution of the range, R_n, and midrange, M_n, are given by the following theorem (see Galambos 1978)).

Theorem 3.13. (Asymptotic distributions of range and midrange). *Under the assumptions of theorem 3.10, we have*

$$(3.54) \qquad \lim_{n \to \infty} P(R_n < a_n - c_n + b_n x) = 1 - \int_{-\infty}^{\infty} L_{c_2}(y - x)\, dH_{c_1}(y)$$

and

$$(3.55) \qquad \lim_{n \to \infty} P(2M_n < a_n + c_n + b_n x) = \int_{-\infty}^{\infty} L_{c_2}(x - y)\, dH_{c_1}(y)$$

This theorem gives the limit distributions of the range and midrange, and some normalizing sequences to be used.

3.9. Asymptotic Distributions of Maxima of Samples with Random Size

In some cases the sample size is not fixed but random (see section 2.3.3). Then, the limit distribution of maxima and minima depends on the random properties of the sample size. The following theorem (Galambos (1978)) gives the asymptotic distribution of maxima for this case.

Theorem 3.14. (Asymptotic distribution of maxima of samples with random size). *Let $N(n)$ be a positive integer random variable such that $N(n)/n \to \tau$ in probability, where τ is a positive random variable. Assume that*

$$(3.56) \qquad \lim_{x \to \infty} P[Z_n < a_n + b_n x] = H(x); \qquad \text{for all } x$$

Where $H(x)$ is a non-degenerated cdf.
   Then

$$(3.57) \qquad \lim_{n \to \infty} P[Z_{N(n)} < a_n + b_n x] = \int_{-\infty}^{\infty} H^y(x)\, dF(y)$$

where $F(y)$ is the cdf of τ.

3.10.   Approximation of Distribution Functions in Their Tails

In this section we study the problem of approximation of distribution functions in the tails and show that this problem is closely related to that of extremes.

In previous sections we have used the distribution function $F(x)$ in order to study the asymptotic properties of maxima. However, because all distribution functions tend to one when x goes to $\omega(F)$, a comparison of distributions based on $F(x)$ itself leads to some problems, as indicated in section 3.2. For this reason, it seems more natural and reasonable to compare the hazard functions, which become infinitesima when $x \rightarrow \omega(F)$. This is equivalent to comparing the return periods (inverse of the hazard values). In this context, the concept of tail equivalence arises in a very natural way.

Definition 3.3. (Tail equivalence).   Two distribution functions $F(x)$ and $G(x)$ are said to be right tail equivalent if, and only if,

$$\omega(F) = \omega(G)$$

and

$$\lim_{x \rightarrow \omega(F)} \frac{1 - F(x)}{1 - G(x)} = 1$$

Similarly, $F(x)$ and $G(x)$ are said left tail equivalent if, and only if,

$$\alpha(F) = \alpha(G)$$

and

$$\lim_{x \rightarrow \alpha(F)} \frac{F(x)}{G(x)} = 1$$

The following theorem, from Resnick (1971a), shows the clear connection between tail equivalence and domains of attraction.

Theorem 3.15. (Tail equivalence and domains of attraction).   *Let $F(x)$ and $G(x)$ be two right tail equivalent distribution functions. If*

$$\lim_{n \rightarrow \infty} F^n(a_n + b_n x) = H(x); \qquad \text{for all } x$$

then

$$\lim_{n \rightarrow \infty} G^n(a_n + b_n x) = H(x); \qquad \text{for all } x$$

Note that the theorem states that: (a) if two distributions are tail equivalent and one belongs to some domain of attraction, the other must also belong to the same domain of attraction, and (b) the sequences of normalizing constants coincide.

This has very important practical implications, as will be seen in following chapters, because a distribution $F(x)$ can be replaced by a tail equivalent distribution $G(x)$ without altering either the domain of attraction or the set of admissible constants. If $G(x)$ is a von-Mises distribution, the problem becomes considerably simpler. In other words, from a practical point of view, we fit a von-Mises distribution to the tail of a given distribution and use it for extreme value purposes. This will be the basis for many methods to be presented in chapters 4, 5 and 6.

Some detailed analyses of the relation between tail equivalence and domains of attraction can be seen in Resnick (1971) and Castillo et al. (1980).

3.11. The Penultimate Form of Approximation to Extremes

One important problem with practical implications is the speed of convergence of the sequence $F^n(a_n + b_n x)$ to the limit distribution $H(x)$. Fisher and Tippett (1928) mention that the convergence of $\Phi^n(a_n + b_n x)$ ($\Phi(x)$ is the cdf of the standard normal distribution) to $H_{3,0}(x)$ is extremely slow and indicate that $\Phi^n(a_n + b_n x)$ is closer to a suitably chosen Weibull distribution.

Gomes (1984) proves that a penultimate approximation of Weibull or Frechet type to Gumbel type distributions is much closer than the ultimate itself, for normal random variables and Cohen (1982) proves that the error in approximating $\Phi^n(a_n + b_n)$ by a Weibull sequence of distributions is uniformly of order $(\log n)^{-2}$ instead of order $(\log n)^{-1}$ when Gumbel sequences are utilized.

Note that in the classical (ultimate) approach, the distribution of maxima $H_n(x)$ is approximated by $H_0((x - a_n)/b_n)$, where $H_c(x)$ is the von-Mises form, defined in (3.28), and that a penultimate approximation is of the form $H_{c_n}((x - a_n)/b_n)$, i.e. it includes one more sequence $\{c_n\}$. This new degree of freedom explains the better quality of the latter approximation.

The above authors only use the penultimate approximations for distributions with Gumbel type domains of attraction. However, it can also be utilized for distributions in a Weibull or Frechet type domain of attraction. Instead of approximating $H_n(x)$ by $H_c((x - a_n)/b_n)$ with c constant, as with the ultimate approach, an approximation by $H_{c_n}((x - a_n)/b_n)$ (penultimate) with c_n dependent upon n, can be done. The new degree of freedom (c_n) explains, as above, the better approximation.

Example 3.21. (Weibull distribution for maxima in the left tail). The Weibull distribution for maxima

$$F(x) = \exp(x): \qquad -\infty < x < 0$$

belongs to the domain of attraction for minima of the Gumbel distribution

$$L_{3,0}(x) = 1 - \exp[-\exp(x)]$$

with a possible choice of normalizing sequences given by

$$c_n = -\log n$$

$$d_n = 1$$

This implies that the exact distribution of minima

$$L_n(x) = 1 - [1 - \exp(x)]^n$$

can be approximated by (ultimate approach)

$$L\left(\frac{x - c_n}{d_n}\right) = 1 - \exp[-\exp(x + \log n)]$$

However, as an alternative, we can choose the penultimate approximation (Frechet type)

$$G_n(x) = 1 - \exp\left[-\left(\frac{\delta_n}{\lambda_n - x}\right)^{\beta_n}\right]$$

where δ_n, λ_n and β_n can be estimated by the percentile method (see expressions (5.60) $-$ (5.64)), i.e. by equating three percentiles of $L_n(x)$ and $G_n(x)$. If the percentiles are p_i ($i = 1, 2, 3$) we get

$$\lambda_n - \delta_n[-\log(1 - p_i)]^{-1/\beta_n} = \log[1 - (1 - p_i)^{1/n}]; \qquad i = 1, 2, 3$$

from which λ_n, δ_n and β_n can be obtained.

In table 3.2., where, following Gomes (1984), $k_{0,n}$ and $k_{1,n}$ denote the Kolmogorov-Smirnov distances between $L_n(x)$ and $L((x - c_n)/d_n)$ and $L_n(x)$ and $G_n(x)$, respectively, some numerical comparisons show the superiority of the penultimate approximation. They were obtained for $p_1 = 0.2$, $p_2 = 0.5$ and $p_3 = 0.8$. ∎

n	$k_{0,n}$	$k_{1,n}$
10	0.28×10^{-1}	0.34×10^{-2}
10^2	0.27×10^{-2}	0.33×10^{-3}
10^3	0.27×10^{-3}	0.33×10^{-4}
10^4	0.27×10^{-4}	0.33×10^{-5}
10^5	0.27×10^{-5}	0.33×10^{-6}
10^6	0.27×10^{-6}	0.33×10^{-7}
10^7	0.27×10^{-7}	0.33×10^{-8}

TABLE 3.2. Kolmogorov-Smirnov distances of ultimate and penultimate approximations to the exact distribution of maximum

Example 3.22. (Uniform distribution). The ultimate approximation to the distribution of maxima from a sample of size n from a uniform distribution is

$$H\left(\frac{x - a_n}{b_n}\right) = \exp[n(x - 1)], \qquad -\infty < x \le 1$$

However, the exact distribution

$$H_n(x) = x^n; \qquad 0 \le x \le 1$$

can also be approximated by the penultimate distribution

$$G_n(x) = \exp\left[-\left(\frac{\lambda_n - x}{\delta_n}\right)^{\beta_n}\right]$$

where, as in the previous example, δ_n, λ_n and β_n can be estimated by the percentile method, i.e. by means of the system of equations

$$p_i^{1/n} = \lambda_n - \delta_n(-\log p_i)^{1/\beta_n}; \qquad i = 1, 2, 3$$

Numerical results, similar to those above, show the superiority of the penultimate approach with respect to the ultimate. ∎

The most important practical implication of all the above is that Weibull and Frechet distributions are justified to approximate distributions belonging to a Gumbel type domain of attraction and that the shape parameter can be dependent upon n.

Chapter 4

Shortcut Procedures: Probability Papers and Least-Squares Methods

4.1. Introduction

Under the heading of "shortcut procedures" we include in this chapter some methods to analyze, interpret, estimate and extrapolate data by quick techniques such as probability paper and least-squares methods.

Graphic methods are important tools used by engineers and scientists in daily practice. Because of their simplicity and usefulness, probability paper methods have been used for many years in the statistical area.

The basic idea of probability paper, of a given parametric family of distributions, is to modify the random variable, X, and probability, P, scales in such a way that the plot against X of any cdf, $F(x)$, belonging to that family, appears as a straight line, such that no other cdf satisfies this property. This implies that the drawing of any cdf on such paper allows one to decide whether or not it belongs to that family and if the answer is affirmative, to estimate its parameters.

However, the situation described above is only a theoretical one, and does not appear in practice. In this case, the cdf associated with the random variable under consideration is not known, but a sample obtained from it is. So, the normal policy consists of drawing the empirical cdf on probability paper as an approximation of the actual cdf. But due to the random character of the sample, even if it comes from a population for which the probability paper was designed, the trend of the empirical cdf changes from sample to sample and in most cases does not appear as a perfect straight line. This complicates the problem considerably. However, the expected departures from the straight line trend, in this case, are smaller than if the sample were to come from a different population, and a decision can be made either subjectively or based on acceptance regions.

In order to present some of the problems associated with probability paper and graphic techniques, we introduce them by means of one practical example, taken from Castillo et al. (1982).

Let us assume that we give to 4 different engineers the data of maximum yearly significant wave height measured in Myken-Skomvaer (Norway) in the period 1949-1976 and published by Houmb et al (1978). Let us assume also that each one of them decides to fit the above data by using a different kind of paper as follows:

Engineer 1: Arithmetic paper

Engineer 2: Half-log paper

Engineer 3: Double-log paper

Engineer 4: Gumbel probability paper

According to the usual policy, they represent the empirical cdf of the sample on their paper and get as results figures 4.1a. to 4.1d. In addition, because the data appears to be as very close to a linear trend they decide to visually fit a straight line, as shown in the same figures, and to use it for design purposes.

Figure 4.2. shows the same 4 lines all in arithmetic scale (note that the three curves coming from papers different from the arithmetic one are not straight lines any more).

The four design curves are close in probability interval $(0.1, 0.9)$, but differ very much in the right tail $(0.9, 1)$ which is the only interval used for design purposes. This becomes more serious if we take into account that the design cost is roughly proportional to the square of the design wave height, and wave heights in the neighbourhoods of 12.3 and 17.5m. would be selected by engineers 3 and 4 respectively. Consequently, the cost associated with engineer 4 would be twice that for engineer 3.

This introductory example is the basis for raising some important questions as: In order to accept a sample as coming from a given parametric family, which decision criteria, based on the probability plot of the empirical cdf on its corresponding probability paper must be used? Is the visual fitting of a straight line sufficient? Which are the analytical criteria to be used for the fitting? Must all data points be given the same weight? Must any data point be excluded from the analysis? Is the purpose of the fitting related to the method to be used? What is to be done when there are large differences in the tails but small ones on the intermediate range as in the example above? These and other questions of practical importance will be answered in the following sections.

In order to illustrate the methods in this chapter, we apply them to several sets of data which are included and explained in appendix C. We assume that an extrapolation is the aim of the analysis, or, in other words, that the engineer

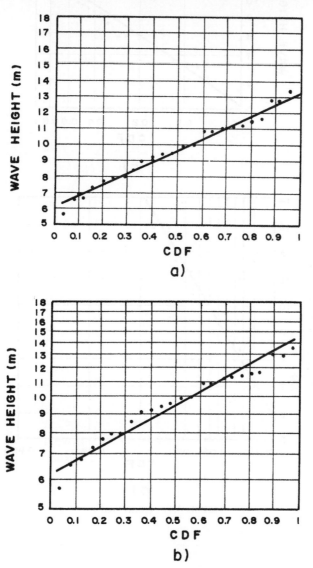

FIGURE 4.1.   Maximum yearly significant wave height data drawn on four different papers

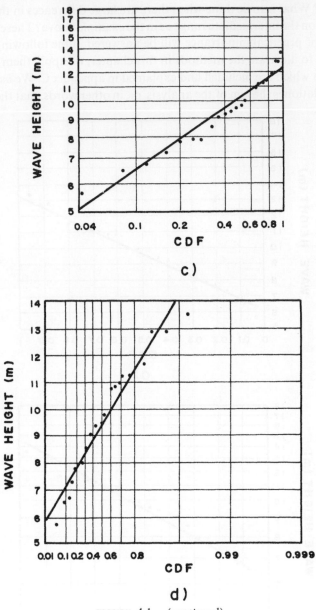

c)

d)

FIGURE 4.1. (continued)

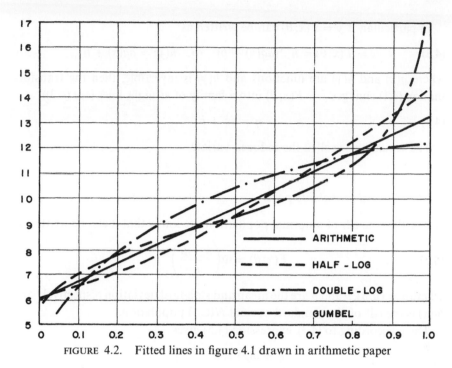

FIGURE 4.2.   Fitted lines in figure 4.1 drawn in arithmetic paper

or scientist is willing to determine the asymptotic distribution or the distribution of extremes of samples of sizes much larger than the sample itself.

4.2.   The Theoretical Bases of Probability Paper

As mentioned in the introduction, probability paper is simply paper in which the scales have been changed in such a manner that the cdfs associated with a given family of distributions, when represented graphically on that paper, become straight lines.

Let $F(x; \theta)$ be a parametric family of cdfs, where θ is the vector parameter. We look for some transformation

$$(4.1) \qquad \left.\begin{array}{l} \xi = g(x) \\ \eta = h(y) \end{array}\right\}$$

such that the family of curves of the equation

$$(4.2) \qquad\qquad y = F(x; \theta)$$

when transformed by (4.1) becomes a family of straight lines.

In particular, if $y = F(x; \theta)$ can be written as

$$(4.3) \qquad y = F(x; \theta) = h^{-1}(ag(x) + b) \quad \leftrightarrow \quad h(y) = ag(x) + b$$

where $g(x)$ and $h(y)$ are functions and $h(y)$ is invertible, then the transformation (4.1) changes $y = F(x; \theta)$ into a family of straight lines (see (4.3))

$$(4.4) \qquad \eta = a\xi + b$$

The variable η is called the reduced variate.

4.2.1. Normal probability paper.

If $F(x; \theta)$ is the cdf of a normal population, we know that it can be written as

$$(4.5) \qquad F(x; \theta) = \Phi\left(\frac{x - \mu}{\sigma}\right)$$

where $\theta = (\mu, \sigma)$, μ and σ are the mean and standard deviation respectively and $\Phi(x)$ is the cdf of the standard normal $N(0, 1)$ population.

Then, according to (4.3), expression (4.5) gives

$$(4.6) \qquad \left. \begin{array}{l} \xi = g(x) = x \\ \eta = h(y) = \Phi^{-1}(y) \end{array} \right\}$$

and

$$(4.7) \qquad \left. \begin{array}{l} a = \dfrac{1}{\sigma} \\ \\ b = \dfrac{-\mu}{\sigma} \end{array} \right\}$$

The family of straight lines becomes

$$(4.8) \qquad \eta = a\xi + b = \frac{\xi - \mu}{\sigma}$$

Figure 4.3. shows normal probability paper, where the ordinate axis has been transformed by $\eta = \Phi^{-1}(y)$ and the abscissas axis has undergone no transformation. Note that the probability, Y, and the reduced variate, η, are shown.

Estimation of the parameters μ and σ can be made after fitting a straight line to the data by noting that by setting $\eta = 0$ and $\eta = 1$ in (4.8) we get

$$(4.9) \qquad 0 = \frac{\xi - \mu}{\sigma} \quad \rightarrow \quad \xi = \mu$$

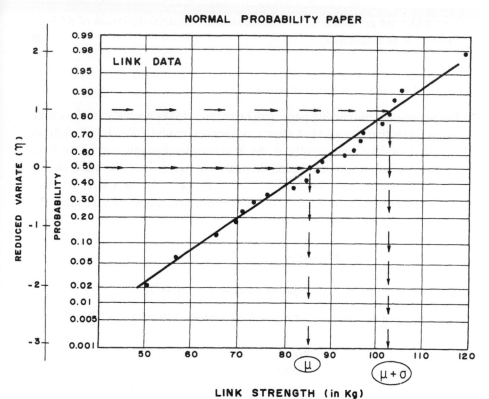

FIGURE 4.3.   Link data drawn on normal probability paper

$$(4.10) \qquad\qquad 1 = \frac{\xi - \mu}{\sigma} \quad \rightarrow \quad \xi = \mu + \sigma$$

Consequently, the value of ξ (or X) corresponding to a value of the reduced variate $\eta = 0$ is the estimated mean and the difference between the value of ξ associated with $\eta = 1$. In addition, the mean is the estimator of the standard deviation σ. (see figure 4.3.).

Example 4.1. (Link data).   In figure 4.3., the link data has been plotted on normal probability paper. Because the trend appears as a straight line, there is no reason to reject the assumption that the data comes from a normal population. Thus, the line shown in the figure was visually fitted to the data. The estimation method described above led to

$$\eta = 0 \quad \rightarrow \quad \xi = \mu = 85 \text{ Kgs.}$$

$$\eta = 1 \quad \rightarrow \quad \xi = \mu + \sigma = 102.5 \text{ Kgs.}$$

from which we get estimates for the mean and standard deviation of 85 Kgs. and 17.5 Kgs. respectively.   ■

4.2.2. Log-normal probability paper.

The log-normal probability paper can be reduced to the normal case if we take into account that X is said to follow a log-normal model if $Y = \log(X)$ follows a normal model. Consequently, if the X is transformed into $\log(X)$ we have normal probability paper. Then, the only change is that the scale of the abscissas axis becomes logarithmic (see figure 4.4.). If the mean, μ^*, and the standard deviation, σ^*, of the log-normal distribution are needed, the following formulas must be used

$$(4.11) \qquad \mu^* = \exp\left(\mu + \frac{\sigma^2}{2}\right)$$

$$(4.12) \qquad \sigma^{*2} = \exp(2\mu)[\exp(2\sigma^2) - \exp(\sigma^2)]$$

where μ and σ are the values obtained as explained in section 4.2.1.

Example 3.2. (Electrical insulation data). The electrical insulation data plotted on log-normal probability paper, which appears in figure 4.4., shows a linear trend. The estimation method for normal paper gives

$$\eta = 0 \quad \rightarrow \quad \xi = \exp(\mu) = 1075$$

$$\eta = 1 \quad \rightarrow \quad \xi = \exp(\mu + \sigma) = 1265$$

where the exp( ) function is due to the logarithmic scale. Thus, the values of μ and σ become 6.98 and 0.163, respectively. Then, according to (4.11) and (4.12), the mean and standard deviation of the lognormal fitted distribution are

$$\mu^* = \exp\left(6.98 + \frac{0.163^2}{2}\right) = 1089.29 \text{ hours.}$$

$$\sigma^{*2} = \exp(2 \times 6.98)[\exp(2 \times 0.163^2) - \exp(0.163^2)] = 31942 \text{ hour}^2 \quad ■$$

4.2.3. Gumbel probability paper.

The cdf curve of the Gumbel family for maxima is given by

$$(4.13) \quad y = F(x; \lambda, \delta) = \exp\left[-\exp\left(-\frac{x - \lambda}{\delta}\right)\right]; \qquad -\infty < x < \infty$$

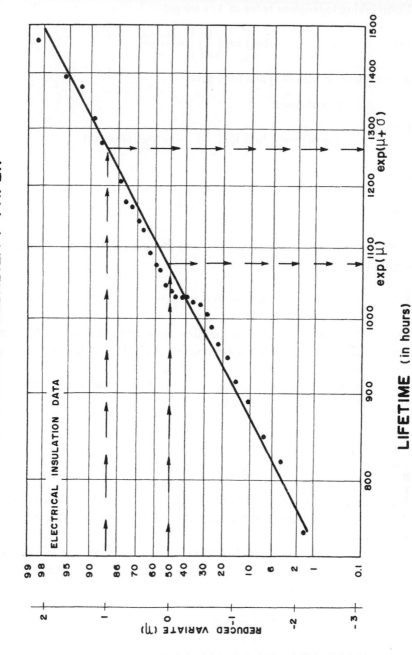

FIGURE 4.4.   Electrical insulation data drawn on lognormal probability paper

Upon taking logarithms twice of $1/y$ we get

(4.14)
$$-\log\left[\log\left(\frac{1}{y}\right)\right] = \frac{x - \lambda}{\delta}$$

and a comparison with (4.1) and (4.3) gives

$$\xi = g(x) = x$$

(4.15)
$$\eta = h(y) = -\log\left[\log\left(\frac{1}{y}\right)\right] = -\log[-\log(y)]$$

and

(4.16)
$$\left.\begin{array}{l} a = 1/\delta \\ b = -\lambda/\delta \end{array}\right\}$$

which shows that the transformation (4.15) changes (4.13) into the family of straight lines

(4.17)
$$\eta = a\xi + b = \frac{\xi - \lambda}{\delta}$$

Figure 4.5. shows a Gumbel probability paper for maxima in which the ordinate axis has been transformed according to (4.15) and the abscissas axis remains unaltered.

The estimation of the two parameters λ and δ can be done by noting (see (4.17)) that for $\eta = 0$ and $\eta = 1$ we get

(4.18)
$$0 = \xi - \lambda \quad \rightarrow \quad \xi = \lambda$$

(4.19)
$$1 = \frac{\xi - \lambda}{\delta} \quad \rightarrow \quad \xi = \lambda + \delta$$

So, once a straight line has been fitted to the data, the abscissas associated with ordinate values 0 and 1 of the reduced variate η, give the values of λ and $\lambda + \delta$, respectively.

In the case of minima instead of maxima, a change of sign of the data permits the use of the above probability paper, because the Gumbel family of cdfs for minima is given by

(4.20)
$$F(x; \lambda, \delta) = 1 - \exp\left[-\exp\left(\frac{x - \lambda}{\delta}\right)\right]; \qquad -\infty < x < \infty$$

and from (4.13) we get (note the change in sign in the x and λ)

(4.21)
$$F(x; \lambda, \delta) = 1 - F(-x; -\lambda, \delta)$$

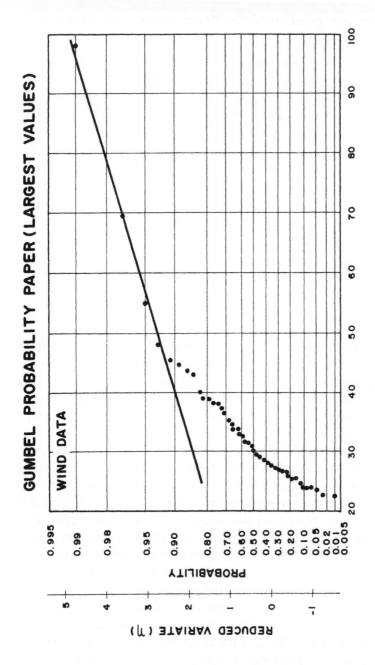

FIGURE 4.5.   Wind data drawn on Gumbel probability paper for maxima

However, the same paper can be directly obtained by a similar process and we get

$$(4.22) \qquad \left. \begin{array}{l} \xi = g(x) = x \\ \eta = h(y) = \log[-\log(1-y)] \end{array} \right\}$$

$$(4.23) \qquad \left. \begin{array}{l} a = \dfrac{1}{\delta} \\ b = -\dfrac{\lambda}{\delta} \end{array} \right\}$$

and

$$(4.24) \qquad \eta = a\xi + b = -\frac{\xi - \lambda}{\delta}$$

which for $\eta = 0$ and $\eta = -1$ gives

$$(4.25) \qquad 0 = \xi - \lambda \quad \rightarrow \quad \xi = \lambda$$

$$(4.26) \qquad 1 = (\xi - \lambda)/\delta \quad \rightarrow \quad \xi = \lambda + \delta$$

Gumbel probability paper becomes the standard starting point when the limit distribution of data (domain of attraction) is not known. The reason for this is that Gumbel type occupies a central position between Weibull and Frechet papers and convex or concave plots, in the tails are indicators of Weibull or Frechet type domains of attraction, respectively. The contrary is true for smallest values.

On the other hand, the plot of data on Weibull or Frechet probability paper requires knowledge of the threshold parameter λ. In fact, if the threshold parameter is not known of inferred from physical considerations, it must be estimated (by trial and error plotting, for example). Thus, the normal procedure starts with the plotting of data on Gumbel probability paper and if the trend cannot be assumed to be linear, then plotting the same data on Weibull or Frechet probability paper, respectively, depending on the convex or concave trend observed. The plot must be done for different values of the threshold parameter, until the linear trend is obtained in the tail of interest.

In the following, according to this procedure, all data sets are plotted on Gumbel probability paper for smallest or largest values depending of the aim of the analysis (see table C.15.), i.e. if the upper order statistics are of interest, the Gumbel paper for largest values will be used. If the lower order statistics play the important part in the problem under study, the Gumbel probability paper for smallest values must be used. We advise the reader to be careful in selecting the proper probability paper. It is not uncommon to see maxima drawn on probability paper for minima and vice versa. Because the interest of our analysis is related to limit distributions, probability papers can be used even though data are not associated with maxima or minima. However, in this case only tail data must be used in the fitting procedure.

Example 4.3. (Wind data). The wind data when plotted on Gumbel probability paper (see figure 4.5), does not show a linear trend in its range, but a concave shape. However, the right tail shows a linear trend and this suggests a Gumbel type limit distribution. Thus, if the data on the right tail is used for fitting a straight line, the line shown in the figure is obtained, which, after the estimation procedure described above, lead to values of λ and δ of -15 and 23.5 miles per hour, respectively.   ■

Example 4.4 (Flood data). The flood data has been plotted in figure 4.6. It can be assumed to follow a Gumbel distribution for maxima, because the trend is a straight line, not only on the right tail but in all of its range. The estimated parameters, based on the visually fitted line in the figure and the process outlined above, are $\lambda = 38.5$ cms. and $\delta = 7.8$ cms.   ■

Example 4.5. (Wave data). Figure 4.7. shows a linear trend for the wave data. Consequently, a straight line can be fitted to the data and values of $\lambda = 11.2$ and $\delta = 5.3$ feet result.   ■

Example 4.6. (Telephone data). The telephone data in figure 4.8. (note that the Gumbel paper is now for smallest values) is clear evidence that it does not come either from a Gumbel distribution for minima or from a distribution belonging to the domain of attractior of Gumbel in the left tail. The curvature of the trend suggests a Weibull limit distributior in the left tail.   ■

Example 4.7. (Epicenter data). The epicenter data in figure 4.9. also suggest a Weibull-type distribution in the left tail due to the curvature of the data trend and its vertical slope in that tail.   ■

Example 4.8. (Link data). The link data in figure 4.10. shows an overall curvature, thus indicating that a Gumbel-type parent is not reasonable. However, the linear trend of the first seven order statistics can suggest a Gumbel limit distribution in that tail.   ■

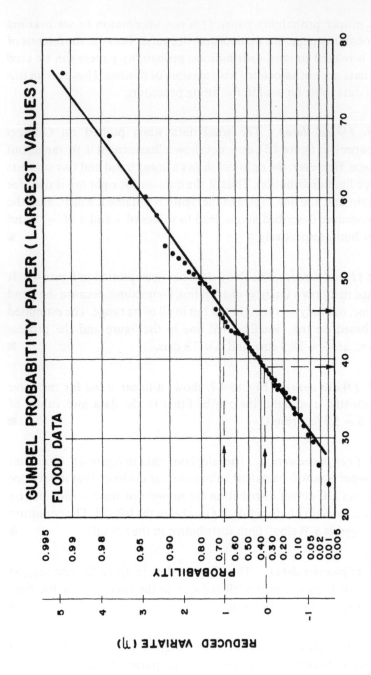

FIGURE 4.6. Flood data drawn on Gumbel probability paper for maxima

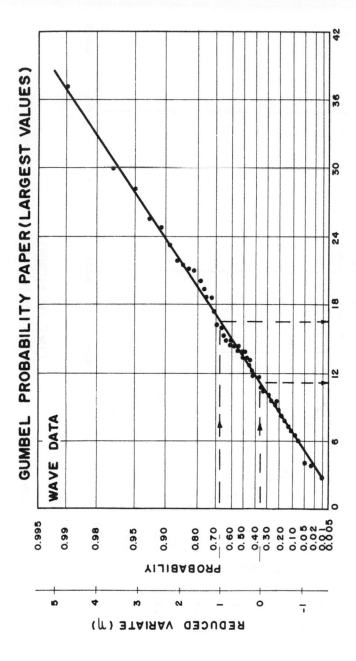

FIGURE 4.7.   Wave data drawn on Gumbel probability paper for maxima

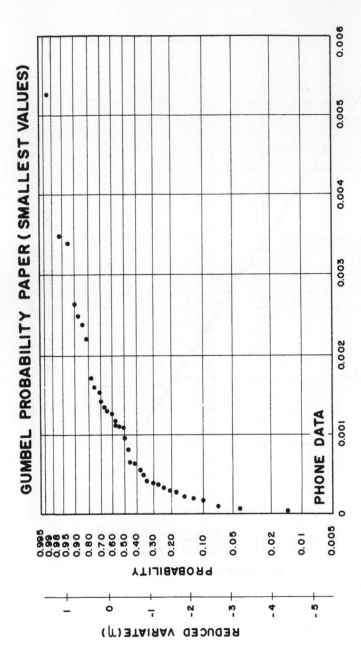

FIGURE 4.8. Phone data drawn on Gumbel probability paper for minima

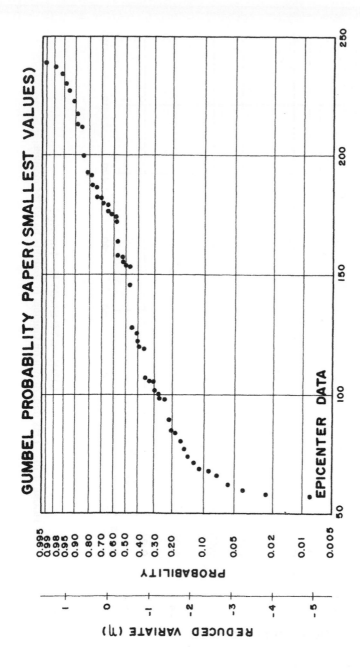

FIGURE 4.9. Epicenter data drawn on Gumbel probability paper for minima

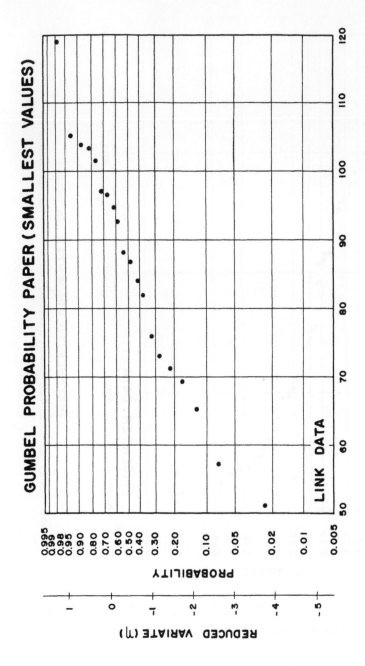

FIGURE 4.10. Link data drawn on Gumbel probability paper for minima

Example 4.9. (Electrical insulation data).   The electrical insulation data in figure 4.11. shows an overall curvature. Consequently, a Gumbel parent should be rejected. However, the almost linear trend shown by the first ten or eleven order statistics indicates a Gumbel-type domain of attraction for minima.   ∎

Example 4.10. (Fatigue data).   The fatigue data in figure 4.12. clearly suggest a Weibull-type limit distribution in the left tail.   ∎

Example 4.11. (Precipitation data).   The precipitation data, shown in figure 4.13., can show either a Weibull type limit distribution in the left tail due to its possible curvature (indicated only by the first order statistic), or a Gumbel type domain of attraction if that is given a small weight.   ∎

Example 4.12. (Houmb data).   The Houmb wave data when drawn on Gumbel probability paper (see figure 4.14.) does not give clear enough information to decide between a Gumbel- or Weibull-type domain of attraction. On one hand, the overall curvature, that can be assumed to extend to the right tail, suggests a Weibull-type domain of attraction. On the other hand, a linear trend assumption, which is also reasonable, points to a Gumbel-type.

                                                                            ∎

Example 4.13. (Ocmulgee river).   The Ocmulgee river data, shown in figures 4.15. and 4.16., for Macon and Hawkinsville, respectively, shows Weibull-type behavior at Macon and a Gumbel-type at Hawkinsville.   ∎

Example 4.14. (Oldest ages at death in Sweden).   The oldest ages at death in Sweden for women and men appear in figures 4.17. and 4.18. respectively. In both cases the curvature indicates a Weibull-type domain of attraction though the women data on the right tail is irregular.   ∎

4.2.4.   Weibull probability paper.

The cdf curves of the Weibull family for maxima are given by

$$(4.27) \qquad y = F(x; \lambda, \beta, \delta) = \exp\left[-\left(\frac{\lambda - x}{\delta}\right)^{\beta}\right]; \qquad -\infty < x \leq \lambda$$

Upon taking logarithms twice we get

$$(4.28) \qquad -\log[-\log(y)] = -\beta \log\left(\frac{\lambda - x}{\delta}\right) = -\beta \log(\lambda - x) + \beta \log \delta$$

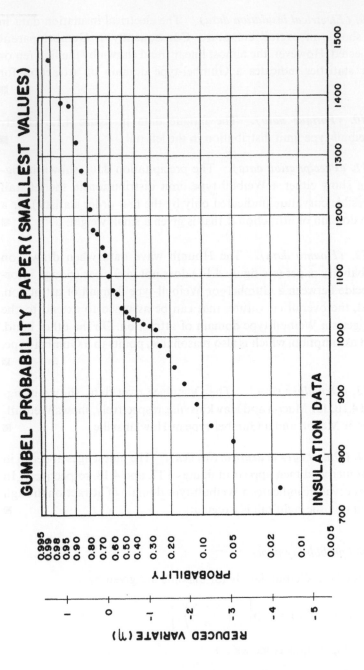

FIGURE 4.11.   Insulation data drawn on Gumbel probability paper for minima

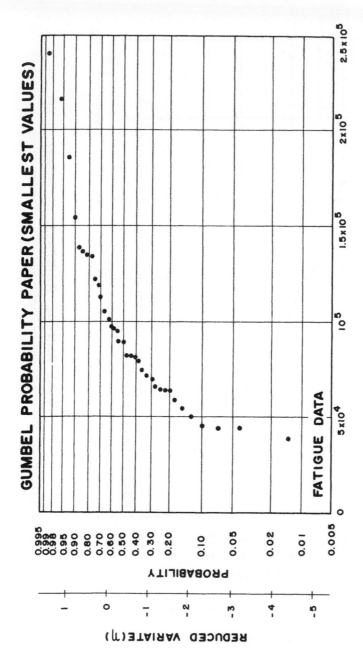

FIGURE 4.12. Fatigue data drawn on Gumbel probability paper for minima

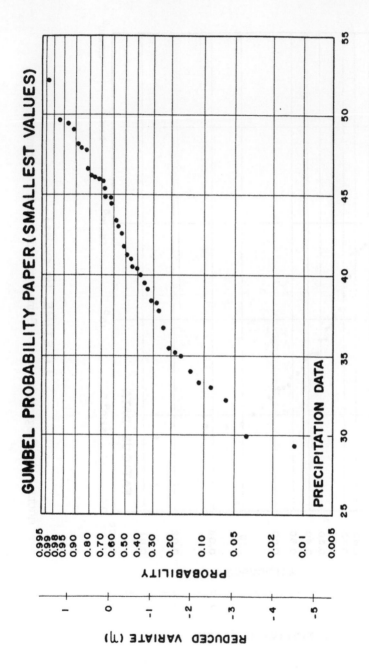

FIGURE 4.13. Precipitation data drawn on Gumbel probability paper for minima

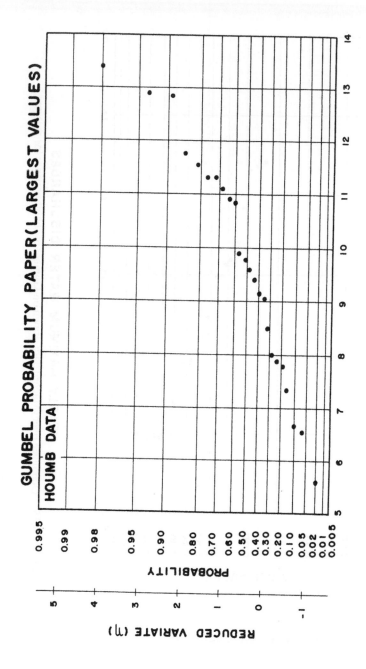

FIGURE 4.14. Houmb data drawn on Gumbel probability paper for maxima

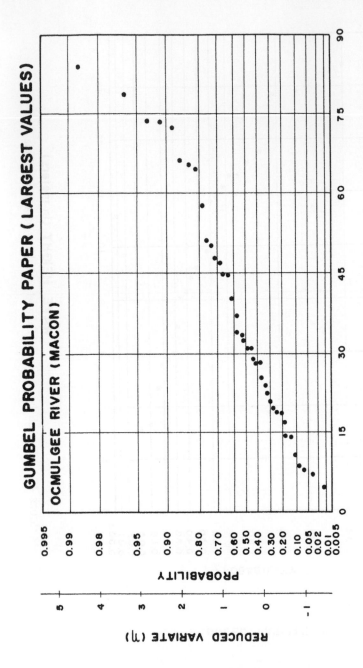

FIGURE 4.15.   Ocmulgee river (Macon) data drawn on Gumbel probability paper for maxima

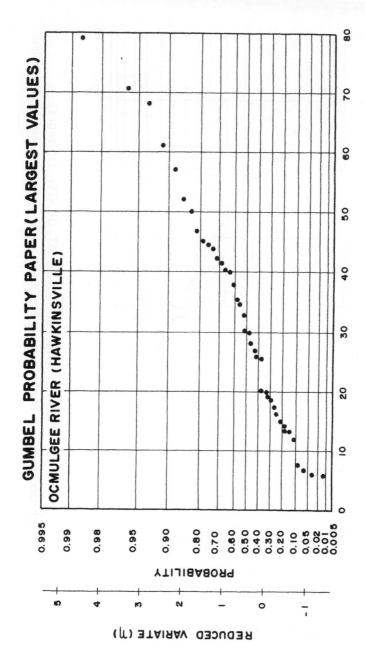

FIGURE 4.16. Ocmulgee river (Hawkinsville) data drawn on Gumbel probability paper for maxima

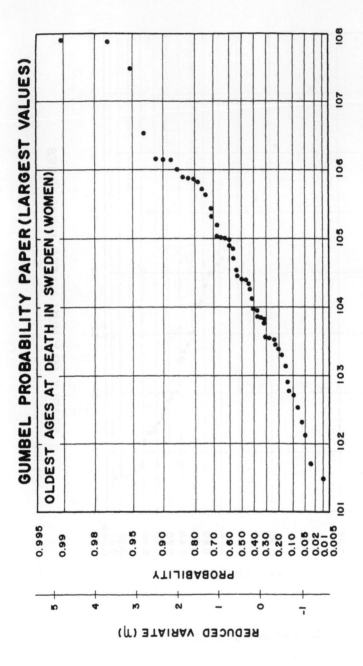

FIGURE 4.17. Oldest ages at death in Sweden (women) data drawn on Gumbel probability paper for maxima.

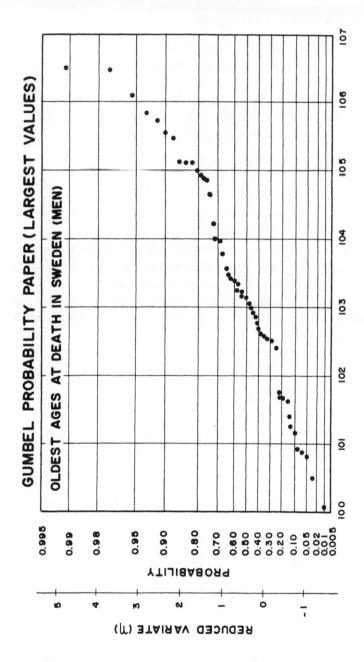

FIGURE 4.18. Oldest ages at death in Sweden (men) data drawn on Gumbel probability paper for maxima

and a comparison with (4.1) and (4.3) gives

(4.29)
$$\left.\begin{array}{l} \xi = g(x) = -\log(\lambda - x) \\ \eta = h(y) = -\log[-\log(y)] \end{array}\right\}$$

and

(4.30)
$$\left.\begin{array}{l} a = \beta \\ b = \beta \log \delta \end{array}\right\}$$

and the family of straight lines becomes

(4.31)
$$\eta = a\xi + b = \beta(\xi + \log \delta)$$

Note that the η-scale coincides with the one in the Gumbel probability paper but the ξ-scale is now logarithmic instead of arithmetic.

Because in this case we have three parameters, if λ is not known, we must proceed by trial and error with λ until a straight line trend is obtained for the cdf, and then proceed with the other two parameters β and δ.

Once λ has been estimated either by trial and error or by any other analytical method, the estimation of β and δ can be done by noting that for $\eta = 0$ and $\eta = 1$ we get

(4.32)
$$0 = \beta(\xi + \log \delta) \quad \rightarrow \quad \xi = -\log \delta$$

(4.33)
$$1 = \beta(\xi + \log \delta) \quad \rightarrow \quad \xi = \frac{1}{\beta} - \log \delta$$

In the case of minima instead of maxima, a change of sign of the data allows the use of the above probability paper, because the Weibull family of cdfs for minima is given by

(4.34)
$$\underline{F}(x; \lambda, \beta, \delta) = 1 - \exp\left[-\left(\frac{x - \lambda}{\delta}\right)^{\beta}\right]; \quad \lambda \le x < \infty$$

From (4.27) we get

(4.35)
$$\underline{F}(x; \lambda, \beta, \delta) = 1 - F(-x; -\lambda, \beta, \delta)$$

which shows that in this case the change in sign of the data reduces the problem to one of maxima. Then the process can be followed by changing λ to $-\lambda$.

Example 4.15. (Telephone data).   Because the telephone data was shown not to reasonably fit a Gumbel distribution in example 4.6, it will be fitted now to a Weibull model. However, as indicated above, the plot on Weibull probability paper requires knowledge of the threshold parameter λ. In this case, due to physical reasons, an obvious candidate for the value of this parameter is zero. Thus, the data can immediately be drawn on Weibull probability paper for smallest values and figure 4.19. is obtained, which shows a linear trend on the right tail with the only exception of the first order statistic. Thus, a Weibull distribution for minima with $\lambda = 0$ can be assumed. The estimation of the parameters δ and β can be done by the method suggested above, i.e.

$$\eta = 0 \quad \rightarrow \quad \xi = -\log \delta = -\log(0.001)$$

$$\eta = -1 \quad \rightarrow \quad \xi = \frac{1}{\beta} - \log \delta = -\log(0.00042)$$

from which $\delta = 0.001$ and $\beta = 1.152$.   ∎

Example 4.16. (Epicenter data).   The epicenter data was plotted in example 4.7. on Gumbel probability paper, and its curvature in the left tail suggested a Weibull type limit distribution. For this reason, the data was plotted (see figure 4.20.) on Weibull probability paper for a zero threshold parameter value and no linear trend in the left tail was obtained Figure 4.9. and 4.20. suggest a threshold value in the neighborhood of 58 miles. However, geological and geotechnical investigations revealed the presence of a fault as the main cause of earthquakes to occur in the area. Thus, due to the fact that the closest distance from the fault to the nuclear power plant's tentative location was 50 miles, this value was taken as the threshold value for the parent distribution of distances from the nuclear plant to epicenters. The associated Weibull probability plot is shown in figure 4.21.   ∎

Example 4.17. (Fatigue data).   In order to plot the fatigue data on Weibull probability paper, a value of the threshold parameter is needed. One obvious lower bound for the number of cycles to failure is zero; thus, taking $\lambda = 0$ the plot in figure 4.22. was obtained. It shows no linear trend at the left tail and suggests a threshold value, λ, larger than zero. So by trial and error technique, different plots can be made with different λ values until linearity in the tail is obtained. A good value is given by $\lambda = 34000$ cycles (see figure 4.23.).   ∎

4.2.5.   Frechet probability paper.

The cdf curves of the Frechet family for maxima are given by

$$(4.36) \qquad y = F(x; \lambda, \beta, \delta) = \exp\left[-\left(\frac{\delta}{x - \lambda}\right)^{\beta}\right]; \qquad \lambda \le x < \infty$$

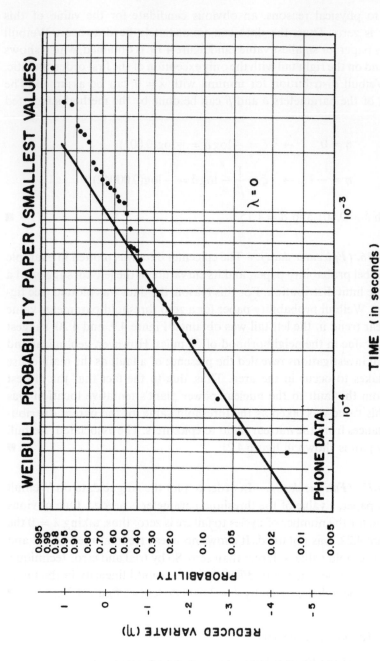

FIGURE 4.19.   Phone data drawn on Weibull probability paper for minima

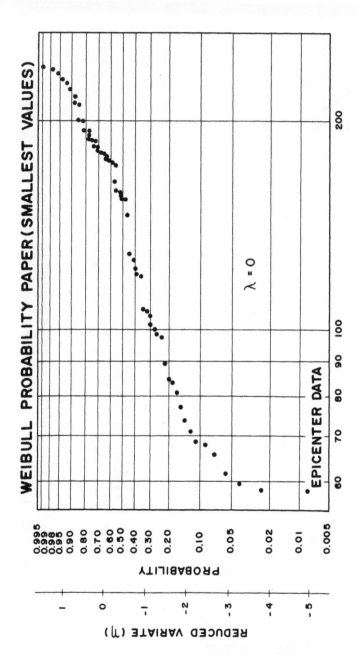

FIGURE 4.20. Epicenter data drawn on Weibull probability paper for minima ($\lambda = 0$)

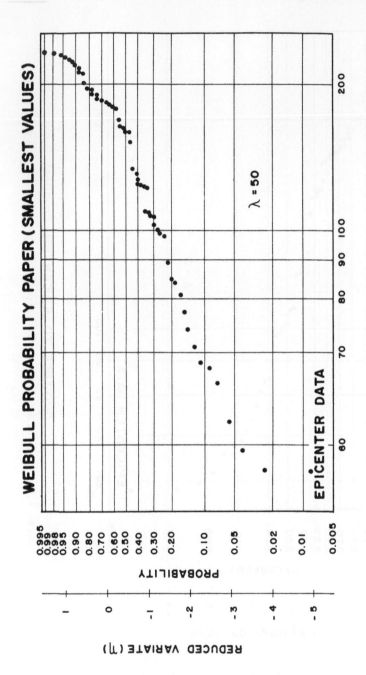

FIGURE 4.21. Epicenter data drawn on Weibull probability paper for minima ($\lambda = 50$)

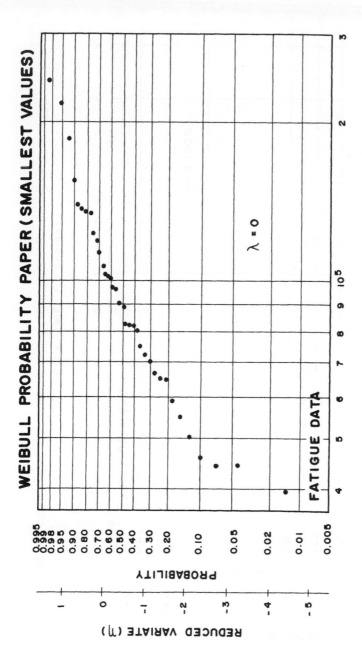

FIGURE 4.22.   Fatigue data drawn on Weibull probability paper for minima ($\lambda = 0$)

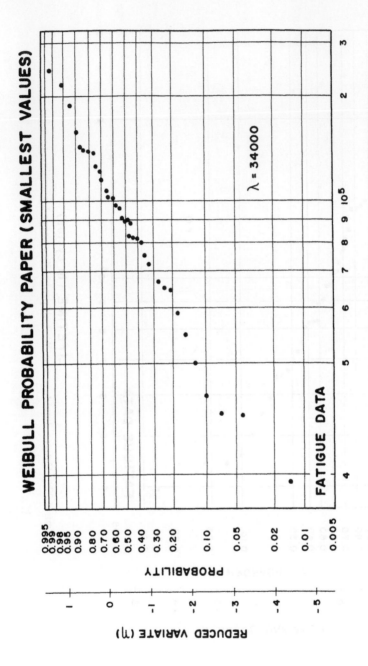

FIGURE 4.23. Fatigue data drawn on Weibull probability paper for minima ($\lambda = 34000$)

Upon taking logarithms twice we get

$$(4.37) \quad -\log[-\log(y)] = -\beta \log\left(\frac{\delta}{x-\lambda}\right) = -\beta \log \delta + \beta \log(x-\lambda)$$

and a comparison with (4.1) and (4.3) gives

$$(4.38) \qquad \left. \begin{array}{l} \xi = g(x) = \log(x-\lambda) \\ \eta = h(y) = -\log[-\log(y)] \end{array} \right\}$$

and

$$(4.39) \qquad \left. \begin{array}{l} a = \beta \\ b = -\beta \log \delta \end{array} \right\}$$

and the family of straight lines becomes

$$(4.40) \qquad \eta = a\xi + b = \beta(\xi - \log \delta)$$

It is interesting to note that the Frechet probability paper coincides with the Weibull probability paper (compare (4.29) and (4.38)) with the only difference being that $\log(x - \lambda)$ must be represented instead of $-\log(\lambda - x)$ on the abscissas axis

Example 4.18. (Wind data).   The wind data, which was shown to have a Frechet type right tail (see example 4.3.) is represented now on Frechet probability paper for maximum values (see figure 4.24.). A value of $\lambda = 0$ was used because zero is a natural lower bound for wind speed. The resulting plot shows linearity in the right tail.   ■

4.3.   The Problem of Plotting Positions

In the introduction to this chapter we mentioned that one of the steps when using probability papers is the representation of the data on the paper. More precisely, the empirical cdf must be drawn. It is well known that this is a step function with steps taking the values $0, 1/n, 2/n, \ldots, 1$. However, the two extremes 0 and 1, when transformed by the probability paper transformation (4.1) become $-\infty$ and ∞, respectively, for many probability paper families. This makes it impossible to draw these steps. Because the step at the $x_{(i)}$ order statistic jumps from $(i - 1)/n$ to i/n, a compromise solution to this problem consists of using the value $(i - 1/2)/n$ and then, the following set of points can

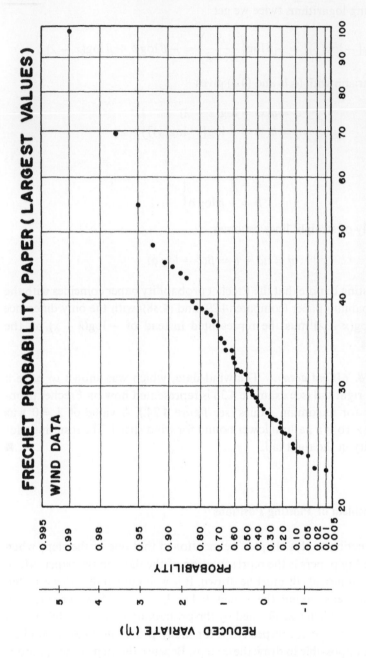

FIGURE 4.24. Wind data drawn on Frechet probability paper for maxima

be plotted on the paper:

$$(4.41) \qquad \left(x_{(i)}, \frac{i - \frac{1}{2}}{n}\right); \qquad i = 1, 2, \ldots n$$

This plotting position formula comes from Hazen.

If the cdf $F(x)$ were known, every value x would have only one associated $p = F(x)$ value. In this case there is no doubt about which point should be plotted on the probability paper because there is only one possibility. However, due to the random character of the sample to the i-th order statistic, $x_{(i)}$, a whole set of values (generally infinitely many) of p can occur. If $F(x)$ is fixed, $x_{(i)}$ becomes a random variable and so is the value $U_{(i)} = F(x_{(i)})$, which has been denoted by $U_{(i)}$ because it is the i-th order statistic of a sample of size n taken from a uniform $U(0, 1)$ population. Consequently, any value between 0 and 1 is admissible and then, any value in that range could be initially associated with the plotting position. However, the pdf of the i-th order statistic for the uniform $U(0, 1)$ governs its random behavior, and some central or characteristic value, as the mode, mean or median seems to be more justified.

Because the mean value of $U_{(i)}$ is $i/(n + 1)$ (see 2.41)) one of the recommended plotting position formulas is

$$(4.42) \qquad \left(x_{(i)}, \frac{i}{n + 1}\right); \qquad i = 1, 2, \ldots, n$$

Similarly, for the mode and median of $U_{(i)}$, two more plotting positions can be obtained.

However, as pointed out by Chernoff and Lieberman (1954,1956), the problem of plotting positions is closely related to the purpose for which the graph is drawn, and no general plotting position formula can be given with optimum characteristics unless this purpose is clearly stated. Kimball (1960) has also indicated the importance of having in mind the purpose of the probability paper, which generally falls in one of the following three, possibly overlapping, categories:

1. To test whether or not the sample comes from a given family of distributions.
2. To estimate the parameters of the family.
3. To graphically extrapolate at one of the extremes.

This last is the purpose most commonly served by plotting data from an extreme value inverse which is the case with which we are mainly concerned.

At this point, we can say that the optimum plotting formula depends on the purpose of the graphic technique and on the kind of paper to be used.

It is interesting to note here that probability papers are thought for visual fitting of the cdf of data by straight lines. One of the most important difficulties in analyzing the optimality of plotting positions is to state in a precise way what is understood by "visual fitting". It can be said without error that the visual fitting technique is clearly dependent on the individual who does it, and that no general equivalent analytical technique exists. This has led to some authors, as Chernoff and Lieberman (1954,1956), to assume that the visual technique can be made equivalent to the least-squares technique. However, it seems perhaps more natural to assume that the visual fitting technique is equivalent to the minimization of the sum of the absolute errors or deviations. Other authors, as Kimball (1960) for example, have criticized this assumption in the case of non symmetrical distributions. However, the analysis of the plotting position formulas requires one assumption and as a starting point or if the visual process is substituted by an analytical least-squares method the assumption of Chernoff and Lieberman can be satisfactory.

The very interesting idea of these two authors consists of choosing the plotting positions in such a way that the estimation of the location and scale parameters of the family by fitting a least-squares straight line leads to either unbiased minimum variance linear estimators or to minimum mean square error linear estimators.

Although this technique appears, in principle, very appealing, some short-comings make it useless in practice, at least in some situations. One of the problems, pointed out by Kimball (1960), is that for some cases the plotted line deviates far from the straight line and this voids the main purpose of probability papers. Another important problem is that the trend of the plotted cdf can be so altered that it can even become a decreasing curve. All this led Blom (1958) to state that one of the conditions to be satisfied by any plotting formula must be the following: "The points $(x_{(i)}, p_i)$ must lie on the average on a line which deviates only very little from a straight line".

One way of avoiding this inconvenience consists of using one of the above mentioned plotting position formulas, (4.41) for example, and then fitting a straight line by weighted least-squares. This process is equivalent to the one given by Chernoff and Lieberman, as they themselves suggest.

Another plotting position formula, given by Blom is

$$(4.43) \qquad \left(x_{(i)}, \frac{i - \frac{3}{8}}{n + \frac{1}{4}} \right); \qquad i = 1, 2, \ldots, n$$

Plotting position formula
$\left(x_{(i)}, \dfrac{i}{n+1}\right)$
$\left(x_{(i)}, \dfrac{i-\frac{3}{8}}{n+\frac{1}{4}}\right)$
$\left(x_{(i)}, \dfrac{i-\frac{1}{2}}{n}\right)$
$\left(x_{(i)}, \dfrac{i-0.44}{n+0.12}\right)$
$\{x_{(i)}, F[E(X_{(i)})]\}$
$[x_{(i)}, F\,(\text{median of } X_{(i)})]$
$[x_{(i)}, F\,(\text{mode of } X_{(i)})]$

TABLE 4.1.   Some plotting position formulas

which is recommended for use with normal probability paper for acceptance tests, estimation purposes and extrapolation.

Finally, the plotting formula given by Gringorten

$$(4.44) \qquad \left(x_{(i)}, \frac{i-0.44}{n+0.12}\right); \qquad i = 1, 2, \ldots, n$$

is designed to be used with Gumbel probability paper for acceptance tests and extrapolation purposes.

All of these formulas are summarized in table 4.1.

The previous paragraphs show that the plotting position formulas can affect the linear trend of the cdf so that a careful selection must be made.

There is another point to be discussed in order to clarify the meaning of visual fitting of the data, even if the least-squares method is accepted. Chernoff and Lieberman suggest measuring the quadratic error in the direction of X, i.e. in the random variable units. They argue for doing so because the $x_{(i)}$ are the random variables. However, it is well known that probability papers are presented with the random variable scale as the ordinate and the probability as the abscissa in some applications. So, it is reasonable to assume that many applied engineers and scientists when told to visually fit the data would use the vertical errors independently of the chosen vertical scale. Thus, in some cases, the so called visual fitting of a line involves the probability errors instead of the random variable errors.

In our opinion, both methods are valid, because they are asymptotically equivalent. The errors in probability have the advantage of being non-dimensional and an analysis can be performed without any knowledge of the random variable under consideration. So, if this knowledge is incomplete, this method is better for defining a weighting technique for the fitting than the other. On the other hand, if knowledge of the random variable is complete some weighting technique can easily be defined in terms of the values of the random variable under study.

4.4. Acceptance Regions

Probability papers are a quick tool with wide acceptance among engineers and scientists. The main reasons for this are their simplicity and the large amount of ready-to-use information contained in them. However, they require a lot of subjectivity because of the plotting position problem, the visual fit and the final acceptance or rejection decision. In the last step, it must be taken into account that the variances of all order statistics are not equal and important differences in variances can occur for some location-scale parametric families. This means that some weighting must be used.

In order to reduce subjectivity, acceptance regions can be added to the empirical cdf plot. These regions allow an objective decision to be made as to whether or not the sample can be accepted as coming from the family under consideration.

4.5. Some Recommendations for the Use of Probability
Papers in Extreme Value Problems

This section is mainly concerned with the use of probability paper for extreme value problems. Whether the sample is known to come from a population from the given family or whether this condition is satisfied only asymptotically makes a large difference.

In the case of extreme value problems, the sample is known not to follow the prescribed distribution, i.e. the distribution for which the paper was designed, but only asymptotically.

Due to the fact that the only part of the cdf governing the behavior of the extremes (maximum or minimum) or the upper or lower order statistics are the tails (right or left tails respectively), the rest of the cdf information is not needed. This must be understood correctly if some important errors are to be avoided. To this end, we must point out here that two cdfs with the same right tail, the same values for the $(0.9, 1)$ range say, and very different values in the $(0, 0.9)$ interval (see figure 4.25.) have exactly the same extreme (maximum) limit distribution, and the use of the cdf in the left tail only disturbs the estimation process. The same argument can, of course, be used for the left tail when minima or lower order statistics are involved. Consequently, when the data are known not to exactly follow the associated probability paper distribution, some weights must be used in the fitting procedure and these weights can be even zero. In other words, some points must be excluded from the fitting procedure and the rest of the data points must be given different weights. The theory on which most of our recommendations are based is included in section 3.10.

In selecting the weights to be used in the fitting procedure, two important confronted facts must be taken into consideration. On one hand, estimated percentiles in the tail of interest have the largest variances. On the other hand, they approach more closely the theoretical limit distribution values. So, both facts must be balanced before taking a decision about the weights. However, one thing is true: the points in the other tail must be given small or null weights because for them both facts act in the same direction.

Note that, because only tail data is used in the analysis, we do not need a sample of maxima or minima, as is required by the classical approach, that fits the whole sample with a straight line. The proposed method (based on tail

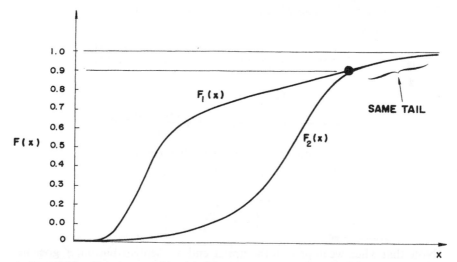

FIGURE 4.25. Two parent cdfs with the same extreme value distribution for maxima

data) is also applicable to samples from the parent population with cdf $F(x)$, the only difference being that a smaller fraction of the tail data will be used in the fitting procedure for the latter than would be used for the former. Note that in the former, assuming independence, the sample has cdf $F^n(x)$, and in the latter it has a cdf $F(x)$.

In order to illustrate graphically the role of tails in the extreme value behavior of a distribution we include in figures 4.26. to 4.29. the cdfs of several known distributions on Gumbel and Weibull probability paper. Because none of them is Gumbel or Weibull, all show non-linearity. Figure 4.26. shows the linearity of the right tail of the exponential distribution when plotted on Gumbel probability paper for largest values. This confirms that it belongs to that domain of attraction. Note also the curvatures of uniform and Cauchy distributions and the vertical and horizontal slopes when approaching their upper ends, thus suggesting Weibull and Frechet domains of attraction, respectively. Figure 4.27. shows the near linearity of the left tail for normal distribution when plotted on Gumbel probability paper for minima and suggests a Weibull-type domain of attraction for exponential and uniform distribution, and Frechet-type for Cauchy distribution. Note the verticallity of slopes of exponential and uniform cdfs close to their lower ends. Similarly, figures 4.28. and 4.29. confirm the Weibull type domain of attraction for both tails in the uniform case and for the left tail in the case of exponential distribution.

In the following section, two results of basic importance in connection with extreme value applications are proved.

If we represent the Von-Mises cdf $H_c(x)$ for largest values on Gumbel probability paper we have

$$\eta = -\log\left\{-\log\left[\exp\left[-\left(1 + \frac{c(\xi - \lambda)}{\delta}\right)^{-1/c}\right]\right]\right\}$$

$$= \frac{\log\left(1 + \frac{c(\xi - \lambda)}{\delta}\right)}{c}; \quad 1 + \frac{c(\xi - \lambda)}{\delta} \geq 0$$

where η and ξ are the ordinate and abscissa values (see (4.1))

Upon taking derivatives with respect to ξ we get

$$\eta' = \frac{1}{\left[\delta\left(1 + \frac{c(\xi - \lambda)}{\delta}\right)\right]}$$

Note that when we approach the upper end of the distribution, η' goes to infinity if $c < 0$ and η' goes to zero if $c > 0$. This fact is very useful in the

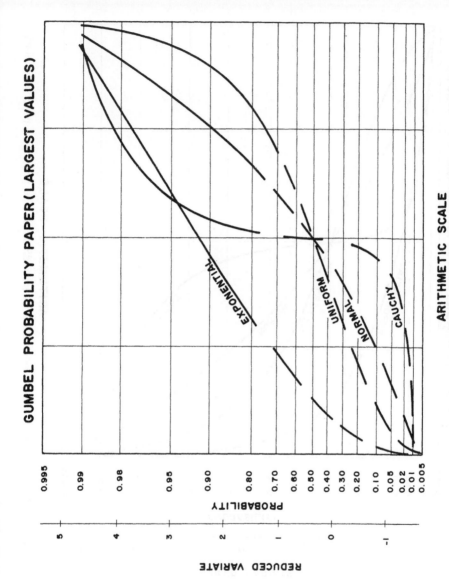

FIGURE 4.26. Exponential, uniform, normal and Cauchy cdfs on Gumbel probability paper for maxima

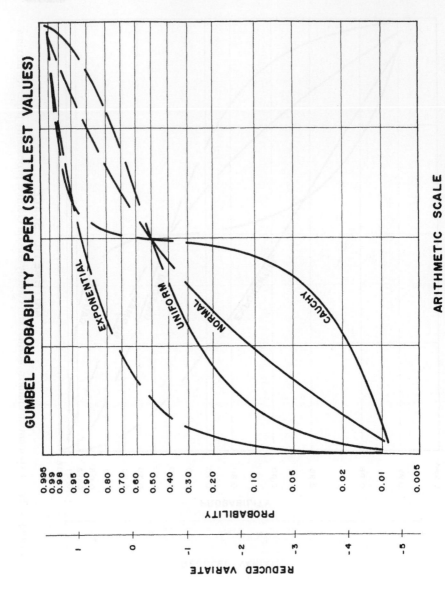

FIGURE 4.27.   Exponential, uniform, normal and Cauchy cdfs on Gumbel probability paper for minima

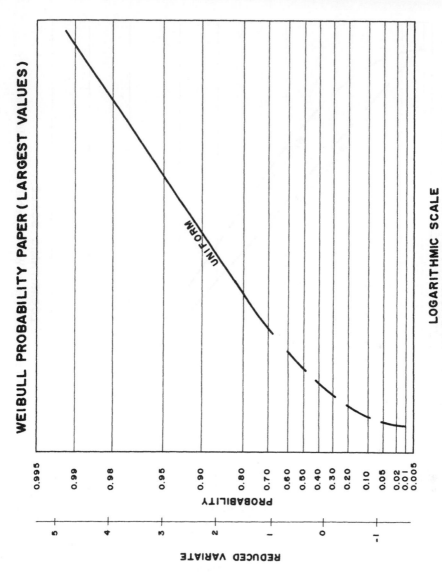

FIGURE 4.28.   Uniform cdf on Weibull probability paper for maxima

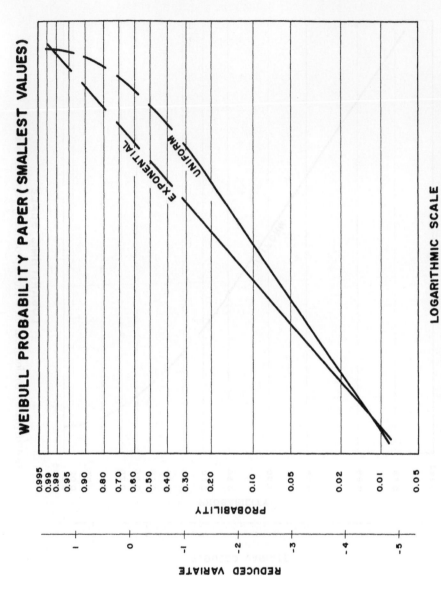

FIGURE 4.29. Exponential and uniform cdfs on Weibull probability paper for minima

identification of Weibull and Frechet type distributions, which show vertical and horizontal slopes, respectively, at the end of interest. Note, however, that η' can tend to zero or infinity for Gumbel type distributions if $\delta \to \infty$ or $\delta \to 0$, respectively.

Upon taking derivatives once more with respect to ξ we get

$$\eta'' = -\frac{c}{\left[\delta^2\left(1 + \frac{c(\xi - \lambda)}{\delta}\right)^2\right]}$$

which shows that

$$\eta'' > 0 \qquad \text{if, and only if } c < 0$$

$$\eta'' = 0 \qquad \text{if, and only if } c = 0$$

$$\eta'' < 0 \qquad \text{if, and only if } c > 0$$

If instead of $H_c(x)$ we use $L_c(x)$, the symbols "$<$", and "$>$" must be reversed.

Consequently, on Gumbel probability paper for largest (smallest) values, the Weibull distributions appear as convex (concave) curves, the Frechet distributions as concave (convex) curves, and the Gumbel distributions as straight lines. This property holds for all the range, in particular for the right tail.

This property is also useful in testing the domain of attraction of the cdf of a given population when only a sample from it is known.

The use of weights considerably complicates the fitting procedure, because visual fitting can be done only by highly trained people. For this reason, we recommend the use of probability papers for only the first of the purposes indicated by Kimball (1960), i.e. for testing whether or not the sample comes from a family of distributions with tails of given types. However, if some analytical methods, now easily available with the help of computers, are used, the extrapolation can also be done by means of probability paper.

In order to help in the selection of these analytical methods, we include the following section where some least-squares methods are discussed.

4.6. Weighted Least-Squares Methods

In the previous section we pointed out the need of some weighting technique to be used when fitting the data on probability paper. In this section we give some criteria for fixing the weights and their corresponding physical meaning. The methods given here are based on absolute and relative error measures of

Method	Function to be Minimized	Domain of Application
Least-squares probability absolute error	$\sum_{i=k}^{n} [p_i - F(x_{(i)}; \theta)]^2$	Not recommended
Least-squares probability relative error or least-squares return period relative error or tail equivalence	$\sum_{i=k}^{n} \frac{1}{(1 - p_i)^2} [p_i - F(x_{(i)}; \theta)]^2$	Right tail
	$\sum_{i=1}^{k} \frac{1}{p_i^2} [p_i - F_{(i)}; \theta)]^2$	Left tail
Least-squares return period absolute error	$\sum_{i=k}^{n} \frac{1}{(1 - p_i)^4} [p_i - F(x_{(i)}; \theta)]^2$	Right tail not recommended
	$\sum_{i=1}^{k} \frac{1}{p_i^4} [p_i - F(x_{(i)}; \theta)]^2$	Left tail not recommended
Standard weighted least-squares	$\sum_{i=k}^{n} \frac{1}{p_i(1 - p_i)} [p_i - F(x_{(i)}; \theta)]^2$	Both tails not recommended
General weighted least-squares	$\sum_{i=k}^{n} w_i [p_i - F(x_{(i)}; \theta)]^2$	Both tails

Note: If samples are maxima (or minima) of other samples of size m, p_i (or $(1 - p_i)$) must be substituted by $p_i^{1/m}$ (or $(1 - p_i)^{1/m}$).

TABLE 4.2.   Weighted least-squares methods for Extreme-value problems

probability and return periods and one of them is based on the concept of tail equivalence. The methods are given for fitting the right tail, i.e. when maxima or upper order statistics are of interest. However, table 4.2., which summarizes all methods, gives the corresponding methods for minima or lower order statistics.

All of the methods described assume minimization of errors in probability. Similar methods can be described for errors on the random variable scale.

4.6.1.   The least-square probability absolute error method.

This method is based on the minimization, with respect to θ, of the distance

$$(4.45) \qquad G_1 = \sum_{i=k}^{n} [p_i - F(x_{(i)}; \theta)]^2$$

where $p_i(i = k, k + 1, \ldots, n)$ are assumed here to be equal to $i/(n + 1)$, though similar distances could be given for p_i equal to the values associated with the order statistic $x_{(i)}$ for the selected plotting position formula, and $F(x_{(i)}; \theta)$ is given by

$$(4.46) \qquad F(x_{(i)}; \theta) = h^{-1}[ag(x_{(i)}) + b]$$

Note that the lower value of index i is k, i.e. the $k - 1$ first order statistics in the sample are not used in the estimation of the θ parameters. However, k can be made equal to 1 if the sample is known to come from that family of populations.

This method gives the same weight to all data points for $i \geq k$. However, it must be pointed out that the contributions to the total distance G_1 of the different order statistics is measured in terms of the square of the difference $[p_i - F(x_{(i)}; \theta)]$ and this means that a difference, of 0.01 say, in the neighbourhood of the 0.5 percentile is considered equivalent to the same difference in the 0.98 percentile, say. It is well known by engineers that this measuring error criteria is not acceptable because the return periods change from 2 to 2.04 in the first case and from 50 to 100 in the second. In other words, one error in the probability estimation of 0.01 in the tails is much worse that the same error in the center of the distribution.

4.6.2.   *The least-squares probability relative error method.*

In this case the distance to be minimized is

$$G_2 = \sum_{i=k}^{n} \left(\frac{p_i - F(x_{(i)}; \theta)}{1 - p_i} \right)^2$$

$$(4.47) \qquad = \sum_{i=k}^{n} \frac{1}{(1 - p_i)^2} [p_i - F(x_{(i)}; \theta)]^2$$

Note that the method is similar to the one above, the only difference being that now some weights, $1/(1 - p_i)^2$, have been added. The weight increases with i and in this way, some of the inconveniences of the above method are avoided. Note that now a difference of 0.01 in the neighbourhood of 0.98 is equivalent to a difference of 0.25 in the neighbourhood of 0.5, which in terms of return periods means that a change from 50 to 100 is equivalent to a change from 2 to 4.

If in (4.47) we assume $k = 1$, we have that all data points are given a weight other than zero. However, the relative weight of a small portion of the tail tends to 1 when n goes to infinity. More precisely, let $r(n)$ be a function of n, that represents the portion of the tail, such that $r(n) \to \infty$ and $r(n)/n \to 0$ (small portion), then we have

$$\lim_{n \to \infty} \frac{\sum\limits_{i=n-r(n)}^{n} \dfrac{1}{(1-p_i)^2}}{\sum\limits_{i=1}^{n} \dfrac{1}{(1-p_i)^2}} = \lim_{n \to \infty} \frac{\sum\limits_{i=n-r(n)}^{n} \dfrac{(n+1)^2}{(n-i+1)^2}}{\sum\limits_{i=1}^{n} \dfrac{(n+1)^2}{(n-i+1)^2}}$$

$$= \lim_{n \to \infty} \frac{1 + \dfrac{1}{2^2} + \dfrac{1}{3^2} + \cdots + \dfrac{1}{[r(n)+1]^2}}{1 + \dfrac{1}{2^2} + \dfrac{1}{3^2} + \cdots + \dfrac{1}{n^2}} = 1$$

This means that as n goes to infinity, the negligible proportion $r(n)/n$ of the data at the right tail concentrates almost all the weight, and proves that this method is clearly a fit of the right tail only, as it is convenient for extreme value purposes.

4.6.3.   The least-squares return period absolute error method.

The distance to be minimized by this method is

$$G_3 = \sum_{i=k}^{n} \left(\frac{1}{1-p_i} - \frac{1}{1-F(x_{(i)};\theta)} \right)^2$$

(4.48)
$$= \sum_{i=k}^{n} \frac{1}{(1-p_i)^2[1-F(x_{(i)};\theta)]^2} [p_i - F(x_{(i)};\theta)]^2$$

Because the term $[1 - F(x_{(i)};\theta)]^2$ depends on the sample, it can be substituted with $(1 - p_i)^2$.

Note that the name of the method comes from the fact that the two terms appearing in the first summation sign are the return periods associated with probabilities p_i and $F(x_{(i)};\theta)$, respectively.

In this case the weights also increase with i and the increase is much larger than in the previous case.

This method seems to give excessive weight to the data in the right tail. Every engineer or scientist knows that errors of 1 year in the return period in the neighborhood of 100 years and 1 year say, are not equivalent.

4.6.4.   The least-squares return period relative error method.

As the name of the method implies, the function to be minimized by this method is the sum of squares of the relative errors of return periods, i.e. the measure

$$G_4 = \sum_{i=k}^{n} \frac{\left(\dfrac{1}{1-p_i} - \dfrac{1}{1 - F(x_{(i)};\theta)}\right)^2}{\left(\dfrac{1}{1 - F(x_{(i)};\theta)}\right)^2}$$

(4.49)
$$= \sum_{i=k}^{n} \frac{1}{(1-p_i)^2} [p_i - F(x_{(i)};\theta)]^2$$

which coincides with G_2.

4.6.5. *The tail equivalence method.*

This method is based on the concept of tail equivalence. Two cdfs, $F(x)$ and $G(x)$, are said to be right tail equivalent if (see definition 3.3)

(4.50)
$$\lim_{x \to \infty} \frac{1 - F(x)}{1 - G(x)} = 1$$

Then, the tail equivalence method minimizes the measure

(4.51) $$G_5 = \sum_{i=k}^{n} \left(\frac{1 - F(x_{(i)};\theta)}{1 - p_i} - 1\right)^2 = \sum_{i=k}^{n} \frac{1}{(1-p_i)^2}[p_i - F(x_{(i)};\theta)]^2$$

which also coincide with G_2. With this, three different physical interpretations can be given to the minimization of function G_2.

4.6.6. *The probability paper method.*

If a straight line is fitted by the least-squares method to some data on probability paper the function which is minimized is

(4.52) $$G_v = \sum[h(p_i) - h(F(x_{(i)};\theta))]^2 = \sum w_i[p_i - F(x_{(i)};\theta)]^2$$

where the weights, w_i are determined by the $h(p)$ function and, consequently are dependent upon the type of paper used.

4.6.7. *The general weighted least-squares method.*

All the methods described in sections 4.6.1 to 4.6.5 are particular cases of the weighted least-squares method that minimizes the function

(4.53) $$G = \sum_{i=1}^{n} w_i[p_i - F(x_{(i)};\theta)]^2$$

with appropriate weights $w_i(i = 1, 2, \ldots, n)$.

One more fact must be remembered when selecting the weights to be used in the analysis. It is that the variance of $\hat{F}(x)$ is given by

$$(4.54) \qquad \operatorname{Var}[\hat{F}(x)] = \frac{F(x)[1 - F(x)]}{n}$$

So, if the usual weights are to be used, these must be proportional to the inverse of the variances, i.e.

$$(4.55) \qquad w_i = \frac{1}{F(x_{(i)}; \theta)[1 - F(x_{(i)}; \theta)]} = \frac{1}{p_i(1 - p_i)}$$

Note that these weights give smaller relative importance to data on the tails than the weights in expressions (4.48) and (4.49). After consideration of the asymptotic convergence of the empirical cdf to the associated probability paper cdf, the fact that points closer to the tails give better approximation and the property given in section 4.6.2, the weights in expression (4.49) are recommended for the fitting procedure.

Finally, as one example of a fitting method using the random variable scale error we include the function

$$(4.56) \qquad G_H = \sum_{i=1}^{n} [g(x_{(i)}) - g(F^{-1}(p_i; \theta))]^2$$

which minimizes the sum of squares of errors in the random variable measured in the probability paper units.

A listing of a computer program for the least-squares fit of data by the above methods is included in Appendix B.

Example 4.19. (Gumbel parameters estimated by least-squares methods). Table 4.3. shows the values of the Gumbel parameters, λ and δ, estimated by means of the least-squares probability absolute error, the least-squares probability relative error and the standard least-squares method, for the data sets that revealed Gumbel behavior in the previous examples. Note that the first and the third methods give very similar results, and that some differences are observed with the second method. These differences are especially important for the scale parameter δ in the cases of the electrical insulation data, the Houmb data, the precipitation data, the oldest ages at death in Sweden (men and women) and the Ocmulgee river (Macon and Hawkinsville). Thus revealing a tail trend different from the overall trend. Consequently, the second method must be used for design when extreme values are of interest, and extrapolation to large n is needed.

Data	Fitted Model	Least-Squares Probability Absolute Error		Least-Squares Probability Relative Error		Standard Least-Squares		Maximum Likelihood	
		λ	δ	λ	δ	λ	δ	λ	δ
Flood	Gumbel (largest)	38.97	6.83	38.42	7.46	38.92	6.99	38.60	7.77
Wave	Gumbel (largest)	11.34	5.59	11.33	5.56	11.33	5.62	11.30	5.60
Link	Gumbel (smallest)	93.73	15.85	91.75	12.24	93.59	14.79	94.29	15.55
Electrical Insulation	Gumbel (smallest)	1130.80	142.44	1099.20	92.82	1134.30	132.70	1170.50	174.10
Rain Data	Gumbel (smallest)	44.10	5.50	42.98	3.64	44.14	5.06	44.18	5.03
Houmb Data	Gumbel (largest)	8.84	2.05	9.39	1.19	8.77	1.86	8.66	1.99
Ocmulgee (Macon)	Gumbel (largest)	26.07	19.67	28.48	14.78	26.51	18.11	26.71	17.61
Ocmulgee (Hawkins)	Gumbel (largest)	23.49	17.06	25.73	13.44	23.87	15.70	23.71	15.06
Ages Death (women)	Gumbel (largest)	103.94	1.25	104.14	0.90	103.91	1.20	103.81	1.34
Ages Death (men)	Gumbel (largest)	102.56	1.45	102.82	0.94	102.53	1.37	102.47	1.45

TABLE 4.3. Gumbel parameters estimated by different methods

As mentioned in previous examples this can also be due to Weibull or Frechet type tails (curvature in the tail of interest when the empirical cdf is plotted on Gumbel probability paper).   ∎

Example 4.20. (Weibull and Frechet parameters estimated by least-squares methods).   Table 4.4. shows the values of the Weibull or Frechet parameters δ and β, estimated by the same methods as those in example 4.15. for the data that revealed Weibull- or Frechet type behavior in the previous examples. Note that the differences between the methods are much smaller than in example 4.19., due to the fact that the data fit to straight lines is much better.
                                                                                         ∎

If instead of parent distribution data, maximum values of samples of size m are used, the following weights are recommended

$$w_i = \frac{1}{(1 - p_i^{1/m})^2}$$

These are based on the fact that if the parent has cdf $F(x)$, the maximum of samples of size m has cdf $F^m(x)$, and the same values of x must lead to the same weights.

4.6.8.   Tests of hypotheses based on least-squares methods.

The least-squares methods allow point estimation of the parameters of any given family, $F(x; \theta)$, of cdfs, according to different criteria. In particular, the statistic in (4.47) allows one to fit the Von-Mises family to the right tail. However, quality of the fit can not be tested while distributional properties of the statistics on which the fit was made remain unknown.

In this section we give some of these properties for the statistics G_1 and G_2.

Under the assumption that the sample comes from a population with cdf $F(x; \theta)$, and because p_i was chosen as $E[U_{(i)}] = i/(n + 1)$, the statistic (4.47) can be written

$$(4.57) \qquad\qquad G = \sum_{i=1}^{n} w_i \{ U_{(i)} - E[U_{(i)}] \}^2$$

where $U_{(i)}(i = 1, 2, \ldots, n)$ are the order statistics of a sample from the uniform population.

Taking expectations and variances in (4.57) we get

Data	Fitted Model	Least-Squares Probability Absolute Error		Least-Squares Probability Relative Error		Standard Least-Squares		Maximum Likelihood	
		δ	β	δ	β	δ	β	δ	β
Wind	Frechet (largest) ($\lambda = 0$)	28.34	4.81	27.83	4.30	28.57	5.10	28.74	5.10
Phone	Weibull (smallest) ($\lambda = 0$)	1.24×10^{-3}	1.01	1.09×10^{-3}	1.30	1.24×10^{-3}	1.06×10^{-3}	1.25	1.08
Epicenter	Weibull (smallest) ($\lambda = 0$)	166.3	2.50	160.5	3.84	163.4	2.93	162.53	2.98
Fatigue	Weibull (smallest) ($\lambda = 0$)	104953	2.54	95107	4.19	105770	2.77	111842	2.25
Fatigue	Weibull (smallest) ($\lambda = 14000$)	90586	2.12	82327	3.17	91322	2.26	96069	1.95
Fatigue	Weibull (smallest) ($\lambda = 34000$)	69891	1.50	68372	1.59	70213	1.48	71482	1.44
Epicenter	Weibull (smallest) ($\lambda = 50$)	113.5	1.51	110.9	1.57	106.8	1.75	105.3	1.68

TABLE 4.4. Weibull and Frechet parameters estimated by different methods

Statistic	Sample Size (n)	$E[G]$	$\text{Var}[G]$
	10	0.1515	0.01836
	20	0.1587	0.02016
	30	0.1613	0.02081
	40	0.1626	0.02115
G_1	50	0.1633	0.02136
	60	0.1639	0.02150
	70	0.1643	0.02160
	80	0.1646	0.02168
	90	0.1648	0.02174
	100	0.1650	0.02178
	10	1.8515	6.4268
	20	2.5251	12.3750
	30	2.9326	16.2445
	40	3.2243	18.9794
G_2	50	3.4511	21.0446
	60	3.6366	22.6791
	70	3.7934	24.0180
	80	3.9293	25.1433
	90	4.0491	26.1086
	100	4.1561	26.9500

TABLE 4.5.   Expected values and variances of statistics G_1 and G_2

$$E[G] = \sum_{i=1}^{n} w_i \, \text{Var}[U_{(i)}] = \sum_{i=1}^{n} \frac{w_i p_i q_i}{n+2}$$

(4.58)
$$= \sum_{i=1}^{n} \frac{w_i i(n-i+1)}{(n+1)^2(n+2)}$$

and

(4.59) $$E[G^2] = \sum_{i=1}^{n} w_i^2 E[U_{(i)} - p_i]^4 + 2 \sum_{i<j} w_i w_j E[(U_{(i)} - p_i)^2 (U_{(j)} - p_j)^2]$$

where $E[(U_{(i)} - p_i)^2 (U_{(j)} - p_j)^2]$ can be obtained from (2.65). Table 4.5. shows the values of $E[G]$ and Var $[G]$ for the statistics G_1 and G_2. With these values at hand and calculating the value of $G_i (i = 1$ or $2)$, for a given $F(x, \theta_0)$, a test of the hypothesis that the sample comes from a parent population with such cdf can be made.

Chapter 5   The Gumbel, Weibull and Frechet Distributions

5.1.   Introduction

From chapter three we know that the only possible asymptotic distributions for the i.i.d. case are Gumbel, Weibull and Frechet. This gives these three families an important role in the extreme value problem. This chapter is dedicated to these families of distributions. Their description and the estimation of their parameters will occupy most of the chapter.

A very important point to be made here is that, contrary to many publications dealing with Gumbel, Weibull and Frechet distributions, our concern is mainly related to these families only as asymptotic distributions, i.e. as approximate distributions. In some particular cases the parent distribution could be one of the above, but this is generally the exception. All this has important implications when data is to be analyzed. In fact, only data in the tails of interest must be used or, if all data is incorporated into the analysis, weights must give adequate compensation. In this context, censored data arise in a natural way, and the associated techniques become the prime techniques for extreme value problems. In spite of this, for the sake of completion, we include here some estimation methods which do not satisfy these requirements. We advise the reader to use these methods only for initial estimation purposes.

183

5.2. The Gumbel Distribution

The Gumbel cdf for maxima and minima are, respectively, given by

$$(5.1) \qquad G(x) = \exp\left[-\exp\left(\frac{-(x - \lambda)}{\delta}\right)\right]; \qquad -\infty < x < \infty; \delta > 0$$

and

$$(5.2) \qquad \underline{G}(x) = 1 - \exp\left[-\exp\left(\frac{-(\lambda - x)}{\delta}\right)\right]; \qquad -\infty < x < \infty; \delta > 0$$

where λ and δ are constants known as the location and scale parameters, respectively. Their associated pdf's are

$$(5.3) \qquad g(x) = \frac{\exp\left(\frac{-(x - \lambda)}{\delta}\right)\exp\left[-\exp\left(\frac{-(x - \lambda)}{\delta}\right)\right]}{\delta}; \qquad -\infty < x < \infty$$

and

$$(5.4) \qquad \underline{g}(x) = \frac{\exp\left(\frac{-(\lambda - x)}{\delta}\right)\exp\left[-\exp\left(\frac{-(\lambda - x)}{\delta}\right)\right]}{\delta}; \qquad -\infty < x < \infty$$

Change of X to $-X$ and λ to $-\lambda$ leads from $G(x)$ to $\underline{G}(x)$ or vice versa. Consequently, any property or estimation method, derived for $G(x)$, can be immediately translated into $\underline{G}(x)$ and vice versa.

The importance of the Gumbel distribution in practice is due to its extreme value behavior. It has been applied, either as the parent distribution or as an asymptotic approximation, to describe extreme wind speeds (Thom (1968a,b), Simiu et al. (1975,1976,1977,1978,1979,1982), Grigoriu (1984b)), sea wave heights (Longuet-Higgins (1952,1975), Dattatri (1973), Borgman (1970,1973), Battjes (1977), Losada (1978)), floods (Shane and Lynn (1964), Benson (1968), Reich (1970), Todorovic (1970, 1971, 1978, 1979), North (1980)), rainfall (Hershfield (1962), Reich (1970)), age at death (Gumbel and Goldstein (1964)), minimum temperature, rainfall during droughts, electrical strength of materials, air pollution problems (Singpurwalla (1972b), Barlow and Singpurwalla (1974)), geologic problems, naval engineering, etc.

The main parameters of the Gumbel distribution are summarized in table 5.1.

Parameter	Distribution	
	$$G(x) = \exp\left[-\exp\left(\frac{-(x-\lambda)}{\delta}\right)\right]$$ (Maxima)	$$\underline{G}(x) = 1 - \exp\left[-\exp\left(\frac{-(\lambda-x)}{\delta}\right)\right]$$ (Minima)
Mean	$\lambda + 0.57772\,\delta$	$\lambda - 0.57772\,\delta$
Median	$\lambda + 0.3665\,\delta$	$\lambda - 0.3665\,\delta$
Mode	λ	λ
Variance	$\dfrac{\pi^2\delta^2}{6}$	$\dfrac{\pi^2\delta^2}{6}$
Kurtosis coefficient	1.1396	-1.1396

TABLE 5.1. Some parameters of the Gumbel distribution

5.2.1. Parameter estimation.

There are many available methods for estimating the parameters of the Gumbel distribution. Some of them are described in the following sections. Even though the method of moments is known to be inefficient and inadequate for extreme value problems, it is given below because it is useful in order to obtain some starting values for other iterative methods. In what follows, only methods for distribution (5.2) or (5.4) will be given. However, as indicated above, change of X into $-X$ and λ into $-\lambda$ leads to the associated methods for (5.1) or (5.3).

5.2.1.1. The method of moments. As is well known, the method of moments consists in stating the equality of as many moments as parameters for population and sample. Because in this case only two parameters exist, the equality of mean and variance will suffice. In other words, the moment equations become

$$(5.5) \qquad \bar{x} = \lambda - 0.5772\,\delta$$

$$(5.6) \qquad s_x^2 = \frac{\pi^2\delta^2}{6}$$

where $\bar{x}$ and s_x^2 are the sample mean and quasi-variance, respectively. From

(5.5) and (5.6) we get the following moment estimates

(5.7)
$$\delta = \frac{s_x \sqrt{6}}{\pi}$$

(5.8)
$$\lambda = \bar{x} + 0.5772\,\delta$$

Example 5.1. (Data examples). Table 5.2. gives the moment estimates of λ and δ for those data sets included in appendix C that revealed Gumbel-type extreme value behavior.   ■

5.2.1.2. The Maximum-likelihood method.

The maximum likelihood method, as its name indicates, is based on the maximization of the likelihood function with respect to the parameters to be estimated.

In the following section, in order to include censoring to the left or to the right (note that this is the natural situation for extreme value problems), we maximize, with respect to the parameters, the likelihood function of the $n - r_1 - r_2$ order statistics

$$X_{r_1 + 1:n}, X_{r_1 + 2:n}, \ldots, X_{n - r_2:n}$$

of a sample coming from a parent population with cdf (5.2), which, according to (2.11) becomes

$$L = \frac{n!}{r_1! r_2!} \left(\frac{1}{\delta}\right)^{n - r_2 - r_1} \exp\left\{ \sum_{i = r_1 + 1}^{n - r_2} \left(\frac{\lambda - x_i}{\delta}\right) - \sum_{i = r_1 + 1}^{n - r_2} \exp\left[\left(\frac{\lambda - x_i}{\delta}\right)\right] \right.$$

(5.9)
$$\left. - r_2 \exp\left[\left(\frac{\lambda - x^{**}}{\delta}\right)\right] \right\} \cdot \left\{ 1 - \exp\left\{ -\exp\left[-\left(\frac{\lambda - x^*}{\delta}\right)\right] \right\} \right\}^{r_1}$$

Note that this likelihood function remains valid for complete data and right and left type II censoring, i.e. when only some central, left or right order statistics are observed, by setting $x^* = x_{r_1 + 1:n}$ and $x^{**} = x_{n - r_2:n}$.

In the case of type I censoring, i.e. if the sample is restricted to values in the interval (y_0, y_1), the same likelihood function is still valid if x^* and x^{**} are substituted with the censoring values at the left, y_0, and the right, y_1, tails, respectively.

Maximization of L is equivalent to maximization of its logarithm less its constant terms

$$\text{Log}\,L = (r_2 + r_1 - n)\log \delta - \sum_{i = r_1 + 1}^{n - r_2} \left[\frac{\lambda - x_i}{\delta} + \exp\left(\frac{-(\lambda - x_i)}{\delta}\right) \right]$$

(5.10)
$$- r_2 \exp\left(\frac{-(\lambda - x^{**})}{\delta}\right) + r_1 \log\left\{ 1 - \exp\left[-\exp\left(\frac{-(\lambda - x^*)}{\delta}\right)\right] \right\}$$

TABLE 5.2. Estimates of Gumbel parameters for some examples in Appendix C

Data	Fitted Model	Percentile Method		Method of Moments		Maximum Likelihood		Covariance Matrix	
		λ	δ	λ	δ	λ	δ		
Flood Data	Gumbel (largest)	38.77	7.18	38.66	7.32	38.60	7.77	1.12 −0.254	0.565
Wave Data	Gumbel (smallest)	11.33	5.58	11.36	5.47	11.30	5.60	0.696 −0.162	0.381
Link Data	Gumbel (smallest)	94.39	15.12	93.65	13.46	94.29	15.55	13.55 −3.15	6.85
Insulation Data	Gumbel (smallest)	1160.6	136.4	1159.2	131.8	1170.5	174.1	1150.3 −271.3	528.8
Precipitation Data	Gumbel (smallest)	44.27	5.07	43.98	4.51	44.18	5.03	0.7058 −0.1675	0.378
Houmb Data	Gumbel (largest)	8.67	1.84	8.77	1.63	8.66	1.99	0.186 −0.0435	0.093
Ocmulgee (Macon)	Gumbel (largest)	26.38	18.67	27.18	17.06	26.71	17.61	8.586 −2.05	4.97
Ocmulgee (Hawkins)	Gumbel (largest)	23.36	15.92	24.10	14.44	23.71	15.06	6.22 −1.49	3.57
Oldest Ages (women)	Gumbel (largest)	103.83	1.20	103.87	1.12	103.81	1.34	0.037 −0.0085	0.018
Oldest Ages (men)	Gumbel (largest)	102.49	1.34	102.53	1.25	102.47	1.45	0.043 −0.0102	0.022

One way of calculating maximum likelihood estimates is the following. Upon setting the partial derivatives of $\log L$ with respect to the parameters to zero, one gets

$$\delta \frac{\partial \log L}{\partial \delta} = r_2 + r_1 - n - \sum_{i=r_1+1}^{n-r_2} (\lambda - x_i) \frac{\left[\exp\left(\frac{-(\lambda - x_i)}{\delta} \right) - 1 \right]}{\delta}$$

$$- \frac{r_2 \exp\left(\frac{-(\lambda - x^{**})}{\delta} \right)(\lambda - x^{**})}{\delta}$$

$$+ r_1 \left\{ 1 - \exp\left[-\exp\left(\frac{-(\lambda - x^*)}{\delta} \right) \right] \right\}^{-1}$$

$$(5.11) \qquad \cdot \frac{\left\{ \exp\left[-\exp\left(\frac{-(\lambda - x^*)}{\delta} \right) \right] \right\} \exp\left(\frac{-(\lambda - x^*)}{\delta} \right)(\lambda - x^*)}{\delta} = 0$$

$$\delta \frac{\partial \log L}{\partial \lambda} = r_2 + r_1 - n + \sum_{i=r_1+1}^{n-r_2} \exp\left(\frac{x_i - \lambda}{\delta} \right)$$

$$+ r_2 \exp\left(\frac{-(\lambda - x^{**})}{\delta} \right) - r_1 \left\{ 1 - \exp\left[-\exp\left(\frac{-(\lambda - x^*)}{\delta} \right) \right] \right\}^{-1}$$

$$(5.12) \qquad \cdot \exp\left[-\exp\left(\frac{-(\lambda - x^*)}{\delta} \right) \right] \exp\left(\frac{-(\lambda - x^*)}{\delta} \right) = 0$$

The non-linear system (5.11)-(5.12), when solved by numerical procedures, gives the maximum likelihood estimates.

If $r_1 = 0$ equations (5.11) and (5.12) simplify considerably and can be written as

$$(5.13) \qquad \lambda = \delta \log\left[\frac{\sum_{i=1}^{n-r_2} \exp\left(\frac{x_i}{\delta} \right) + r_2 \exp\left(\frac{x^{**}}{\delta} \right)}{n - r_2} \right]$$

$$(5.14) \qquad \frac{\sum_{i=1}^{n-r_2} x \exp\left(\frac{x_i}{\delta} \right) + r_2 x^{**} \exp\left(\frac{x^{**}}{\delta} \right)}{\sum_{i=1}^{n-r_2} \exp\left(\frac{x_i}{\delta} \right) + r_2 \exp\left(\frac{x^{**}}{\delta} \right)} - \delta - \sum_{i=1}^{n-r_2} \frac{x_i}{n - r_2} = 0$$

Note that equation (5.14) depends on δ only. Thus, it can be used to obtain the estimator δ. Then, equation (5.13) gives λ.

Another way of calculating maximum likelihood estimates is by the direct maximization of (5.10). A wide range of methods for such calculations are available in the literature, as the steepest ascent, the Powell, the Fletcher-Powell-Davidson, the secant, the Newton-Raphson or the random search methods, for example. All these methods start with some initial estimates, as those above, and they iteratively improve them by modifying the parameter values in such a way that the log-likelihood function is increased in every step. The process is repeated until some condition (gradient bound, precision requirement, etc.) is satisfied.

A computer program for the estimation by the maximum likelihood procedure using one of the methods described above is included in Appendix B.

A rapid rate of convergence of many of the above iterative method depends upon a good initial guess or estimate of the parameters. For this reason, initial estimates play an important role in any iterative estimation procedure. One good method, which from now on will be called the percentile method, is the following.

From (5.2) we get

(5.15) $$x_i = \lambda + \delta u_{p_i}$$

where

(5.16) $$u_{p_i} = \log[-\log(1 - p_i)]$$

If now $p_i = \underline{G}(x_{(i)})$, is approximated by $(i - 1/2)/n$, we can write

(5.17) $$\sum_{i=s_1}^{s_2} x_i = \lambda(s_2 - s_1 + 1) + \delta \sum_{i=s_1}^{s_2} \log\left[-\log\left(1 - \frac{i - \frac{1}{2}}{n}\right)\right]$$

for any pair, (s_1, s_2), of integers smaller than n.

Upon taking two pairs contained in the interval $[r_1 + 1, n - r_2]$, a linear system of two equations can be obtained, from which the starting values λ_0 and δ_0 can be derived.

Note that obtaining an initial estimate based on a freely selected fraction of the sample is possible by this method. In particular, this fraction can be in one of the tails, as required by extreme value or tail problems. On the other hand, the method of moments requires the full sample.

Example 5.2. (Data examples). Some estimates of λ and δ, obtained by the percentile method with the two pairs of values of s_1 and s_2: $(1, n/2)$ and $(n/2 + 1, n)$ are given in Table 5.2. ∎

An estimation of the asymptotic covariance matrix of the maximum likelihood estimates λ and δ can be obtained by inverting the local Fisher information matrix evaluated at (λ, δ), i.e.

$$(5.18) \qquad \begin{pmatrix} \text{Var}(\lambda) & \text{Cov}(\lambda, \delta) \\ \text{Cov}(\lambda, \delta) & \text{Var}(\delta) \end{pmatrix} = \begin{pmatrix} -\dfrac{\partial^2 L}{\partial \lambda^2} & -\dfrac{\partial^2 L}{\partial \lambda \partial \delta} \\ -\dfrac{\partial^2 L}{\partial \lambda \partial \delta} & -\dfrac{\partial^2 L}{\partial \delta^2} \end{pmatrix}^{-1}_{(\lambda, \delta)}$$

Due to the asymptotic normality of maximum likelihood estimates, confidence bounds of the parameters can be easily obtained by means of standard normal tables and the variances above.

An asymptotically normal estimator for the $100p$-th percentile is

$$(5.19) \qquad Y_p = \lambda + \delta u_{p_i}$$

and its variance is given by

$$(5.20) \qquad \text{Var}[Y_p] = \delta^2 [\text{Var}(\lambda) + u_p^2 \, \text{Var}(\delta) + 2u_p \, \text{Cov}(\lambda, \delta)]$$

When dealing with extreme value problems, a critical step in the estimation process is the selection of the r_1 and r_2 values or, in other words, selection of the fraction of the sample to be used. The main reason for using a small fraction of the sample is that the parent population generally does not follow an extreme value model. However, it is known that the tail behaves as one extreme value tail, asymptotically. Because of this asymptotic property, two factors, sample size and speed of convergence must be considered before any selection can be made. If the problem is related to smallest values, in order to have data from the left tail, r_1 must be small and r_2, large. A good choice for medium and high speed of convergence would be

$$r_1 = 0; \qquad r_2 = n - 2\sqrt{n}$$

Similarly, for largest values

$$r_1 = n - 2\sqrt{n}; \qquad r_2 = 0$$

For slower speeds of convergence the portion of the sample used should be reduced even more.

Example 5.3. (Data examples—complete sample). The maximum likelihood estimates of λ and δ, based on complete samples of some data sets in Appendix C are given in table 5.2. Their associated covariance matrices are given in the same table. We remind the reader that a complete sample must not be used for extreme value purposes unless the parent distribution is exactly Gumbel, Weibull or Frechet. ■

Example 5.4. (Data examples—tail data). Table 5.3. shows the maximum likelihood estimates of λ and δ, based on a small fraction $(2\sqrt{n})$ of the data in the tail of interest for some data sets in Appendix C. Note, by a comparison of tables 5.2. and 5.3., that estimates for the full sample and tail data differ substantially in the cases showing tails with slopes very different

Data	Fitted Model	Maximum Likelihood		Covariance Matrix	
		λ	δ		
Flood Data	Gumbel (largest)	40.64	6.44	7.28 3.408	2.574
Wave Data	Gumbel (largest)	13.48	4.63	3.592 1.711	1.425
Link Data	Gumbel (smallest)	89.73	12.07	28.05 12.45	15.74
Insulation Data	Gumbel (smallest)	1103.09	94.32	1673.8 792.3	795.8
Precipitation Data	Gumbel (smallest)	42.18	3.78	2.51 1.18	1.06
Houmb Data	Gumbel (largest)	10.22	0.90	0.150 0.0695	0.0799
Ocmulgee (Macon)	Gumbel (largest)	31.54	16.03	44.98 21.22	19.08
Ocmulgee (Hawkinsville)	Gumbel (largest)	30.20	11.73	24.74 11.82	10.54
Age at Death (women)	Gumbel (largest)	104.82	0.674	0.0824 0.0389	0.0302
Age at Death (men)	Gumbel (largest)	103.76	0.748	0.0989 0.0464	0.0365

TABLE 5.3.   Estimates of Gumbel parameters for some examples in Appendix C (Tail data)

from the overall slope of the cdf's, as in the cases of insulation data. Houmb data and age at death data, for example. Note also the larger variance values for tail estimates due to the fact that only a fraction of the sample is utilized. ∎

5.2.1.3. Best linear unbiased estimators (BLUES). One very simple esti-mation procedure for small samples is by means of the best linear unbiased estimators, which were described in section 2.3.7.1. The BLUES for esti-mating the parameters of (5.2) are given by

(5.21)
$$\lambda^* = \sum_{i=r_1}^{n-r_2} a_i(n, r_1, r_2) X_{(i)}$$

i	$n-r_2$	n	$a_i(n,0,r_2)$	$b_i(n,0,r_2)$	i	$n-r_2$	n	$a_i(n,0,r_2)$	$b_i(n,0,r_2)$
1	2	4	-0.706	-0.869	2	2	4	1.706	0.869
1	3	5	-0.210	-0.434	2	3	5	-0.086	-0.364
3	3	5	1.296	0.799	1	3	6	-0.316	-0.447
2	3	6	-0.203	-0.389	3	3	6	1.519	0.835
1	4	6	-0.087	-0.286	2	4	6	-0.028	-0.265
3	4	6	0.065	-0.186	4	4	6	1.050	0.737
1	3	7	-0.404	-0.455	2	3	7	-0.301	-0.406
3	3	7	1.705	0.861	1	4	7	-0.146	-0.294
2	4	7	-0.094	-0.276	3	4	7	-0.007	-0.210
4	4	7	1.247	0.780	1	5	7	-0.039	-0.211
2	5	7	-0.004	-0.206	3	5	7	0.046	-0.169
4	5	7	0.113	-0.099	5	5	7	0.884	0.686
1	3	8	-0.479	-0.461	2	3	8	-0.385	-0.418
3	3	8	1.864	0.879	1	4	8	-0.198	-0.300
2	4	8	-0.150	-0.283	3	4	8	-0.068	-0.227
4	4	8	1.416	0.811	1	5	8	-0.078	-0.217
2	5	8	-0.047	-0.213	3	5	8	-0.00009	-0.180
4	5	8	0.064	-0.123	5	5	8	1.062	0.733
1	3	9	-0.546	-0.466	2	3	9	-0.458	-0.428
3	3	9	2.003	0.893	1	4	9	-0.243	-0.314
2	4	9	-0.990	-0.290	3	4	9	-0.122	-0.240
4	4	9	1.564	0.834	1	5	9	-0.112	-0.222
2	5	9	-0.085	-0.217	3	5	9	-0.040	-0.189
4	5	9	0.021	-0.139	5	5	9	1.216	0.767
1	6	9	-0.045	-0.171	2	6	9	-0.024	-0.172
3	6	9	0.006	-0.155	4	6	9	0.044	-0.122
5	6	9	0.092	-0.072	6	6	9	0.926	0.692

TABLE 5.4. Coefficients of the BLUES for the Gumbel distribution (5.2)(Part one)

$$(5.22) \qquad \delta^* = \sum_{i=r_1}^{n-r_2} b_i(n, r_1, r_2) X_{(i)}$$

where $a_i(n, r_1, r_2)$ and $b_i(n, r_1, r_2)$ are coefficients which can be obtained by means of the methods in section 2.3.7.1. Some values, for $r_1 = 0$ and $r_2 > n - 2\sqrt{n}$, are given in table 5.4. For more complete tables with $2 \le n - r_2 \le n \le 25$ see Mann (1967a).

The covariance matrix is

$$(5.23) \qquad \begin{pmatrix} \mathrm{Var}(\lambda^*) & \mathrm{Cov}(\lambda^*, \delta^*) \\ \mathrm{Cov}(\lambda^*, \delta^*) & \mathrm{Var}(\delta^*) \end{pmatrix} = \delta^{*2} \begin{pmatrix} A(n, r_1, r_2) & C(n, r_1, r_2) \\ C(n, r_1, r_2) & B(n, r_1, r_2) \end{pmatrix}$$

where $A(n, r_1, r_2)$, $B(n, r_1, r_2)$ and $C(n, r_1, r_2)$ are values that can be obtained by

i	$n - r_2$	n	$a_i(n, 0, r_2)$	$b_i(n, 0, r_2)$	i	$n - r_2$	n	$a_i(n, 0, r_2)$	$b_i(n, 0, r_2)$
1	3	10	−0.605	−0.469	2	3	10	−0.522	−0.435
3	3	10	2.127	0.904	1	4	10	−0.283	−0.307
2	4	10	−0.242	−0.294	3	4	10	−0.169	−0.250
4	4	10	1.694	0.852	1	5	10	−0.143	−0.225
2	5	10	−0.117	−0.221	3	5	10	−0.075	−0.195
4	5	10	−0.017	−0.152	5	5	10	1.352	0.793
1	6	10	−0.069	−0.175	2	6	10	−0.051	−0.175
3	6	10	−0.023	−0.160	4	6	10	0.014	−0.131
5	6	10	0.060	−0.088	6	6	10	1.068	0.729
1	3	11	−0.658	−0.472	2	3	11	−0.580	−0.441
3	3	11	2.238	0.913	1	4	11	−0.319	−0.310
2	4	11	−0.281	−0.298	3	4	11	−0.211	−0.259
4	4	11	1.811	0.867	1	5	11	−0.170	−0.228
2	5	11	−0.147	−0.223	3	5	11	−0.106	−0.201
4	5	11	−0.051	−0.162	5	5	11	1.474	0.814
1	6	11	−0.091	−0.178	2	6	11	−0.074	−0.178
3	6	11	−0.048	−0.164	4	6	11	−0.013	−0.138
5	6	11	0.031	−0.101	6	6	11	1.194	0.758
1	3	12	−0.706	−0.475	2	3	12	−0.633	−0.446
3	3	12	2.339	0.921	1	4	12	−0.351	−0.312
2	4	12	−0.316	−0.301	3	4	12	−0.249	−0.265
4	4	12	1.916	0.878	1	5	12	−0.195	−0.230
2	5	12	−0.173	−0.226	3	5	12	−0.134	−0.205
4	5	12	−0.082	−0.171	5	5	12	1.584	0.832
1	6	12	−0.111	−0.180	2	6	12	−0.096	−0.180
3	6	12	−0.070	−0.167	4	6	12	−0.037	−0.144
5	6	12	0.005	−0.111	6	6	12	1.308	0.781

TABLE 5.4. Coefficients of the BLUES for the Gumbel distribution (5.2)(Part two)

$n-r_2$	n	$A(n,0,r_2)$	$B(n,0,r_2)$	$C(n,0,r_2)$	$n-r_2$	n	$A(n,0,r_2)$	$C(n,0,r_2)$	$B(n,0,r_2)$
3	5	0.529	0.417	0.235	3	6	0.653	0.333	0.432
4	6	0.324	0.270	0.102	3	7	0.788	0.417	0.443
4	7	0.373	0.280	0.157	5	7	0.235	0.050	0.198
3	8	0.927	0.450	0.489	4	8	0.431	0.205	0.288
5	8	0.258	0.205	0.086	3	9	1.066	0.553	0.456
4	9	0.494	0.293	0.248	5	9	0.287	0.118	0.211
6	9	0.198	0.161	0.051	3	10	1.204	0.611	0.461
4	10	0.559	0.298	0.286	5	10	0.321	0.146	0.215
6	10	0.214	0.166	0.073	3	11	1.339	0.662	0.464
4	11	0.626	0.301	0.321	5	11	0.357	0.172	0.219
6	11	0.234	0.169	0.094	3	12	1.471	0.709	0.468
4	12	0.693	0.304	0.352	5	12	0.395	0.196	0.222
6	12	0.256	0.172	0.113					

TABLE 5.5. Coefficients to calculate the variances and covariances of the BLUES

the same methods (see (2.123)). Some of these values are tabulated in table 5.5.

Two sided exact 100α confidence intervals for λ^* and δ^*, if $r_1 = 0$, are given by

$$\lambda^* - \frac{\delta^*\left[C + t\left(\dfrac{1+\alpha}{2}, n, n - r_2\right)\right]}{1 + B}, \quad \lambda^* - \frac{\delta^*\left[C + t\left(\dfrac{1-\alpha}{2}, n, n - r_2\right)\right]}{1 + B}$$

(5.24)

(5.25) $$\frac{\delta^*}{(1 + B)w\left(\dfrac{1+\alpha}{2}, n, n - r_2\right)}, \quad \frac{\delta^*}{(1 + B)w\left(\dfrac{1-\alpha}{2}, n, n - r_2\right)}$$

where the functions $t(\alpha, n, r)$ and $w(\alpha, n, r)$ are given in tables 5.6. and 5.7.

The BLUE estimator for the $100p$-th percentile $Y_p = \lambda + u_p\delta$ is

(5.26) $$Y_p^* = \lambda^* + u_p\delta^*$$

and its variance is

(5.27) $$\text{Var}[Y_p^*] = \delta^{*2}[A(n, r_1, r_2) + u_p B(n, r_1, r_2) + 2u_p C(n, r_1, r_2)]$$

These estimators have a distribution which is approximately normal $N(Y_p^*, \text{Var}(Y_p^*))$.

Though the BLUES have good statistical properties comparable to maximum likelihood estimates, their use in extreme value problems is highly restricted by the lack of tables of coefficients for sample sizes larger than 25. Note that $n = 25$ is a rather small size for extreme value problems. However, the future appearance of such tables for estimation from tail order statistics would make this method very attractive.

5.2.1.4.   Best linear invariant estimators (BLIES).   These estimators were discussed in section 2.3.7.2 in a general context. According to (2.124), (2.125), (5.21) and (5.22), they become

(5.28) $$\lambda^{**} = \sum_{i=r_1}^{n-r_2}\left\{a_i(n, r_1, r_2) - \frac{b_i(n, r_1, r_2)C(n, r_1, r_2)}{1 + B(n, r_1, r_2)}\right\}X_{(i)}$$

(5.29) $$\delta^{**} = \sum_{i=r_1}^{n-r_2}\frac{b_i(n, r_1, r_2)X_{(i)}}{1 + B(n, r_1, r_2)}$$

and their mean square errors are given by (2.127)–(2.129).

		α							
n	r	0.02	0.05	0.10	0.25	0.75	0.90	0.95	0.98
5	3	−9.35	−5.22	−3.04	−1.22	0.40	0.86	1.20	1.76
6	3	−10.54	−6.12	−3.72	−1.56	0.33	0.75	1.02	1.39
6	4	−3.69	−2.39	−1.59	−0.67	0.38	0.76	1.03	1.42
7	3	−13.00	−7.39	−4.45	−1.87	0.26	0.68	0.90	1.20
7	4	−4.67	−2.95	−1.94	−0.84	0.32	0.66	0.89	1.20
7	5	−2.48	−1.59	−1.10	−0.48	0.34	0.66	0.89	1.21
8	3	−14.36	−8.15	−5.01	−2.14	0.24	0.67	0.88	1.12
8	4	−5.34	−3.30	−2.18	−0.99	0.30	0.64	0.83	1.07
8	5	−2.78	−1.86	−1.25	−0.56	0.32	0.62	0.82	1.07
9	3	−15.68	−9.12	−5.64	−2.38	0.20	0.66	0.86	1.06
9	4	−6.31	−3.78	−2.47	−1.08	0.28	0.61	0.79	1.00
9	5	−3.19	−2.10	−1.40	−0.63	0.30	0.58	0.76	0.98
9	6	−2.01	−1.38	−0.94	−0.41	0.30	0.57	0.76	0.99
10	3	−17.45	−9.98	−6.05	−2.58	0.17	0.66	0.87	1.07
10	4	−6.54	−4.17	−2.70	−1.22	0.27	0.60	0.77	0.96
10	5	−3.56	−2.37	−1.56	−0.73	0.28	0.56	0.72	0.93
10	6	−2.21	−1.51	−1.03	−0.48	0.28	0.54	0.71	0.93
11	3	−18.52	−10.68	−6.42	−2.76	0.13	0.65	0.87	1.07
11	4	−7.26	−4.57	−2.95	−1.37	0.24	0.58	0.75	0.92
11	5	−4.00	−2.58	−1.75	−0.81	0.26	0.54	0.69	0.88
11	6	−2.45	−1.67	−1.16	−0.53	0.26	0.52	0.66	0.85
12	3	−19.08	−11.23	−6.92	−3.03	0.10	0.64	0.88	1.10
12	4	−7.44	−4.81	−3.17	−1.47	0.21	0.58	0.75	0.92
12	5	−4.17	−2.72	−1.88	−0.89	0.24	0.53	0.68	0.84
12	6	−2.63	−1.83	−1.27	−0.60	0.25	0.50	0.64	0.81

TABLE 5.6.   Values of $t(\alpha, n, r)$

Two sided exacted 100α confidence intervals for λ^{**} and δ^{**} are

$$(5.30) \quad \left[\lambda^{**} - \delta^{**}t\left(\frac{1+\alpha}{2}, n, n - r_2\right), \quad \lambda^{**} - \delta^{**}t\left(\frac{1-\alpha}{2}, n, n - r_2\right) \right]$$

$$(5.31) \quad \left[\frac{\delta^{**}}{w\left(\frac{1+\alpha}{2}, n, n - r_2\right)}, \quad \frac{\delta^{**}}{w\left(\frac{1-\alpha}{2}, n, n - r_2\right)} \right]$$

5.2.1.5.   Estimation by means of probability paper.   As described in chapter 4, once a straight line has been fitted to the tail (see section 3.10) and drawn on

n	r	α							
		0.02	0.05	0.10	0.25	0.75	0.90	0.95	0.98
5	3	0.09	0.14	0.21	0.37	0.94	1.32	1.59	1.93
6	3	0.09	0.14	0.21	0.36	0.93	1.32	1.59	1.92
6	4	0.18	0.25	0.32	0.49	1.01	1.33	1.55	1.84
7	3	0.08	0.14	0.20	0.35	0.92	1.30	1.56	1.92
7	4	0.17	0.24	0.31	0.48	1.01	1.32	1.54	1.82
7	5	0.25	0.32	0.40	0.56	1.05	1.33	1.52	1.75
8	3	0.08	0.13	0.19	0.35	0.92	1.31	1.58	1.95
8	4	0.16	0.23	0.31	0.47	1.00	1.33	1.55	1.83
8	5	0.23	0.31	0.39	0.55	1.05	1.33	1.52	1.76
9	3	0.08	0.13	0.19	0.34	0.92	1.31	1.58	1.92
9	4	0.16	0.23	0.31	0.47	1.00	1.33	1.55	1.84
9	5	0.23	0.31	0.39	0.54	1.04	1.33	1.52	1.76
9	6	0.30	0.38	0.45	0.60	1.06	1.31	1.48	1.70
10	3	0.08	0.13	0.19	0.34	0.93	1.31	1.59	1.92
10	4	0.16	0.23	0.30	0.46	1.00	1.33	1.57	1.86
10	5	0.23	0.30	0.38	0.54	1.04	1.33	1.53	1.77
10	6	0.29	0.37	0.45	0.60	1.06	1.32	1.49	1.71
11	3	0.08	0.13	0.19	0.34	0.92	1.31	1.60	1.96
11	4	0.15	0.22	0.30	0.46	1.00	1.34	1.58	1.87
11	5	0.22	0.30	0.38	0.54	1.04	1.34	1.54	1.82
11	6	0.28	0.36	0.44	0.60	1.07	1.33	1.52	1.73
12	3	0.08	0.13	0.19	0.34	0.92	1.30	1.56	1.87
12	4	0.16	0.22	0.30	0.46	1.00	1.33	1.55	1.82
12	5	0.23	0.30	0.38	0.54	1.04	1.33	1.53	1.78
12	6	0.29	0.36	0.44	0.60	1.06	1.33	1.49	1.72

TABLE 5.7.   Values of $w(\alpha, n, r)$

probability paper, the method explained in section 4.2.3 can be used to obtain some estimates of λ and δ. The main advantage of this method lies in its simplicity and the fact that no computer is required for the calculations. One important shortcoming is that no confidence limits can be given.

Application of this method to the flood and wave data can be seen in examples 4.4 and 4.5.

5.2.1.6. Other methods of estimation.   There are many other methods for the estimation of parameters of the Gumbel distribution. Among them, the Blom nearly best estimators (see Blom (1958,1962), David (1983)) and the methods based on a few order statistics (see Chan and Kabir (1969)) are perhaps some of the most important. However, we advise the reader, once more, to use the methods based on tail data only.

5.3.   The Weibull Distribution

The Weibull cdf for maxima and minima are given, respectively, by

$$
(5.32) \qquad G(x) = \begin{cases} \exp\left[-\left(\dfrac{\lambda - x}{\delta} \right)^{\beta} \right]; & x \leq \lambda \\ 1 & \text{otherwise} \end{cases}
$$

and

$$
(5.33) \qquad \underline{G}(x) = \begin{cases} 1 - \exp\left[-\left(\dfrac{x - \lambda}{\delta} \right)^{\beta} \right]; & x \geq \lambda \\ 0 & \text{otherwise} \end{cases}
$$

where λ, δ and β are constants known as the location, scale and shape parameters, respectively and such that $\delta > 0$ and $\beta > 0$. Their associated pdfs are

$$
(5.34) \qquad g(x) = \begin{cases} \dfrac{\beta\left(\dfrac{\lambda - x}{\delta} \right)^{\beta - 1} \exp\left[-\left(\dfrac{\lambda - x}{\delta} \right)^{\beta} \right]}{\delta}; & x \leq \lambda \\ 0 & \text{otherwise} \end{cases}
$$

and

$$
(5.35) \qquad \underline{g}(x) = \begin{cases} \dfrac{\beta\left(\dfrac{x - \lambda}{\delta} \right)^{\beta - 1} \exp\left[-\left(\dfrac{x - \lambda}{\delta} \right)^{\beta} \right]}{\delta}; & x \geq \lambda \\ 0 & \text{otherwise} \end{cases}
$$

Changing of X to $-X$ and λ to $-\lambda$ leads from $G(x)$ to $\underline{G}(x)$ or vice versa. Consequently, any property or estimation method, derived for $\underline{G}(x)$, can be immediately translated into $G(x)$.

The importance of the Weibull distribution in practice is due to its extreme value behavior. It has been applied to fatigue strength of materials (Gumbel (1957b,1962c), Phoenix (1975), Tierney (1982), Castillo et al. (1985), Castillo and Galambos (1985)), lifetime of vacuum tubes (Kao (1959)), electrical insulations (Nelson (1982)), ball bearings (Lieblein and Zelen (1956)), human reliability (Hannaman (1985)), semiconductor devices, motors, capacitors,

Parameter	Distribution	
	$G(x) = \exp\left[-\left(\dfrac{\lambda - x}{\delta}\right)^{\beta}\right]$ (Maxima)	$\underline{G}(x) = 1 - \exp\left[-\left(\dfrac{x - \lambda}{\delta}\right)^{\beta}\right]$ (Minima)
Mean	$\lambda - \delta\,\Gamma\left(1 + \dfrac{1}{\beta}\right)$	$\lambda + \delta\,\Gamma\left(1 + \dfrac{1}{\beta}\right)$
Median	$\lambda - \delta\,0.693^{1/\beta}$	$\lambda + \delta\,0.693^{1/\beta}$
Mode	$\lambda - \delta\left(\dfrac{\beta - 1}{\beta}\right)^{1/\beta};\quad \beta > 1$ $\lambda \qquad\quad ;\quad \beta \leq 1$	$\lambda + \delta\left(\dfrac{\beta - 1}{\beta}\right)^{1/\beta};\quad \beta > 1$ $\lambda \qquad\quad ;\quad \beta \leq 1$
Variance	$\delta^{2}\left[\Gamma\left(1 + \dfrac{2}{\beta}\right) - \Gamma^{2}\left(1 + \dfrac{1}{\beta}\right)\right]$	$\delta^{2}\left[\Gamma\left(1 + \dfrac{2}{\beta}\right) - \Gamma^{2}\left(1 + \dfrac{1}{\beta}\right)\right]$

TABLE 5.8.   Some parameters of the Weibull distribution

photoconductive cells, corrosion resistance, leakage failure of batteries, etc.

The main parameters of the Weibull distribution are summarized in table 5.8.

5.3.1.   Parameter estimation.

5.3.1.1.   The two-parameter Weibull distribution.   Here we refer to the Weibull distribution (5.32) or (5.33) when the location parameter λ is known.

Estimation of the two-parameter Weibull distribution can be reduced to the estimation of the Gumbel distribution if we take into account that the transformation

$$(5.36) \qquad\qquad Y = \log(X - \lambda)$$

transforms the X Weibull distribution (5.33) into the Y Gumbel distribution (5.2) with the following relation between parameters

$$(5.37) \qquad\qquad \lambda_G = \log(\delta_W)$$

$$(5.38) \qquad\qquad \delta_G = \frac{1}{\beta_W}$$

where the subindexes G and W refer to Gumbel and Weibull parameters respectively.

Data	Fitted Model	Percentile Method		Method of Moments		Maximum Likelihood		Covariance Matrix	
		δ	β	δ	β	δ	β		
Wind Data	Frechet (largest) ($\lambda = 0$)	28.64	4.57	28.54	4.49	28.74	5.10	0.706 −0.149	0.355
Phone Data	Weibull (smallest) ($\lambda = 0$)	1.27×10^{-3}	1.061	1.22×10^{-3}	1.176	1.25×10^{-3}	1.078	$4.31E-8$ $9.41E-6$	0.020
Epicenter Data	Weibull (smallest) ($\lambda = 0$)	164.9	2.665	161.1	3.02	162.53	2.98	55.51 0.734	0.097
Epicenter Data	Weibull (smallest) ($\lambda = 50$)	107.1	1.44	106.03	1.49	105.3	1.68	71.1 0.455	0.034
Fatigue Data	Weibull (smallest) ($\lambda = 0$)	110614	2.649	108991	2.89	111842	2.249	$7.97E+7$ 826.5	0.076
Fatigue Data	Weibull (smallest) ($\lambda = 34000$)	70952	1.48	70353	1.54	71482	1.44	$7.85E+7$ 526.3	0.035

TABLE 5.9. Estimates of Weibull and Frechet parameters for some examples in Appendix C

Similarly, the transformation

(5.39) $$Y = -\log(\lambda - x)$$

transforms the X Weibull distribution (5.32) into the Y Gumbel distribution (5.1) with the following relation between parameters

(5.40) $$\lambda_G = -\log(\delta_W)$$

(5.41) $$\delta_G = \frac{1}{\beta_W}$$

Thus, any of the methods in section 5.2 are immediately applicable to the Weibull distribution by means of the transformations (5.36) and (5.37)–(5.38).

If this method is used, the covariance matrix of estimators must also be transformed (see, for example, Nelson (1982) pag. 374). As one example, the covariance matrix (5.18) for the Weibull parameters as a function of the Gumbel parameters becomes

(5.42) $$\begin{pmatrix} \text{Var}(\delta_W) & \text{Cov}(\delta_W, \beta_W) \\ \text{Cov}(\delta_W, \beta_W) & \text{Var}(\delta_W) \end{pmatrix}$$

$$= \begin{pmatrix} \exp(2\lambda_G)\text{Var}(\lambda_G) & -\dfrac{\exp(\lambda_G)\text{Cov}(\lambda_G, \delta_G)}{\delta_G^2} \\ -\dfrac{\exp(\lambda_G)\text{Cov}(\lambda_G, \delta_G)}{\delta_G^2} & \dfrac{\text{Var}(\delta_G)}{\delta_G^4} \end{pmatrix}$$

for the transformation (5.36) with (5.37)–(5.38).

Example 5.5. (Data examples—complete sample).   Table 5.9. gives the moment and percentile method estimates of some data examples included in Appendix C. These estimates were obtained by transforming the data to Gumbel type by means of expression (5.36), and then using relations (5.37) and (5.38).   ∎

Example 5.6. (Data examples—tail data).   Table 5.10. shows the maximum likelihood estimates of δ and β, based on a small fraction $(2\sqrt{n})$ of the data in the tail of interest for some data sets in Appendix C.   ∎

In the following section we include some direct methods, which give estimates different from those obtained through the above mentioned transformation.

Data	Fitted Model	Maximum Likelihood		Covariance Matrix		
		δ	β			
Wind Data	Frechet (largest) ($\lambda = 0$)	29.42	4.81	6.298 2.365	1.544	
Phone Data	Weibull (smallest) ($\lambda = 0$)	0.708 $\times 10^{-3}$	1.643	3.46E $-$ 8 -6.3E $-$ 5	0.225	
Epicenter Data	Weibull (smallest) ($\lambda = 0$)	136.8	3.62	242.8 -10.77	0.786	
Epicenter Data	Weibull (smallest) ($\lambda = 50$)	126.8	1.25	1756.0 -10.16	0.096	
Fatigue Data	Weibull (smallest) ($\lambda = 0$)	88584.8	4.15	8.28E $+$ 7 -7619.9	1.396	
Fatigue Data	Weibull (smallest) ($\lambda = 34000$)	67865.3	1.55	3.51E $+$ 8 -5957.4	0.199	

TABLE 5.10.   Estimates of Weibull and Frechet parameters for some examples in Appendix C (Tail data)

The method of moments. Upon establishing the equality of mean and variance of population and sample we get

$$(5.43) \qquad \bar{x} = \lambda + \delta\,\Gamma\left(1 + \frac{1}{\beta}\right)$$

$$(5.44) \qquad s_x^2 = \delta^2\left[\Gamma\left(1 + \frac{2}{\beta}\right) - \Gamma^2\left(1 + \frac{1}{\beta}\right)\right]$$

from which

$$(5.45) \qquad \delta = \frac{\bar{x} - \lambda}{\Gamma\left(1 + \dfrac{1}{\beta}\right)}$$

$$(5.46) \qquad s_x^2 = (\bar{x} - \lambda)^2 \left[\frac{\Gamma\left(1 + \dfrac{2}{\beta}\right)}{\Gamma^2\left(1 + \dfrac{1}{\beta}\right)} - 1 \right]$$

Equation (5.46) can be used to iteratively obtain β. Then expression (5.45) gives δ.

Best linear unbiased estimators (BLUES) and best linear invariant estimators (BLIES). Due to the logarithmic transformation, a linear combination of the Gumbel order statistics is not linear on the Weibull transformed values. Consequently, the transformation method does not give either the BLUES or the BLIES for the Weibull case. Thus, in order to obtain the coefficients of both, the theory in sections 2.3.7.1. and 2.3.7.2. must be used to obtain these estimates.

$$(5.47) \qquad \delta^* = \sum_{i=r_1}^{n-r_2} a_i(n, r_1, r_2) X_{(i)}$$

$$(5.48) \qquad \beta^* = \sum_{i=r_2}^{n-r_2} b_i(n, r_1, r_2) X_{(i)}$$

where $a_i(n, r_1, r_2)$ and $b_i(n, r_1, r_2)$ are coefficients which can be obtained by the method in section 2.3.7.1.

The covariance matrix is

$$(5.49) \qquad \begin{pmatrix} \mathrm{Var}(\delta^*) & \mathrm{Cov}(\delta^*, \beta^*) \\ \mathrm{Cov}(\delta^*, \beta^*) & \mathrm{Var}(\beta^*) \end{pmatrix} = \beta^{*2} \begin{pmatrix} A(n, r_1, r_2) & C(n, r_1, r_2) \\ C(n, r_1, r_2) & B(n, r_1, r_2) \end{pmatrix}$$

where $A(n, r_1, r_2)$, $B(n, r_1, r_2)$ and $C(n, r_1, r_2)$ can be calculated by (2.123).

The BLIES can be obtained by (2.124) and (2.125).

Estimation by means of probability paper. Section 4.2.4. explains how Weibull parameter estimates can be easily obtained with the help of Weibull probability paper. We remind the reader that a straight line must be fitted to the data in the tail of interest, because the Weibull distribution is an asymptotic distribution and generally gives an approximation for that tail only.

Application of this method to the telephone data is illustrated in Example 4.15.

5.3.1.2. The three-parameter Weibull distribution. We include here some methods for estimating the location, scale and shape parameters of the Weibull distribution.

The maximum likelihood method.   The likelihood function of the $n - r_1 - r_2$ order statistics

$$X_{r_1+1:n}, X_{r_1+2:n}, \ldots, X_{n-r_2:n}$$

from the Weibull parent (5.35) is given by (see (2.11)):

$$L = \left(\frac{\beta}{\delta}\right)^{n-r_1-r_2} \left[\prod_{i=r_1+1}^{n-r_2} \left(\frac{x_i - \lambda}{\delta}\right)^{\beta-1}\right] \exp\left[\sum_{i=r_1+1}^{n-r_2} \left(\frac{x_i - \lambda}{\delta}\right)^{\beta}\right]$$

$$(5.50) \qquad \cdot \exp\left[-r_2\left(\frac{x^{**} - \lambda}{\delta}\right)^{\beta}\right]\left\{1 - \exp\left[-\left(\frac{x^* - \lambda}{\delta}\right)^{\beta}\right]\right\}^{r_1} \frac{n!}{r_1! r_2!}$$

where

$$(5.51) \qquad\qquad\qquad x^* = x_{r_1+1:n}$$

$$(5.52) \qquad\qquad\qquad x^{**} = x_{n-r_2:n}$$

Its logarithm, apart from constant terms, becomes

$$\log L = (n - r_1 - r_2)\log\left(\frac{\beta}{\delta}\right) + (\beta - 1)\left\{\sum_{i=r_1+1}^{n-r_2} \log\left(\frac{x_i - \lambda}{\delta}\right) - \sum_{i=r_1+1}^{n-r_2} \left(\frac{x_i - \lambda}{\delta}\right)^{\beta}\right.$$

$$(5.53) \qquad \left. -r_2\left(\frac{x^{**} - \lambda}{\delta}\right)^{\beta} + r_1 \log\left\{1 - \exp\left[-\left(\frac{x^* - \lambda}{\delta}\right)^{\beta}\right]\right\}\right\}$$

Upon setting to zero its partial derivatives with respect to δ, β and λ we get

$$\frac{\partial \log L}{\partial \delta} = -(n - r_1 - r_2)\frac{\beta}{\delta} + \beta\sum_{i=r_1+1}^{n-r_2} \frac{(x_i - \lambda)^{\beta}}{\delta^{\beta+1}} + \frac{r_2\beta(x^{**} - \lambda)^{\beta}}{\delta^{\beta+1}}$$

$$(5.54) \qquad -\frac{r_1\beta \exp\left[-\left(\frac{x^* - \lambda}{\delta}\right)^{\beta}\right](x^* - \lambda)^{\beta}\delta^{-1-\beta}}{\left\{1 - \exp\left[-\left(\frac{x^* - \lambda}{\delta}\right)^{\beta}\right]\right\}} = 0$$

$$\frac{\partial \log L}{\partial \beta} = (n - r_1 - r_2)\left(\frac{1}{\beta} - \log \delta\right) + \sum_{i=r_1+1}^{n-r_2} \log(x_i - \lambda)$$

$$-\sum_{i=r_1+1}^{n-r_2} \left(\frac{x_i - \lambda}{\delta}\right)^{\beta}\log\left(\frac{x_i - \lambda}{\delta}\right) - r_2\left(\frac{x^{**} - \lambda}{\delta}\right)^{\beta}\log\left(\frac{x^{**} - \lambda}{\delta}\right)$$

$$(5.55) \qquad +\frac{r_1 \exp\left[-\left(\frac{x^* - \lambda}{\delta}\right)^{\beta}\right](x^* - \lambda)^{\beta}\log\left(\frac{x^* - \lambda}{\delta}\right)\delta^{-\beta}}{\left\{1 - \exp\left[-\left(\frac{x^* - \lambda}{\delta}\right)^{\beta}\right]\right\}} = 0$$

$$\frac{\partial \log L}{\partial \lambda} = (1 - \beta) \sum_{i=r_1+1}^{n-r_2} (x_i - \lambda)^{-1} + \frac{\beta}{\delta^\beta} \sum_{i=r_1+1}^{n-r_2} (x_i - \lambda)^{\beta-1} + \frac{r_2 \beta (x^{**} - \lambda)^{\beta-1}}{\delta^\beta}$$

$$(5.56) \qquad - \frac{r_1 \beta \exp\left[-\left(\frac{x^* - \lambda}{\delta}\right)^\beta\right](x^* - \lambda)^{\beta-1}\delta^{-\beta}}{\left\{1 - \exp\left[-\left(\frac{x^* - \lambda}{\delta}\right)^\beta\right]\right\}} = 0$$

The system (5.54)-(5.56) must be solved iteratively, with the help of a computer. However, we need to point out here that, in some cases, a solution cannot be found, i.e. a relative maximum does not exist. It is also worthwhile mentioning that the Weibull pdf presents a discontinuity at $\beta = 1$, and for $\beta < 1$ and $\beta > 1$ this family of distributions shows different behavior at the lower finite end. For $\beta < 1$, the likelihood function can be made infinity if λ is made equal to the minimum of the sample values. For this reason and for the reason that some regularity problems occur for $1 < \beta < 2$, the maximum likelihood method is not appropriate for Weibull populations with values of $\beta < 2$. Another alternative is to use the multinomial discretization method to be described below.

Some initial estimates of the parameters can be obtained by the method given by Dubey (1966)

$$(5.57) \qquad \lambda = \frac{X_{(1)}X_{(k)} - X_{(j)}^2}{X_{(1)} + X_{(k)} - 2X_{(j)}}$$

$$(5.58) \qquad \beta = \frac{2.99}{\log(X_{[0.9737n]+1} - \lambda) - \log(X_{[0.1673n]+1} - \lambda)}$$

$$(5.59) \qquad \delta = \frac{X_{[0.1673n]+1} - \lambda}{0.183^{1/\beta}}$$

where $X_{(k)}$ and $X_{(j)}$ are order statistics such that $X_{(1)} < X_{(j)} < (X_{(1)}X_{(k)})^{1/2}$, and $[x]$ is the integer part of x.

Another method of obtaining quick initial estimates of the parameters consists of equating three percentiles of sample and population, i.e.

$$(5.60) \qquad x_{[p_i]} = \lambda + \delta u_{p_i}^{1/\beta}; \qquad i = 1, 2, 3$$

where

$$(5.61) \qquad u_{p_i} = -\log(1 - p_i)$$

and u_{p_i} is the sample $100p_i$-th percentile.

From (5.60) we get

$$(5.62) \qquad \frac{x_{[p_2]} - x_{[p_1]}}{x_{[p_3]} - x_{[p_2]}} = \frac{u_{p_2}^{1/\beta} - u_{p_1}^{1/\beta}}{u_{p_3}^{1/\beta} - u_{p_2}^{1/\beta}}$$

which depends on β only and can be solved by an iterative procedure or with the help of a small computer.

If the solution of (5.62) is less than zero, the validity of the Weibull distribution for the data under consideration must be carefully tested.

Once β has been obtained, δ and λ can be calculated by

$$(5.63) \qquad \delta = \frac{x_{[p_2]} - x_{[p_1]}}{u_{p_2}^{1/\beta} - u_{p_1}^{1/\beta}}$$

$$(5.64) \qquad \lambda = x_{[p_1]} - \delta u_{p_1}^{1/\beta}$$

Widely distant percentiles are recommended for this formula. If the data is known to follow the Weibull distribution exactly, the following values are good choices

$$p_1 = 0.10; \qquad p_2 = 0.5; \qquad p_3 = 0.90$$

If the data is known to follow the Weibull distribution only asymptotically, the following values can be chosen

$$p_1 = \frac{a}{n}; \qquad p_2 = \frac{b}{n}; \qquad p_3 = \frac{c}{n}$$

where a, b and c are constants such that

$$k \log n > a > b > c > 2$$

and k is a constant. Note that this implies that only values in the left tail are used and their associated percentiles go to zero as n goes to infinity. So it is a true tail fit.

The multinomial discretization method.   This method is based on the following: Let $-\infty = a_1 < a_2 < \ldots < a_{k+1} = \infty$ be a set of extended real numbers. Let $F(x; \theta)$ be the cdf of a random variable X that belongs to a θ-parametric family. Then, the experiment consisting of counting the number of elements, in a random sample of size n, in each of the subintervals (a_1, a_{i+1}) $(i = 1, 2, \ldots, k)$, is a multinomial experiment, and its associated random variable has pmf given by

$$(5.65) \qquad P(n_1, n_2, \ldots, n_k) = \frac{n! p_1^{n_1}(\theta) p_2^{n_2}(\theta) \ldots p_k^{n_k}(\theta)}{[n_1! n_2! \ldots n_k!]}$$

where

$$(5.66) \qquad p_i(\theta) = F(a_{i+1};\theta) - F(a_i;\theta)$$

The maximization of (5.65) with respect to θ leads to a maximum likelihood multinomial approximation of the family $F(x;\theta)$ of cdfs. If $F(x:\theta)$ is the Weibull family, some instability problems of the standard maximum likelihood method are avoided, because now the likelihood function is always finite.

Note that an adequate selection of the a_i values allows the estimation for type I and type II censoring.

5.4. The Frechet Distribution

The Frechet cdf for maxima and minima are given, respectively, by

$$(5.67) \qquad F(x) = \begin{cases} \exp\left[-\left(\dfrac{\delta}{x-\lambda}\right)^{\beta}\right]; & x \geq \lambda \\ 0 & \text{otherwise} \end{cases}$$

and

$$(5.68) \qquad \underline{F}(x) = \begin{cases} 1 - \exp\left[-\left(\dfrac{\delta}{\lambda-x}\right)^{\beta}\right]; & x \leq \lambda \\ 1 & \text{otherwise} \end{cases}$$

where λ, δ and β are constants known as the location, scale and shape parameters, respectively such that $\delta > 0$ and $\beta > 0$. Their associated pdfs are

$$(5.69) \qquad f(x) = \begin{cases} \dfrac{\beta}{\delta}\left(\dfrac{\delta}{x-\lambda}\right)^{\beta+1} \exp\left[-\left(\dfrac{\delta}{x-\lambda}\right)^{\beta}\right]; & x \geq \lambda \\ 0 & \text{otherwise} \end{cases}$$

and

$$(5.70) \qquad \underline{f}(x) = \begin{cases} \dfrac{\beta}{\delta}\left(\dfrac{\delta}{\lambda-x}\right)^{\beta+1} \exp\left[-\left(\dfrac{\delta}{\lambda-x}\right)^{\beta}\right]; & x \leq \lambda \\ 0 & \text{otherwise} \end{cases}$$

Changing X to $-X$ and λ to $-\lambda$ leads from $F(x)$ to $\underline{F}(x)$ or vice versa. Consequently, any property, estimation method, etc. derived for $F(x)$, can be immediately translated to $\underline{F}(x)$.

The importance of the Frechet distribution in practice is due also to their

Parameter	Distribution	
	$G(x) = \exp\left[-\left(\dfrac{\delta}{x-\lambda}\right)^{\beta}\right]$ (Maxima)	$\underline{G}(x) = 1 - \exp\left[-\left(\dfrac{\delta}{\lambda-x}\right)^{\beta}\right]$ (Minima)
Mean	$\lambda + \delta\Gamma\left(1 - \dfrac{1}{\beta}\right);\quad \beta > 1$	$\lambda - \delta\Gamma\left(1 - \dfrac{1}{\beta}\right);\quad \beta > 1$
Median	$\lambda + \delta\,0.693^{-1/\beta}$	$\lambda - \delta\,0.693^{-1/\beta}$
Variance	$\delta^2\left[\Gamma\left(1 - \dfrac{2}{\beta}\right) - \Gamma^2\left(1 - \dfrac{1}{\beta}\right)\right]$ $\beta > 2$	$\delta^2\left[\Gamma\left(1 - \dfrac{2}{\beta}\right) - \Gamma^2\left(1 - \dfrac{1}{\beta}\right)\right]$ $\beta > 2$

TABLE 5.11.   Some parameters of the Frechet distribution

extreme value behavior. It has been applied to sea waves (Thom (1967,1973)), wind speeds (Thom (1967,1968a,b)), etc.

The main parameters of the Frechet distribution are summarized in table 5.11.

5.4.1.   Parameter estimation.

5.4.1.1.   The two-parameter Frechet distribution. Here we refer to the Frechet distribution (5.67) or (5.68) when the location parameter λ is known.

Estimation of the two-parameter Frechet distribution can be reduced to the estimation of the Gumbel distribution if we take into account that the transformation

$$(5.71) \qquad\qquad Y = \log(x - \lambda)$$

transforms the X Frechet distribution (5.67) into the Y Gumbel distribution (5.1) with the following relation between parameters

$$(5.72) \qquad\qquad \lambda_G = \log(\delta_F)$$

$$(5.73) \qquad\qquad \delta_G = \frac{1}{\beta_F}$$

where the subindexes G and F refer to Gumbel and Frechet parameters respectively.

Similarly, the transformation

$$(5.74) \qquad\qquad Y = -\log(\lambda - x)$$

transforms the X Frechet distribution (5.68) into the Y Gumbel distribution (5.2) with the following relation between parameters

$$(5.75) \qquad\qquad \lambda_G = -\log(\delta_F)$$

$$(5.76) \qquad\qquad \delta_G = \frac{1}{\beta_F}$$

If this method is used, the covariance matrix of estimators must be accordingly transformed. For example, the covariance matrix (5.18) for the Frechet parameters, as a function of the Gumbel parameters, becomes

$$\begin{pmatrix} \mathrm{Var}(\delta_F) & \mathrm{Cov}(\delta_F, \beta_F) \\ \mathrm{Cov}(\delta_F, \beta_F) & \mathrm{Var}(\delta_F) \end{pmatrix}$$

$$(5.77) \qquad = \begin{vmatrix} \exp(2\lambda_G)\,\mathrm{Var}(\lambda_G) & \dfrac{-\exp(\lambda_G)\mathrm{Cov}(\lambda_G, \delta_G)}{\delta_G^2} \\ \dfrac{-\exp(\lambda_G)\,\mathrm{Cov}(\lambda_G, \delta_G)}{\delta_G^2} & \dfrac{\mathrm{Var}(\delta_G)}{\delta_G^4} \end{vmatrix}$$

Example 5.7. (Wind data—complete sample). Table 5.9 shows the moment, percentile and maximum likelihood estimates of δ and β for the wind data. These values were obtained using the full sample through transformation (5.71) and relations (5.72) and (5.73). Similarly, the covariance matrix was obtained with the help of (5.77). ∎

Example 5.8. (Wind data—tail data). In an analogous manner, the maximum likelihood estimates, based on the $2\sqrt{n}$ largest order statistics appear in Table 5.10. The similarity of estimates is due to the similarity of the right tail and overall slopes of the cdf's. ∎

5.4.1.2.   The three parameter Frechet distribution. For the three-parameter Frechet distribution, a maximum likelihood system of equations can be derived in a similar way as was done for the Weibull distribution. Direct methods for the maximization of the log-likelihood function are also applicable. The main difference with respect to the Weibull case is that the likelihood function is now bounded and the associated problems do not appear here.

Chapter 6

Selection of Limit Distributions from Data

6.1. Statement of the Problem

One of the first steps when dealing with a problem of extremes is to determine the domain of attraction of the parent distribution under study.

In chapter 3 we have given some theorems that allow determination of the domain of attraction of a given cdf $F(x)$. This is the case of theorems 3.3, 3.4 and 3.6 to 3.9, which are all written in terms of $F(x)$ or $F^{-1}(x)$. However, the most common problem faced in daily practice appears in a completely different form: a sample comes to the engineer or scientist and some ignorance (complete or partial) about the parent cdf exists. In this case, the basic data is the sample and not the cdf. Consequently, the above theorems cannot be used, at least in the form stated. Thus, some alternatives become necessary. At this point the new problem can be stated as: Determine the domain of attraction of a parent distribution when a sample from it is known. The aim of this chapter is to give some tools to make this analysis possible.

In order to avoid some confusion, we want to point out here that this problem is different from the problem of testing whether or not a sample comes from a Weibull, Gumbel or Frechet population. In other words, most of the known methods for this test are not valid for the problem under consideration. The reasons for this have, in some way, been indicated before.

Nevertheless, due to their practical implications, we expose here them more clearly. The domain of attraction of a given cdf, $F(x)$, is only determined by its tail (right or left, depending on whether the upper or lower order statistics, respectively, are of interest). Thus, only the tail contains this information and only upper, for right tail interest, and lower, for left tail interest, order statistics in the sample can be used to this end. On the contrary, all the range has information about the parent distribution. Thus, if one sample is to be tested for a Gumbel parent say, all the order statistics can be used, because all contain information about the parent. In conclusion, it is one thing to test whether $F(x)$ lies in the domain of attraction of $H_c(x)$ or $L_c(x)$ and a very different thing to test whether it is exactly equal to $H_c(x)$ or $L_c(x)$ for some specified value of c.

6.2.   Methods for Determining the Domain of Attraction of a Parent Distribution from Samples

In this section we give several methods for determining the domain of attraction of a parent distribution when only a sample from it is known.

For many years, graphic methods have been used by applied statisticians. Among them probability paper has played a very important role in the area of extremes. Following in this tradition, we start this section with one graphic and very simple method, based on probability paper.

Some recent publications as those of Pickands (1975), Hill (1975), Boos (1984), Weissman (1978,1980), Galambos (1980), Mason (1981,1982), DuMouchel (1983), Davis and Resnick (1984), etc. deal with the present or related problems. From them we have selected two methods, from Pickands and Galambos.

Because the domain of attraction of a given cdf, $F(x)$, is completely known as soon as the value of the parameter c, in the von-Mises family, is known, the problem of identifying this domain of attraction can be made equivalent to that of estimating the value of the parameter c. We also know that positive values of c are associated with Frechet type domains of attraction, negative values with Weibull types, and zero values with Gumbel types.

With this in mind, and using data in the tails only, estimation methods can also be useful in solving the present problem.

Finally, we indicate here that we assume, throughout this chapter, that the mentioned cdfs lie in the domain of attraction of any of the $H_c(x)$ or $L_c(x)$ cdfs in the corresponding tail.

6.2.1.   The probability paper method.

We have shown in section 4.5 that when a cdf, $F(x)$, is represented graphically on Gumbel probability paper its concavity in the associated tail is dependent on the type of domain of attraction in which it lies. More precisely, we have proven that distributions lying in a Weibull type domain of attraction for maxima (for minima) show convexity (concavity) in the right (left) tail, distributions in a Frechet type domain of attraction for maxima (minima) show concavity (convexity) in the right (left) tail, and, finally, distributions in a Gumbel type domain of attraction appear as straight lines in the corresponding tail.

We have also shown that cdfs which lie in the domain of attraction of a Weibull type show a vertical tangent slope at their associated ends (upper or lower). Similarly, cdfs in some domain of attraction of Frechet type show a horizontal slope, instead.

Thus, at the end of interest we have either

 (i) A horizontal slope (Frechet type domain of attraction),
 (ii) A vertical slope (Weibull type domain of attraction) or
(iii) An intermediate slope (Gumbel type domain of attraction).

Consequently, the method, which obviously follows from all the above, consists of the following steps

 (i) Draw the empirical cdf on Gumbel probability paper.
(ii) Observe the curvature (concavity or convexity) and the slopes of the cdf in the tail of interest. If the convexity is not negligible and the slopes go to zero or infinity, then reject the hypothesis that $F(x)$ lies in a Gumbel type domain of attraction. Depending on the curvature and the slope, decide in favor of Weibull or Frechet according to the rules above.

The main drawback of this method is its subjectivity. Note that no precise criteria is given in order to know what negligible convexity means or what is exactly meant by tail. However, it has been proven very useful in practical applications because of its simplicity and accurate results.

It must be remembered that a Gumbel type cdf can be approximated as accurately as desired by Weibull and Frechet type cdfs (see section 3.11). Thus, from a practical point of view, the wrong rejection of a Gumbel type distribution can be corrected in the estimation process that usually follows this decision.

All the above, justifies the use of this method and we recommend its use as a complement to any of the following methods.

Example 6.1. (Data examples in Appendix C).   Application of this method to the data in Appendix C, which was drawn in figures 4.5. to 4.18., can lead to the domains of attraction shown in table 6.1. Note that in some cases two

	Type of Domain of Attraction		
Data Set	Weibull	Gumbel	Frechet
Wind Data		X	X
Flood Data		X	
Wave Data		X	
Telephone Data	X		
Epicenter Data	X		
Link Data		X	
Insulation Data		X	
Fatigue Data	X		
Precipitation Data	X	X	
Houmb Data	X		
Ocmulgee (Macon)	X	X	
Ocmulgee (Hawk)	X		
Oldest Ages (women)	X	X	
Oldest Ages (men)	X	X	

TABLE 6.1.   Domain of attraction of examples in Appendix C, according to the probability paper method

possible decisions are included because the above arguments are not clear enough to lead to a unique decision.   ■

6.2.2.   The Pickands III method.

Suppose that the random variable X has a continuous cdf, $F(x)$. Then Pickands III (1975) has demonstrated that the assumption that $F(x)$ lies in the domain of attraction for maxima of an extreme value distribution is equivalent to

$$\lim_{u \to \omega(F)} P[X > x + u / X > u] = \frac{F(u + x)}{1 - F(u)} = \left(1 + \frac{cx}{a}\right)^{-1/c}$$

(6.1) $a > 0, -\infty < c < \infty; 1 + \dfrac{cx}{a} \geq 0$

where for $c = 0$, the right hand side of (6.1) must be given a limiting sense, i.e. for $c = 0$

(6.2) $\lim_{u \to \infty} P[X > x + u / X > u] = \exp\left(-\frac{x}{a}\right),$

where if $c > 0, c < 0$ or $c = 0$, $F(x)$ lies in the domain of attraction for maxima of the Frechet, Weibull or Gumbel distributions, respectively.

Expression (6.1) shows that the conditional cdf, $G(x)$, of $X - u$ given $X \geq u$, for u large enough, is approximately of the form

(6.3) $$G(x) = 1 - \left(1 + \frac{cx}{a}\right)^{-1/c}$$

This fact allows the estimation of c from a sample. Based on this value, a decision on domains of attractions can be made.

Pickands (1975) gives the following procedure for this estimation. Let us assume that $X_{(1)}, X_{(2)}, \ldots, X_{(n)}$ are the descending order statistics of a sample of size n. Then for $s = 1, 2, \ldots, [n/4]$, where $[x]$ means the integer part of x, we compute

(6.4) $$d_s = \max_x |F_s(x) - G_s(x)|$$

where $F_s(x)$ is the empirical distribution of $X - X_{(4s)}$, given that $X \geq X_{(4s)}$, and $G_s(x)$ is the distribution (6.3) with a and c replaced by the estimators

(6.5) $$c = \frac{\log\left(\dfrac{X_{(s)} - X_{(2s)}}{X_{(2s)} - X_{(4s)}}\right)}{\log(2)}$$

(6.6) $$a = \frac{c(X_{(2s)} - X_{(4s)})}{2^c - 1}$$

Finally, we choose M the smallest integer solution of

(6.7) $$d_M = \min_{1 \leq s \leq \left[\frac{n}{4}\right]} d_s$$

and take as values of c and a those associated with $s = M$ in (6.5) and (6.6). In other words, the idea of Pickands III is to fit the distribution (6.3) to data for several u ($u = X_{(4s)}$) values and then selecting the value of u leading to the best fit in the sense of expression (6.7). Note that for any s, the quality of the fit is measured by (6.4).

Pickands proves that estimation of tail probabilities of $F_s(x)$ by means of $G_s(x)$, with c and a given by the above method, leads to consistent estimates.

The method of Pickands can be used and a decision on domain of attraction types be made based on the c estimates. In order to be able to obtain significance levels for testing the hypothesis that $F(x)$ belongs to a Gumbel-type domain of attraction rather than a Weibull- or Frechet-type, we have simulated this algorithm by the Monte Carlo method (5000 replications) for samples sizes of $n = 10, 20, 40, 60, 80$ and 100, assuming a Gumbel parent population. The resulting cdf of c are shown in table 6.2. This table allows

	Sample Size					
cdf	10	20	40	60	80	100
0.01	-2.540	-1.040	-0.756	-0.608	-0.536	-0.520
0.02	-1.976	-0.908	-0.664	-0.540	-0.484	-0.452
0.05	-1.348	-0.736	-0.544	-0.456	-0.412	-0.404
0.10	-0.888	-0.572	-0.460	-0.388	-0.348	-0.340
0.20	-0.412	-0.368	-0.332	-0.288	-0.272	-0.260
0.30	-0.124	-0.200	-0.232	-0.220	-0.212	-0.209
0.50	0.476	0.112	-0.040	-0.072	-0.093	-0.112
0.70	1.128	0.536	0.232	0.140	0.076	0.038
0.80	1.608	0.884	0.448	0.288	0.212	0.156
0.90	2.308	1.500	0.772	0.568	0.424	0.330
0.95	2.888	2.048	1.140	0.820	0.604	0.544
0.98	3.660	3.024	1.744	1.256	0.868	0.760
0.99	4.180	3.488	2.232	1.704	0.980	0.916

TABLE 6.2.   Simulated cdf of c (Gumbel parent)

approximate critical values of c, for selected values of the type I error probability, or approximate significance levels to be obtained.

However, because this method is based on an asymptotic property and the parameters a and c are estimated by a naive percentile method, it can be expected to give good results for large samples sizes only.

Example 6.2. (Data examples in Appendix C).   The Pickands method, described above, was applied to all data sets in Appendix C, even to those of very small sample size. The estimates of c, their associated significance levels and the domains of attraction types resulting from the test are shown in table 6.3.   ■

Data Set	Sample Size	C Estimates	Significance Level	Domain of Attraction Type
Link Data	20	−0.619	0.087	Gumbel or Weibull
Precipitation Data	40	−0.579	0.038	Weibull
Houmb Data	24	0.823	0.180	Gumbel
Ocmulgee (Macon)	40	−0.595	0.032	Weibull
Ocmulgee (Hawk)	40	−0.162	0.380	Gumbel
Oldest Ages (women)	54	−0.461	0.060	Gumbel or Weibull
Oldest Ages (men)	54	−0.148	0.40	Gumbel
Wind Data	50	0.086	0.37	Gumbel
Flood Data	60	−0.310	0.178	Gumbel
Wave Data	50	−0.328	0.17	Gumbel
Phone Data	35	−0.538	0.07	Gumbel or Weibull
Epicenter Data	60	−1.390	<0.0002	Weibull
Fatigue Data	35	−0.959	<0.01	Weibull
Insulation Data	30	0.690	0.19	Gumbel

TABLE 6.3.   c estimates, significance levels and domains of attraction types resulting from the Pickands method

In appendix B we include the listing of a computer program, in BASIC, to estimate the values of a and c by the above method.

6.2.3. The Galambos method.

Galambos (1980) suggests a method to test whether or not a given cdf, $F(x)$, lies in the domain of attraction of the Gumbel distribution for largest values. The method is based on the fact that if $F(x)$ lies in that domain of attraction we have (see (6.2))

$$(6.8) \qquad \lim_{u \to \omega(F)} \frac{P[X > x + u/X > u]}{E[X - u/X > u]} = \exp(-x)$$

i.e. the random variable $(X - u)/E[X - u/X > u]$ is unit exponential.

Thus, Galambos gives the following procedure. Choose a number u. Select those X_j from the sample such that exceed u. For each $X_j > u$ compute the transformed value

$$(6.9) \qquad Y_j = \frac{X_j - u}{E^*}$$

where

$$(6.10) \qquad E^* = E[X_j - u/X_j > u] = \sum_{X_j > u} \frac{X_j - u}{m(u)}$$

and $m(u)$ is the number of X_j which exceed u, and finally test if the Y_j follow a unit exponential law.

This can be done by the Kolmogorov-Smirnov test, i.e. by means of the statistic

$$(6.11) \qquad \max_y |F_{m(u)}(y) - 1 + \exp(-y)|$$

where $F_{m(u)}(y)$ is the empirical cdf associated with the exceedances of the level u in the sample.

In order to calculate the significance level, we must take into account that the mean value of the exponential law was estimated from the sample by means of (6.10). Thus, the standard tables for the Kolmogorov-Smirnov test are no longer valid and we must use the tables of H. W. Lilliefors (1969), a summary of which appears in table 6.4.

Sample Size	Significance Level				
	0.20	0.15	0.10	0.05	0.01
5	0.359	0.382	0.406	0.442	0.504
10	0.263	0.277	0.295	0.325	0.380
20	0.188	0.199	0.212	0.234	0.278
25	0.170	0.180	0.191	0.210	0.247
30	0.155	0.164	0.174	0.192	0.226
> 30	$\dfrac{0.86}{\sqrt{n}}$	$\dfrac{0.91}{\sqrt{n}}$	$\dfrac{0.96}{\sqrt{n}}$	$\dfrac{1.06}{\sqrt{n}}$	$\dfrac{1.25}{\sqrt{n}}$

TABLE 6.4.   Critical values of the Kolmogorov-Smirnov statistic for testing a exponential parent when the mean is estimated from the sample.

As in the case of the Pickands method, this method is useful when the sample size is large, because only then the size of the censored sample used to test exponentiality is large enough to get significant results. For small samples, the powers of the test become very small and the null hypothesis (Gumbel type domain of attraction) is hardly ever rejected. However, this can be avoided if the critical significance levels of the test are increased. This can be done in practice when the wrong rejection of the null hypothesis is not a serious problem (Any Gumbel cdf can be obtained as a limit of Weibull or Frechet cdf's), as occurs when parameter estimation follows. In these cases, critical values of 0.25 or even 0.50, instead of the usual values 0.05 or 0.01 can be useful.

6.2.4.   The curvature method.

The method to be described below (Castillo and Galambos (1986b)) has the same appealing geometrical property of the basic idea that was used for the probability paper method, i.e. the statistic upon which a decision will be made is based on the tail curvature.

This curvature can be measured in many different ways, as for example, by the difference, or the quotient of slopes at two points. In addition, any of these two slopes can be measured by utilizing two or more data points. The latter option seems to be better in order to reduce variances.

Here we propose to fit two straight lines, by least-squares, to two tail intervals and to use the quotient of their slopes to measure the curvature. More precisely, we use the statistic

$$(6.12) \qquad S = \frac{S_{n_1, n_2}}{S_{n_3, n_4}}$$

where $S_{i,j}$ is the slope of the least-squares straight line fitted on Gumbel probability paper, to the r-th order statistics with $i \le r \le j$. Thus, we can write

$$(6.13) \qquad S = \frac{m \Sigma_{11} - \Sigma_{10} \Sigma_{01}}{m \Sigma_{20} - \Sigma_{10} \Sigma_{10}}$$

where

$$(6.14) \qquad m = n_j - n_i + 1$$

$$(6.15) \qquad \Sigma_{10} = \sum_{k=n_i}^{n_j} - \log\left[-\log\left(\frac{k - 0.5}{n} \right) \right]$$

$$(6.16) \qquad \Sigma_{01} = \sum_{k=n_i}^{n_j} x_k$$

$$(6.17) \qquad \Sigma_{11} = \sum_{k=n_i}^{n_j} - x_k \log\left[-\log\left(\frac{k - 0.5}{n} \right) \right]$$

$$(6.18) \qquad \Sigma_{20} = \sum_{k=n_i}^{n_j} \left\{ -\log\left[-\log\left(\frac{k - 0.5}{n} \right) \right] \right\}^2$$

and n is the sample size.

An important property of the least squares slope $S_{i,j}$ is that it is a linear combination of order statistics with coefficients which add up to zero. This property makes the statistic S location and scale invariant.

The selection of n_1, n_2, n_3 and n_4 must be based on the sample size and the speed of convergence to the asymptotic distribution, which sometimes can be inferred from the sample. Apart from speed of convergence considerations, we have selected the following values when the right tail is of interest

$$
\begin{aligned}
n_1 &= n - [2\sqrt{n}\,] \\
(6.19) \qquad n_2 &= n_3 = n - \frac{[2\sqrt{n}\,]}{2} \\
n_4 &= n
\end{aligned}
$$

where $[x]$ means integer part of x.

According to the theory above and with the values in (6.19), if the statistic S is well above 1 we can decide that the domain of attraction is Weibull type. On the contrary, if it is well below 1, the decision is in favor of a Frechet type. However, in order to be able to give significance levels of the test, we need to know the cdf of S. Due to the analytical difficulties associated with this problem, this distribution has been determined by Monte Carlo simulation techniques assuming a Gumbel parent. After 5000 repetitions for samples sizes $n = 10, 20, 40, 60, 80, 100$ and 200, the cdfs for S in table 6.5. were obtained. From this table, critical values associated with given significance levels can be obtained. However, in selecting these values it must be taken into account that a wrong decision in rejecting a Gumbel type domain of attraction can, in many applications, be corrected, if estimation follows this decision (see section 3.11).

In order to check the goodness of the test, three Monte Carlo simulations with 5000 replications were performed assuming uniform, Cauchy and

Sample Size							
cdf	10	20	40	60	80	100	200
0.01	0.120	0.159	0.216	0.265	0.285	0.297	0.360
0.02	0.155	0.209	0.264	0.303	0.336	0.345	0.415
0.05	0.238	0.294	0.362	0.396	0.418	0.433	0.506
0.10	0.340	0.387	0.465	0.485	0.518	0.533	0.592
0.20	0.497	0.543	0.622	0.651	0.658	0.678	0.713
0.30	0.694	0.724	0.768	0.783	0.796	0.805	0.817
0.50	1.135	1.133	1.083	1.085	1.073	1.055	1.046
0.70	1.903	1.782	1.558	1.485	1.481	1.405	1.345
0.80	2.643	2.302	1.955	1.836	1.758	1.657	1.530
0.90	4.333	3.401	2.662	2.427	2.264	2.135	1.895
0.95	6.477	4.770	3.392	3.090	2.877	2.588	2.250
0.98	11.211	6.739	4.673	4.045	3.720	3.153	2.732
0.99	15.723	9.363	5.814	4.990	4.673	3.613	2.969

TABLE 6.5.   Simulated cdf of S (Gumbel parent)

exponential parents. The resulting cdfs for S with values in (6.19), are shown in tables 6.6., 6.7. and 6.8. respectively.

Table 6.9. shows critical values and their associated percentages of erroneous decisions when testing Gumbel against Weibull or Gumbel against Frechet type domains of attractions for different probabilities of the type I error and uniform, Cauchy and exponential parents.

Note that the exponential belongs to the domain of attraction for maxima of the Gumbel distribution and thus, the percentages of wrong decisions can be approximately inferred from the probabilities of type I errors.

As was stated before, if estimation of parameters follows, the erroneous decision to reject a Gumbel type domain of attraction has no serious consequences (see section 3.11). This is why we include in table 6.9. error probabilities as high as 0.20 or 0.50.

	Sample Size						
cdf	10	20	40	60	80	100	200
0.01	0.252	0.469	0.801	1.063	1.219	1.340	1.766
0.02	0.368	0.588	1.008	1.235	1.370	1.486	1.979
0.05	0.554	0.873	1.319	1.613	1.756	1.857	2.343
0.10	0.809	1.176	1.650	2.005	2.172	2.283	2.720
0.20	1.245	1.686	2.210	2.548	2.720	2.830	3.247
0.30	1.691	2.131	2.680	3.086	3.234	3.324	3.715
0.50	2.778	3.281	3.765	4.098	4.288	4.288	4.655
0.70	4.587	4.854	5.319	5.605	5.708	5.593	5.734
0.80	6.361	6.614	6.545	6.720	6.720	6.510	6.614
0.90	10.417	9.881	8.711	8.803	8.651	8.197	7.911
0.95	15.625	13.441	11.111	11.062	10.823	9.843	9.191
0.98	26.316	19.080	14.970	14.124	13.158	11.737	10.965
0.99	37.879	26.320	17.606	16.779	15.625	13.661	12.438

TABLE 6.6.   Simulated cdf of S (Uniform parent)

| Sample Size | | | | | | |
cdf	10	20	40	60	80	100	200
0.01	0.007	0.004	0.003	0.003	0.003	0.003	0.003
0.02	0.012	0.008	0.007	0.006	0.006	0.005	0.004
0.05	0.030	0.022	0.017	0.014	0.014	0.014	0.013
0.10	0.064	0.045	0.033	0.029	0.029	0.027	0.022
0.20	0.140	0.101	0.075	0.064	0.060	0.054	0.045
0.30	0.236	0.161	0.119	0.102	0.094	0.082	0.072
0.50	0.520	0.356	0.239	0.201	0.178	0.162	0.130
0.70	1.056	0.664	0.440	0.360	0.331	0.286	0.222
0.80	1.586	1.015	0.618	0.511	0.454	0.392	0.296
0.90	2.769	1.618	0.992	0.786	0.709	0.582	0.443
0.95	4.554	2.618	1.401	1.124	0.972	0.813	0.587
0.98	7.962	3.834	2.115	1.666	1.368	1.128	0.793
0.99	11.521	6.173	2.612	2.140	1.890	1.361	1.013

TABLE 6.7.   Simulated cdf of S (Cauchy parent)

Example 6.3. (Data examples in Appendix C).   The method above has been applied to all data sets in Appendix C. The resulting values for the statistic c and their associated significance levels are shown in table 6.10., which allows one to make the following decisions

   (i) The Ocmulgee river (Macon), the epicenter and the fatigue data come from parents which belong to the domain of attraction of a Weibull type.
   (ii) The wind data comes from a parent which belongs to the domain of attraction of a Frechet type.
   (iii) The rest of the data sets come from parents which belong to a Gumbel type domain of attraction.

These three conclusions were arrived at using a value for the critical

Sample Size						
cdf	10	20	40	60	80	100	200
0.01	0.091	0.138	0.196	0.239	0.262	0.278	0.344
0.02	0.117	0.178	0.238	0.272	0.309	0.322	0.396
0.05	0.181	0.250	0.322	0.361	0.388	0.402	0.480
0.10	0.263	0.340	0.418	0.448	0.481	0.499	0.564
0.20	0.401	0.468	0.559	0.600	0.616	0.634	0.683
0.30	0.570	0.637	0.697	0.724	0.748	0.757	0.780
0.50	0.939	1.014	0.996	1.009	1.007	0.994	1.005
0.70	1.608	1.607	1.434	1.384	1.397	1.329	1.293
0.80	2.246	2.061	1.812	1.711	1.663	1.560	1.472
0.90	3.709	3.094	2.465	2.294	2.146	2.036	1.833
0.95	5.631	4.409	3.161	2.907	2.747	2.473	2.178
0.98	9.843	6.297	4.378	3.870	3.546	2.994	2.623
0.99	14.205	8.803	5.342	4.726	4.480	3.458	2.874

TABLE 6.8.   Simulated cdf of S (Exponential parent)

significance level of 0.05. However, as the aim of the analysis, as indicated in Appendix C, is the estimation of parameters, large critical values for the significance levels could be chosen and then it could be also reasonable to substitute conclusion (iii) with the two following

(iiia) The precipitation, the oldest ages of both women and men, the flood and the telephone data come from parents which belong to the domain of attraction of a Weibull type.

(iiib) The rest of the data sets come from parents which belong to a Gumbel type domain of attraction.   ∎

6.2.5   Methods based on other shape parameter estimates.

As was mentioned before a decision on domains of attraction can be made based on the c shape parameter of the Von-Mises form. To this end, this

Sample Size	Probability Type I Error	Critical Values		Percentages of Wrong Decisions		
		Weibull	Frechet	Uniform	Cauchy	Exponential
10	0.05	6.477	0.238	80.5	70.0	4.0
	0.10	4.333	0.340	68.0	61.5	8.0
	0.20	2.643	0.497	48.0	51.0	16.0
	0.50	1.135	1.135	17.5	28.5	43.5
20	0.05	4.770	0.294	69.0	55.0	4.0
	0.10	3.401	0.387	52.0	48.0	8.5
	0.20	2.302	0.543	33.0	36.0	16.5
	0.50	1.133	1.133	9.0	17.0	44.0
40	0.05	3.392	0.362	44.0	36.5	4.0
	0.10	2.662	0.465	29.5	28.0	8.0
	0.20	1.955	0.622	14.5	20.0	17.0
	0.50	1.083	1.083	2.5	8.5	45.0
60	0.05	3.090	0.396	30.0	27.5	4.5
	0.10	2.427	0.485	17.5	21.5	8.5
	0.20	1.836	0.651	7.5	14.5	17.0
	0.50	1.085	1.085	1.0	5.5	45.5
80	0.05	2.877	0.418	23.0	22.5	4.5
	0.10	2.264	0.518	11.5	16.5	8.5
	0.20	1.758	0.658	5.0	11.0	17.5
	0.50	1.073	1.073	0.5	4.0	46.0
100	0.05	2.588	0.433	15.5	17.0	4.5
	0.10	2.135	0.533	8.0	12.0	8.5
	0.20	1.657	0.678	3.0	7.5	17.5
	0.50	1.055	1.055	0.4	2.5	46.0
200	0.05	2.250	0.506	4.0	7.5	4.5
	0.10	1.895	0.592	1.5	5.0	8.5
	0.20	1.530	0.713	0.5	3.0	18.0
	0.50	1.046	1.046	0.04	1.0	46.5

TABLE 6.9   Percentages of wrong decisions for different parents and type I error probabilities

Data Set	Sample Size	S Estimates	Significance Level	Domain of Attraction Type
Link Data	20	1.490	0.37	GUMBEL
Precipitation Data	40	2.567	0.10	GUMBEL OR WEIBULL
Houmb Data	24	1.168	0.48	GUMBEL
Ocmulgee (Macon)	40	3.999	0.03	WEIBULL
Ocmulgee (Hawk)	40	1.091	0.49	GUMBEL
Oldest Ages (women)	54	0.629	0.18	GUMBEL
Oldest Ages (men)	54	1.876	0.190	GUMBEL
Wind Data	50	0.366	0.045	FRECHET
Flood Data	60	0.562	0.14	GUMBEL
Wave Data	50	0.750	0.28	GUMBEL
Phone Data	35	2.602	0.11	GUMBEL
Epicenter Data	60	7.075	0.005	WEIBULL
Fatigue Data	35	5.243	0.015	WEIBULL
Insulation Data	30	1.587	0.31	GUMBEL

TABLE 6.10. S estimates, significance levels and domains of attraction types resulting from the curvature method.

parameter can be estimated together with its associated confidence interval by using tail data. If this interval does not contain the value $c = 0$, then a Gumbel type domain of attraction can be rejected. Note that this method makes use of the usual connection between confidence intervals and tests of hypotheses. Thus, any estimation method based on tail data which allows confidence intervals to be determined, leads to an associated test for domains of attraction.

Chapter 7

Limit Distributions of
k-th order Statistics

7.1. Introduction

In previous chapters the problems have been concerned with the limit distributions of maxima and minima of random (i.i.d.) samples. One of the most important result was that all the limit distributions were shown to belong to the von-Mises families (3.28) and (3.29) for maxima and minima, respectively. These families are an aggregate of the three classical Gumbel, Weibull and Frechet families, thus giving an unified approach to the classical three-fold treatment of extremes. Some criteria to identify the domain of attraction of a given cdf were also given.

In this chapter, we concentrate on the problem of limit distributions of k-th order statistics, which also play an important role in applications, as was shown in chapter 2. Questions similar to those stated in the introduction to chapter 3, pertaining to the existence and variety of limit distributions can be stated now for the k-th order statistics. However, due to our knowledge about extremes, we can add new questions such as: Is there any family that includes all the limit distributions for the k-th order statistics as there was for the extremes? What is, if any, the relation between those limit distributions?, etc.

Summarizing, the aim of this chapter is to analyze the existence of limit distributions for the k-th order statistics, to identify them, if they exist, and to find, if any, their relationships to the limit laws for extremes.

7.2.   Statement of the Problem and Previous Definitions

As in the problem of extremes, we assume an i.i.d. sample $X_1, X_2, \ldots, X_n$, coming from a parent with cdf $F(x)$ and we look for sequences of constants a_n, b_n, c_n and d_n such that the limit distributions

(7.1)
$$\lim_{n \to \infty} H_{n-r+1:n}(a_n + b_n x) = H_r(x)$$

(7.2)
$$\lim_{n \to \infty} L_{r:n}(c_n + d_n x) = L_r(x)$$

where $H_{n-r+1:n}(x)$ and $L_{r:n}(x)$ are the cdfs of the $(n - r + 1)$-th and r-th order statistics, respectively, become non-degenerated.

More specifically, we shall study the limit distribution of $H_{r(n):n}(x)$ where $r(n)$ is a function of n. We shall distinguish three cases for the function $r(n)$, which lead to the definition of the so-called upper (lower), moderately upper (lower) and central order statistics.

Definition 7.1. (Upper and lower order statistics).   The order statistic $X_{r:n}$ is called a lower order statistic and the $X_{n-r+1:n}$ an upper order statistic if, when n tends to infinity, $\lim r(n) = k < \infty$, where k is an integer.   ∎

Definition 7.2. (Moderately upper and lower order statistics).   The order statistic $X_{r:n}$ is called a moderately low order statistic and $X_{n-r+1:n}$ is called a moderately upper order statistic if, when n tends to infinity, $\lim r(n) = \infty$ and $\lim r(n)/n = 0$.   ∎

Definition 7.3. (Central order statistics). $X_{r:n}$ and $X_{n-r+1:n}$ are central order statistics if, when n tends to infinity, $\lim r(n)/n = p$ with $0 < p < 1$.   ∎

Our first discussion will be for lower and upper order statistics.

7.3.   Limit Distributions of Upper and Lower Order Statistics

The simplest of the three cases above is that of upper and lower order statistics. Their limit distributions are given by the following theorem (Galambos (1984)).

Theorem 7.1. (Asymptotic distribution of upper order statistics).   *If for sequences a_n and b_n we have*

$$(7.3) \qquad \lim_{n \to \infty} F^n(a_n + b_n x) = H_c(x)$$

then

$$(7.4) \qquad \lim_{n \to \infty} H_{n-r+1:n}(a_n + b_n x) = \begin{cases} H_c(x) \displaystyle\sum_{i=0}^{r-1} \frac{[-\log H_c(x)]^i}{i!}; & H_c(x) > 0 \\ 0 & \text{otherwise} \end{cases}$$

Proof.   By (7.3), we have that for any x such that $0 < H_c(x)$

$$(7.5) \qquad \lim_{n \to \infty} F(a_n + b_n x) = 1$$

because if it is not, then

$$\lim_{n \to \infty} F^n(a_n + b_n x) = 0 \neq H_c(x)$$

contradicting (7.3). Consequently, for n large

$$\log H_c(x) \approx n \log[F(a_n + b_n x)]$$
$$= n \log\{1 - [1 - F(a_n + b_n x)]\}$$
$$(7.6) \qquad \approx -n[1 - F(a_n + b_n x)]$$

where $\approx$ means asymptotically equivalent.

Now from (2.2) we get

$$H_{n-r+1:n}(a_n + b_n x) = F_{X_{n-r+1:n}}(a_n + b_n x)$$
$$= \sum_{k=n-r+1}^{n} \binom{n}{k} F^k(a_n + b_n x)[1 - F(a_n + b_n x)]^{n-k}$$
$$(7.7) \qquad = \sum_{i=0}^{r-1} \binom{n}{i} F^{n-i}(a_n + b_n x)[1 - F(a_n + b_n x)]^i$$

and noting that

$$(7.8) \qquad F^{n-i}(a_n + b_n x) \approx F^n(a_n + b_n x)$$

$$(7.9) \qquad \binom{n}{i}[1 - F(a_n + b_n x)]^i \approx \frac{\{n[1 - F(a_n + b_n x)]\}^i}{i!}$$

and substitution in (7.7), taking into account (7.6) leads to (7.4).   ∎

This theorem has very important implications because

(i) It guarantees the existence of limit distributions of upper order statistics if the limit distribution for the maximum exists,

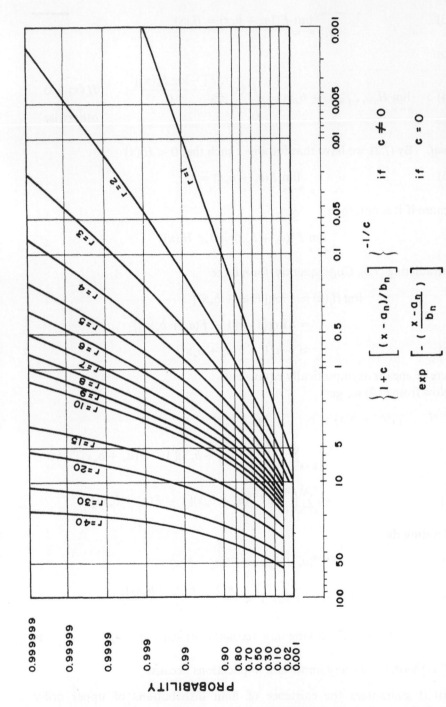

$$\left\{1 + c\left[(x - a_n)/b_n\right]\right\}^{-1/c} \qquad \text{if} \quad c \neq 0$$

$$\exp\left[-\left(\frac{x - a_n}{b_n}\right)\right] \qquad \text{if} \quad c = 0$$

FIGURE 7.1. Normalized limit cdf of the statistics of order n to $n - 9$ and $n - 14$, $n - 19$, $n - 29$ and $n - 39$

(ii) it states that the same sequences of normalizing constants as for maxima can be used, and

(iii) it gives the cdfs of their limit distributions as a function of the limit distribution for maxima.

Substitution in (7.4) of $H_c(x)$ by its value (see (3.28)) leads to

$$H^*_{n-r+1:n}(x) = \exp\left\{-\left[1 + c\left(\frac{x-a_n}{b_n}\right)\right]^{-1/c}\right\} \sum_{i=0}^{r-1} \frac{\left[1 + c\left(\frac{x-a_n}{b_n}\right)\right]^{-i/c}}{i!}$$

(7.10)

where $H^*_{n-r+1:n}(x)$ is the limit distribution of the $(n - r + 1)$-th order statistic. Figure 7.1. shows this cdf for $r = 1$ to 10, 15, 20, 30 and 40.

Example 7.1. (Breakwater wave height design).   It is well known that a rubble mound breakwater fails after being reached by *k* waves of height larger than a given threshold value h_0. For this reason, the design engineer is interested in the probabilities of the $(n - k + 1)$-th order statistic, of the sample of all waves in a given period, to exceed the value *x*. In other words, the designer wants to know the cdf of the $(n - k + 1)$-th order statistic. However, because the number of waves in the usual design periods is very large, an asymptotic solution is usually adequate.

If the wave heights, as is usual, are assumed to follow a Rayleigh distribution, $F(x) = 1 - \exp[-x^2/a^2]$, then the limit distribution of the $(n - k + 1)$-th order statistic, according to theorem 7.1, is

$$H^*(a_n + b_n x) = H(x) \sum_{i=0}^{k-1} \frac{\{-\log H(x)\}^i}{i!}$$

where $H(x)$ is Gumbel and a_n and b_n are given in example 3.4, i.e.

$$a_n = a\sqrt{\log n}; \qquad b_n = \frac{a}{2\sqrt{\log n}}$$

If $a = 3m$, $k = 7$ and the design wave height must be such that the probability of failure after 1000 waves is 0.01, we have

$$a_n = 3\sqrt{\log 1000} = 7.8847$$

$$b_n = \frac{3}{[2\sqrt{\log 1000}]} = 0.5707$$

and from figure 7.1, for $p = 0.99$ and $r = 7$, we get

$$\exp\left(\frac{-(x - a_n)}{b_n}\right) = 2$$

from which $x = 7.49\ m$, i.e. the design wave height is $7.49\ m$.

If the design were for $k = 1$, i.e. for the largest wave, we should have had, from figure 7.1, for $r = 1$ and $p = 0.99$

$$\exp\left(\frac{-(x - a_n)}{b_n}\right) = 0.01$$

from which $x = 10.51\ m$.   ■

The corresponding theorem for lower order statistics is

Theorem 7.2. (Asymptotic distribution of lower order statistics) *If for sequences c_n and d_n we have*

(7.11) $$\lim_{n \to \infty} 1 - [1 - F(c_n + d_n x)]^n = L_c(x)$$

then

$$\lim_{n \to \infty} L_{r:n}(c_n + d_n x)$$

(7.12)
$$= \begin{cases} 1 - [1 - L_c(x)] \displaystyle\sum_{i=0}^{r-1} \frac{\{-\log[1 - L_c(x)]\}^i}{i!}; & L_c(x) < 1 \\ 1 & \text{otherwise} \end{cases}$$ ■

Substitution of $L_c(x)$ by its value (see (3.29)) leads to

(7.13) $$L_{r:n}^*(x) = 1 - \exp\left\{-\left[1 + c\left(\frac{c_n - x}{d_n}\right)\right]^{-1/c}\right\} \sum_{i=0}^{r-1} \frac{\left[1 + c\left(\frac{c_n - x}{d_n}\right)\right]^{-i/c}}{i!}$$

where $L_{r:n}^*(c_n + d_n x)$ is the limit distribution of the r-th order statistic, which for $r = 1$ to 10, 15, 20, 30 and 40 are shown in figure 7.2.

Example 7.2. (Fatigue strength of a cable). For analyzing the fatigue strength of a cable, a model based on the following assumptions is developed (see Castillo et al. (1985))

(i) The cable is made of m parallel elements (wires or strands) of identical length L

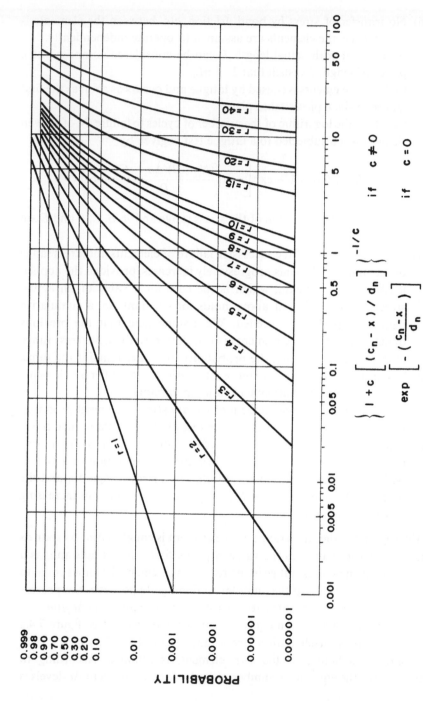

$$\left\{ \begin{array}{l} \left\{ 1 + c \left[(c_n - x) / d_n \right] \right\}^{-1/c} \quad \text{if} \quad c \neq 0 \\[2mm] \exp \left[- \left(\dfrac{c_n - x}{d_n} \right) \right] \quad \text{if} \quad c = 0 \end{array} \right.$$

FIGURE 7.2.   Normalized limit cdf of the statistics of order 1 to 10 and 15, 20, 30 and 40

(ii) No transfer of stress by bond, friction, etc. between the elements is possible, i.e. the elements are assumed to operate independently.

(iii) An element with actual length L can be considered composed of n pieces of length L_0, such that $L = nL_0$.

(iv) Failure of elements is caused by fatigue and occurs as soon as the first failure in the n pieces takes place.

(v) the cdf of the logarithm of the number of cycles to failure, N, of a given element when subjected to a fatigue test is given by

$$E(N; \Delta\sigma) = 1 - \exp\left[-n\left(\frac{(N - B)(\Delta\sigma - C)}{D} + E\right)^A\right]$$

where $\Delta\sigma$ is the logarithm of the stress range and A, B, C, D and E are constants.

(vi) As a consequence of the successive failure of elements, the stress range of the unfailed elements progressively increases, provided the applied loads remain constant, which is the case for a cable of an actual structure. Since the cdf above holds for constant $\Delta\sigma$, a cumulative damage hypothesis is needed. We assume here that the previous damage history in the elements is transferred to the new stress range level in such a way that the probability of failure remains the same.

(vii) The numbers of cycles to failure (strength) for the different elements are assumed to be independent random variables.

(viii) The dynamic effect due to the momentary stress increment produced by the failure of an element on the rest of the elements is neglected.

(ix) The failure of the cable is defined as failure of the k-th weakest element. Figure 7.3. shows the schematic representation of a system with m elements of length L subjected to a stress range $\Delta\sigma_1$ and the number of cycles to failure $N_{(1)}, N_{(2)}, \ldots, N_{(m)}$, ranked in increasing order of magnitude.

At the beginning, the stress range in one element is equal to $\Delta\sigma_1$. As soon as the number of cycles reaches the value $N_{(1)}$, the first element fails and, as a consequence, the new stress range in the remaining elements becomes equal to $m\Delta\sigma/(m - 1)$. According to the cumulative damage hypothesis, the process goes on, as if the cable were subjected to a stress range equal to $m\Delta\sigma_1/(m - 1)$ from the beginning and for an equivalent number of cycles (see figure 7.4.). This process repeats itself until a stress range equal to $m\Delta\sigma_1/(m - k + 1)$, corresponding to failure of the k-th element, is attained. According to assumption (iv), the equivalent number of cycles for two different $\Delta\sigma$-levels is

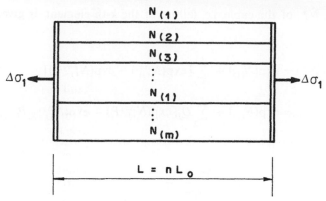

FIGURE 7.3.   Schematic representation of a system with m elements of length L subjected to a stress range $\Delta\sigma_1$

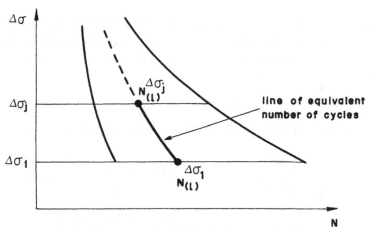

FIGURE 7.4.   Illustration of the concept of equivalent number of cycles

given by the cumulative damage hypothesis, i.e. the probabilities of failure for both levels coincide. This condition, for the s-th failure, becomes

$$(N_{(s)}^{\Delta\sigma_1} - B)(\Delta\sigma_1 - C) = (N_{(s)}^{\Delta\sigma_j} - B)(\Delta\sigma_j - C)$$

where $N_{(s)}^{\Delta\sigma_1}$ is the r-th order statistic of the logarithm of the number of cycles to failure of the elements when subjected to a stress range $\Delta\sigma$ and the pairs $(N_{(s)}^{\Delta\sigma_1}, \Delta\sigma_1)$ and $(N_{(s)}^{\Delta\sigma_j}, \Delta\sigma_j)$ correspond to the same damage state.

The life, N_k^*, of the cable, i.e. failure of the k-th element, is given by (see figure 7.5.)

$$\exp(N_k^*) = \exp(N_{(1)}) + \sum_{j=2}^{k} [\exp(N_{(j)}^{\Delta\sigma_j}) - \exp(N_{(j-1)}^{\Delta\sigma_j})]$$

$$= \exp(N_{(1)}) + \sum_{j=2}^{k} Q_j [\exp(N_{(j)}^{\Delta\sigma_1} P_j) - \exp(N_{(j-1)}^{\Delta\sigma_1} P_{j-1})]$$

where

$$P_j = \frac{\Delta\sigma_1 - C}{\Delta\sigma_j - C}$$

$$Q_j = \exp[B(1 - P_j)]$$

$$\Delta\sigma_j = \log\left(\frac{m}{m - j + 1}\right) + \Delta\sigma_1$$

This equation allows one to calculate the cdf of N_k^* only by numerical or simulation techniques. Since these can be cumbersome, it would be of interest, from a practical point of view, to give upper and lower bounds for N_k^*. These bounds are given by (see figure 7.5)

$$N_{(k)}^{\Delta\sigma_k} \le N_k^* \le N_{(k)}^{\Delta\sigma_1}$$

The cdfs of the k-th order statistics $N_{(k)}^{\Delta\sigma_1}$ and $N_{(k)}^{\Delta\sigma_k}$ are given by (see expression (2.2))

$$G_k(N; \Delta\sigma_p) = \sum_{j=k}^{m} \binom{m}{j} \left\{ 1 - \exp\left[-n\left(\frac{(N - B)(\Delta\sigma_p - C)}{D} + E \right)^A \right] \right\}^j$$

$$\cdot \exp\left[(j - m)n\left(\frac{(N - B)(\Delta\sigma_p - C)}{D} + E \right)^A \right]$$

for $p = 1$ and k, respectively. They give upper and lower bounds for the cdf of N_k^*.

It is interesting to analyze the asymptotic cases. Because of the presence of two parameters m and n, the following asymptotic cases can be considered:

(1) Very long cables $(n \to \infty)$
(2) Very large number of elements $(m \to \infty)$

In the case of very long cables, all the elements can be assumed to fail at the endurance limit $N = B$.

If there is a large number of elements, the upper and lower bounds above

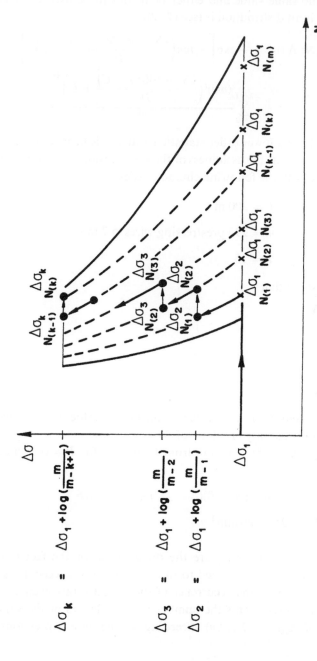

FIGURE 7.5. Evolution of stress ranges and equivalent number of cycles during failure

converge to the same value and either of them can be used for a practical solution. The limit distribution is (see (7.12))

$$G_k(N; \Delta\sigma) = 1 - \exp\left[-nm\left(\frac{(N - B)(\Delta\sigma - C)}{D} + E\right)^A\right]$$

$$\cdot \sum_{t=0}^{k-1} \frac{(nm)^t\left(\frac{(N - B)(\Delta\sigma - C)}{D} + E\right)^{At}}{t!}$$

Note that we have a lower order statistic because k is held constant.

In order to illustrate the usefulness of the above results, we apply them to the case of a cable with the following characteristics

$$L = 200 \text{ m.}$$

$$m = 295 \text{ prestressing wires } \phi \text{ 7 mm.}$$

$$k = 15$$

We also assume that

$A = 4.09$

$B = 9.97$

$C = 5.48$

$D = 1.274$

$E = -0.085$

for a reference length of $L_0 = 1.96$ m, and take a value $N = 10^7$ cycles for design, with a probability of failure of 0.05.

Thus, we get $n = 200/1.96$ and, from figure 7.2, for $P = 0.05$ and $r = 15$ we obtain

$$nm[(N - B)(\Delta\sigma - C) + E]^A = 8.6$$

from which $\Delta\sigma = 251.1$ N/mm^2.   ∎

Expressions (7.4) and (7.12) are the direct result of the fact that upper (lower) order statistics are related to the occurrence of exceedances of very high (low) values, i.e. to the occurrence of the so-called rare events.

In fact, if $\underline{m}_n(x)$ represents the number of exceedances of the value x, i.e. the number of $X_i(i = 1, 2, \ldots, n)$ exceeding x, then (see expressions (2.143) and (2.144))

(7.14) $F_{X_{n-r+1:n}}(x) = P[X_{n-r+1} \leq x] = P[\underline{m}_n(x) < r]$

(7.15) $$F_{X_{r:n}}(x) = P[X_{r:n} \le x] = P[\underline{m}_n(x) \le n - r]$$

Further, for the sequences a_n and b_n to satisfy (3.8) we must have

(7.16) $$\lim_{n \to \infty} n[1 - F(a_n + b_n x)] = u(x); \qquad 0 \le u(x) \le \infty$$

but this implies (see theorem 1.1) that

(7.17) $$\lim_{n \to \infty} P[m_n(a_n + b_n x) < r] = \exp[-u(x)] \sum_{i=0}^{r-1} \frac{u^i(x)}{i!}$$

where the right hand side of (7.17) must be interpreted as one or zero for $u(y) = 0$ or $u(y) = \infty$, respectively.

Expression (7.17) shows, as claimed, that $m_n(a_n + b_n x)$ behaves as a Poisson process of intensity $u(x)$.

Note, however, that the level $a_n + b_n x$ is not constant, but, typically, increases with n. For $0 \le u(x) < \infty$, this implies that $\lim F(a_n + b_n x) = 1$, i.e. the level $a_n + b_n x$ is very rare (exceptionally high); thus, the validity of the Poisson law, which, not in vain, is called the law of rare events.

For the particular case of $r = 1$ we have from (7.14) and (7.17)

(7.18) $$\lim_{n \to \infty} F_{X_{n:n}}(a_n + b_n x) = \lim_{n \to \infty} [\underline{m}_n(x) < 1] = \exp[-u(x)] = H_c(x)$$

from which

(7.19) $$u(x) = -\log H_c(x)$$

Substitution now of (7.19) into (7.17) leads to the general case (7.4).

7.3.1.   *Asymptotic joint distribution of upper or lower order statistics.*

In order to obtain a formula for the asymptotic joint distribution of upper or lower order statistics, we generalize theorem 1.1 first (see Leadbetter et al. (1983)).

Theorem 7.3. (Joint distribution of the number of multiple exceedances).   *Let $\{X_n\}$ be a sequence of i.i.d random variables. Assume that the sequences $\{x_n^{(k)}\}$, $k = 1, 2, \ldots, s$, of real numbers satisfy the conditions*

$$x_n^{(1)} \ge x_n^{(2)} \ge \ldots \ge x_n^{(s)}$$

(7.20)

$$\lim_{n \to \infty} n\{1 - F(x_n^{(k)})\} = \tau_k; \qquad k = 1, 2, \ldots, s$$

Then, for non-negative integers $r_1, r_2 \ldots, r_s$ we have

$$\lim_{n \to \infty} P[\underline{m}_n^{(1)}(x_n^{(1)}) = r_1, \underline{m}_n^{(2)}(x_n^{(2)}) = r_1 + r_2, \ldots, \underline{m}_n^{(s)}(x_n^{(s)}) = r_1 + r_2 + \cdots + r_s]$$

$$(7.21) \qquad = \frac{\tau_1^{r_1}(\tau_2 - \tau_1)^{r_2} \cdots (\tau_s - \tau_{s-1})^{r_s} \exp(-\tau_s)}{r_1! r_2! \ldots r_s!}$$

where $\underline{m}_n^{(k)}(x_n^{(k)})$ is the number of exceedances of $x_n^{(k)}$, $k = 1, 2, \ldots, s$ and the right hand side must be taken as 1 or zero if any τ is zero or infinity, respectively. ∎

This theorem, together with the following equality

$$P[X_{n:n} \le x_n^{(1)}, X_{n-1:n} \le x_n^{(2)}, \ldots, X_{n-s+1:n} \le x_n^{(s)}]$$

$$(7.22) \qquad = P[\underline{m}_n^{(1)}(x_n^{(1)}) = 0, \underline{m}_n^{(2)}(x_n^{(2)}) \le 1, \ldots, \underline{m}_n^{(s)}(x_n^{(s)}) \le r - 1]$$

allows the joint distribution of upper order statistics to be obtained.

For lower order statistics we can use the standard technique of changing X into $-X$.

7.4.   Limit Distributions of Other Order Statistics

We turn now to the case of other order statistics, making a distinction between central order statistics and moderately upper or lower order statistics. However, we give first, without proof, the following general theorem (see Leadbetter et al. (1983)).

Theorem 7.4.   *Assume that for some sequence $\{a_n\}$ and $\{b_n > 0\}$, the following holds*

$$(7.23) \qquad \lim_{n \to \infty} \frac{r(n) - n[1 - F(a_n + b_n x)]}{\left[r(n) \left(1 - \dfrac{r(n)}{n} \right) \right]^{1/2}} = \tau(x)$$

with

$$(7.24) \qquad \lim_{n \to \infty} [n - r(n)] = \infty$$

Then, we have

$$(7.25) \qquad \lim_{n \to \infty} P[X_{n-r(n)+1:n} \le a_n + b_n x] = H(x)$$

where

(7.26) $$H(x) = \Phi(\tau(x))$$

and $\Phi(x)$ is the cdf of the standard normal distribution. Conversely, if (7.25) holds for $H(x)$ non-degenerated, then (7.23) and (7.24) hold.

Example 7.3. (Weibull distribution). Assume a sample of size *n* from a population with cdf

$$F(x) = \exp(x); \qquad x \le 0$$

and consider the sequence of order statistics

$$r(n) = n^{1/2}$$

Then

$$\lim_{n \to \infty} (n - r(n)) = n - n^{1/2} = \infty$$

and with

$$a_n = -n^{-1/2}; \qquad b_n = n^{-3/4}$$

we have

$$\lim_{n \to \infty} \frac{r(n) - n[1 - F(a_n + b_n x)]}{\left[r(n)\left(1 - \frac{r(n)}{n} \right) \right]^{1/2}} = \lim_{n \to \infty} \frac{n^{1/2} - n[1 - \exp(-n^{-1/2} + n^{-3/4}x)]}{\left[n^{1/2}\left(1 - \frac{n^{1/2}}{n} \right) \right]^{1/2}}$$

$$= \lim_{n \to \infty} \frac{n^{1/2} - n[n^{-1/2} - n^{-3/4}x]}{(n^{1/2} - 1)^{1/2}} = x$$

Then, according to theorem 7.4,

$$\lim_{n \to \infty} P[X_{n - r(n) + 1:n} \le a_n + b_n x] = \Phi(x) \qquad \blacksquare$$

7.4.1. *Asymptotic distribution of central order statistics.*

For central order statistics we have the following theorem (see Galambos (1978)).

Theorem 7.5. (Asymptotic normality of central order statistics). *Let $F(x)$ be a continuous cdf with associated continuous pdf, $f(x)$. Let $p_1, p_2, \ldots, p_k$ be a set of* *real numbers in the interval* $(0, 1)$ *such that* $f(F^{-1}(p_j)) \neq 0$; $1 \leq j \leq k$. *If* $r_j(n)$ *are such that*

(7.27)
$$\lim_{n \to \infty} n^{1/2} \left(\frac{r_j(n)}{n} - p_j \right) = 0; \qquad 1 \leq j \leq k$$

then the vector

(7.28)
$$\sqrt{n} [X_{r_j:n} - F^{-1}(p_j)]; \qquad 1 \leq j \leq k$$

is asymptotically a k-dimensional normal vector with zero expectation and covariance matrix

(7.29)
$$\frac{p_i(1 - p_j)}{\{f(F^{-1}(p_i)) f(F^{-1}(p_j))\}}; \qquad p_i \leq p_j$$

This theorem states that central order statistics are jointly asymptotically normal if condition (7.27) is satisfied. The particular case $k = 1$ guarantees the asymptotic normality of any central order statistic satisfying (7.27).

Example 7.4. (Fatigue strength of a cable).   Assume now that the failure of the cable in example 7.2 is assumed to take place when the 5% weakest elements fail. Then we have $r(m) = 5m/100$ (central order statistics) where m is the number of elements, and condition (7.27) holds with $p = 0.05$.

If we define the random variable

(7.30)
$$V = n \left(\frac{(N - B)(\Delta\sigma - C)}{D} + E \right)^A$$

we have, according to (7.28) and (7.29), that

$$\sqrt{m} [V_{r:n} - F^{-1}(p)] \quad \rightarrow \quad N \left(0, \frac{p(1 - p)}{f^2(F^{-1}(p))} \right)$$

where $F(x)$ and $f(x)$ are the cdf and pdf of V, which, according the example 7.2, are

$$F(x) = 1 - \exp(-x)$$

$$f(x) = \exp(-x)$$

Thus, we get

$$\sqrt{m} [V_{r:n} + \log(1 - p)] \quad \rightarrow \quad N \left(0, \frac{p(1 - p)}{(1 - p)^2} \right) = N \left(0, \frac{p}{(1 - p)} \right)$$

Consequently, the 0.05 percentile of $V_{r:n}$ is (notice that -1.64 is the 0.05 percentile of the $N(0, 1)$ variable)

$$V_{0.05} = -\log(1 - p) - \frac{1.64[p(1 - p)]^{1/2}}{m^{1/2}} = 0.02938$$

from which, by setting $p = 0.05$, $m = 295$ and back-substitution gives $\Delta\sigma = 251 \ N/mm^2$, which is practically the same as that obtained in example 7.2.

Note that in these two examples, two approaches, one of central order statistics and one of lower order statistics, have been used, both leading to the same practical result.   ■

The following theorem solves some of the stated problems for central order statistics.

Theorem 7.6. (Some feasible limit distributions for central order statistics).   *Let r(n) be such that*

(7.31) $$\lim_{n \to \infty} n^{1/2}\left(\frac{r(n)}{n} - p\right) = 0; \qquad 0 < p < 1$$

and assume that for some sequences $\{a_n\}$ and $\{b_n > 0\}$

(7.32) $$\lim_{n \to \infty} P[X_{n-r(n)+1:n} \leq a_n + b_n x] = H(x)$$

where $H(x)$ is non-degenerated. Then, $H(x)$ can be only one of the following forms

(i)

(7.33) $$H(x) = \begin{cases} 0 & \text{if } x < 0 \\ \Phi(cx^a) & \text{if } x \geq 0 \end{cases}$$

(ii)

(7.34) $$H(x) = \begin{cases} \Phi(-c|x|^a) & \text{if } x < 0 \\ 1 & \text{if } x \geq 0 \end{cases}$$

(iii)

(7.35) $$H(x) = \begin{cases} 0 & \text{if } x < -1 \\ \dfrac{1}{2} & \text{if } -1 \leq x < 1 \\ 1 & \text{if } x \geq 1 \end{cases}$$

(iv)

$$(7.36) \qquad H(x) = \begin{cases} \Phi(-c_1|x|^a) & \text{if } x < 0 \\ \Phi(c_2|x|^a) & \text{if } x \geq 0 \end{cases}$$

where all constans a, c, c_1 and c_2 are positive. Note that only (7.36) is a continuous distribution function.

If condition (7.31) is relaxed, many more limit distributions are possible. The interested reader is referred to Balkema and de Haan (1978a, 1978b).

7.4.2. *Asymptotic distributions of moderately upper and lower order statistics.*

The asymptotic distribution of moderately upper and lower order statistics can be determined with the help of the following theorem (see Galambos (1978)).

Theorem 7.7. *For an arbitrary distribution $F(x)$ and any values n, r and x, we have*

$$(7.37) \qquad \left| P(X_{r:n} < x) - \Phi\left[\frac{nF(x) - r}{\left[r\left(1 - \frac{r}{n}\right)\right]^{1/2}} \right] \right| \leq \frac{3}{\left[r\left(1 - \frac{r}{n}\right)\right]^{1/2}}$$

Example 7.5. (Exponential distribution).   Assume an exponential distribution and take $r(n) = n - \log n$.

If now we make

$$y = \frac{nF(x) - r}{\left[r\left(1 - \frac{r}{n}\right)\right]^{1/2}}$$

upon consideration of $F(x) = 1 - \exp(-x)$, we get

$$x = \log\left[1 - \frac{r}{n} - \frac{y\left(\frac{r}{n}\right)^{1/2}(n - r)^{1/2}}{n} \right]$$

$$= -\log\frac{(n - r)\left[1 - \frac{\left(\frac{r}{n}\right)^{1/2}y}{(n - r)^{1/2}} \right]}{n}$$

$$= -\log\left(\frac{n-r}{n}\right) - \log\left[1 - \frac{\left(\frac{r}{n}\right)^{1/2}y}{(n-r)^{1/2}}\right]$$

$$= -\log\left(\frac{\log(n)}{n}\right) + \frac{\left(\frac{r}{n}\right)^{1/2}y}{(n-r)^{1/2}}$$

$$\approx -\log\log n + \log n + \frac{y}{(\log n)^{1/2}}$$

from which, upon taking into account (7.37), we get

$$P(X_{r:n} < a_n + b_n y) \quad \rightarrow \quad \Phi(y)$$

where

$$a_n = \log n - \log\log n$$

$$b_n = (\log n)^{-1/2} \qquad \blacksquare$$

Finally, we include the following result, from Wu (1966).

Theorem 7.8. (Feasible asymptotic distributions for moderately lower order statistics).   *Let $r(n)$ be non-decreasing and such that*

(7.38) $$\lim_{n\to\infty}\frac{r(n)}{n} = 0; \quad \lim_{n\to\infty} r(n) = \infty$$

and assume that for some sequences $\{a_n\}$ and $\{b_n > 0\}$

(7.39) $$\lim_{n\to\infty} P[X_{r(n):n} \le a_n + b_n x] = H(x)$$

where $H(x)$ is non-degenerated. Then $H(x)$ can be only one of the following forms

(i)

(7.40) $$H(x) = \begin{cases} \Phi(-a\log|x|)) & \text{if } x \le 0 \\ 1 & \text{if } x > 0 \end{cases}$$

(ii)

(7.41) $$H(x) = \begin{cases} 0 & \text{if } x \le 0 \\ \Phi(a\log x) & \text{if } x > 0 \end{cases}$$

(iii)

$$(7.42) \qquad H(x) = \Phi(x) \qquad\qquad \text{if } -\infty < x < \infty$$

where a is a positive constant.

7.5. Asymptotic Distributions of k-th Order Statistics of Samples with Random Sizes

All the previous sections dealt with the case of samples of fixed size. In this section we assume a sample with random size as in section 3.9.

The following theorem (Galambos (1978)) gives the asymptotic distribution in this case.

Theorem 7.9. (Asymptotic distribution of the k-th order statistic from samples of random size). *Let $X_{j,n}$, $1 \le j \le N(n)$, be independent random variables with common cdf $F_n(x)$. Let $N(n)$ be a positive integer-valued random variable, independent of $X_{j,n}$, $1 \le j \le N(n)$. If $\{a_n\}$ and $\{b_n > 0\}$ are sequences such that*

$$(7.43) \qquad \lim_{n \to \infty} F_n(a_n + b_n x) = 1 \qquad \text{for any } x$$

Then, for each k,

$$(7.44) \qquad \lim_{n \to \infty} P_n(X_{N(n)-k+1:n} < a_n + b_n x) = E_k(x)$$

exists if, and only if,

$$(7.45) \qquad \lim_{n \to \infty} P_n\left(N(n) < \frac{u}{1 - F_n(a_n + b_n x)} \right) = U(u; x)$$

exists, and

$$(7.46) \qquad E_k(x) = \sum_{t=0}^{k-1} \int_0^\infty \frac{u^t \exp(-u)\, dU(u; x)}{t!}$$

IV Asymptotic Distribution of Sequences of Dependent Random Variables

Chapter 8   Limit Distributions in the Case of Dependence

8.1. Introduction

Previous chapters were devoted to the problem of limit distributions of order statistics in the case of independence. Nevertheless, in many practical situations this assumption is satisfied neither exactly nor approximately. Just think of the water levels in a given section of a river on successive days, of the strength of contiguous sections of a longitudinal element, of the height of successive waves reaching a given location or of the rainfall on two consecutive days, for example. Thus, questions similar to those following can be immediately formulated: (a) Are the formulas for the independent case still valid for the case of dependence? (b) Under what conditions do they remain valid? (c) What are the limit distributions when they are not valid?

   The aim of this chapter is to analyze the case of dependence and to derive the limit distributions of order statistics under some dependence structures.

   Let us start by saying that the problem of dependence is much more complicated that the independence case. One of the main reasons for this is that, while for the independence case the probabilities of the n-dimensional events $(X_1 \leq x, X_2 \leq x, \ldots, X_n \leq x)$, can be formulated in terms of the population one-dimensional cdf $F(x)$, the analysis of the case of dependence requires partial knowledge of the joint distribution of $(X_1, X_2, \ldots, X_n)$. This means that, while one unidimensional function $F(x)$ is enough to derive such probabilities in the case of independence, a multivariate function is needed to deal with the dependence case.

The fact that only partial knowledge of the joint distribution of order statistics is needed implies that many dependence structures can lead to the same distribution of order statistics. In this context, the so-called exchangeable sequences, which are studied in section 8.2, become important. It is shown there that any sequence of random variables can be replaced with a sequence of exchangeable variables having both the same distributions for their order statistics.

In practice, one faces many kinds of dependence structures. There are many ways of measuring this dependence. Some dependence conditions, which play an important role in extreme value problems are introduced in section 8.3.

Though the theory of extremes for the case of dependence is not completely developed some particular cases are presently well known. In this chapter, we shall study mainly the cases of m-dependence, some types of mixing and some special correlation structures. Although, in general, the correlation structure is only a partial measure of dependence, some results are based on it. These are specially important for the case of normal random variables, which are analyzed in section 8.4.1.3.

Throughout this chapter we distinguish between extremes (maxima and minima) and other order statistics. The case of limit distributions of extremes is analyzed in section 8.4.

A very important practical case is that of stationary sequences, which is studied in section 8.4.1. For this case we shall identify conditions under which the same limit distribution for extremes as for the independent case still hold. Two special cases of the above are the m-dependent sequences, which are discussed in section 8.4.1.1, and the moving-average models of stable distributions, which generalizes the well known ARMA models, frequently used in the time series field. Some limit results for these models are analyzed in section 8.4.1.2.

The last part of this chapter is devoted to limit distributions of k-th extremes. In fact, section 8.5 gives some important results for this problem. In particular, conditions under which some results for the independent case still remain valid are given.

Finally, we want to point out that, contrary to the independent case, for the case of dependence any limit distribution is possible. Thus, some restrictions are needed in order to have a limited set of admissible asymptotic distributions. In particular, minimal conditions leading to the same limit distributions (von-Mises family) as for the case of independence become of special interest.

In order to illustrate this point, assume a sequence $\{X_n\}$ of random variables such that $X_n = X$ for any n. Then, it is clear that

$$X_{n:n} = \max(X_1, X_2, \ldots, X_n) = X_{r:n} = \min(X_1, X_2, \ldots, X_n) = X_{1:n} = X$$

for any r. Consequently, by selecting as the cdf of X any arbitrary cdf, $F(x)$, we have $F(x)$ as the limit cdf of the above sequence.

Galambos (1978, pp. 135-136) gives other examples and some additional restrictions in order to have meaningful theorems.

8.2. Exchangeable Variables

In section 2.4. we derived inequalities and formulas by which is obtained the exact or approximate cdf of all order statistics for finite sample size n. One interesting conclusion was that the value of the cdf of all order statistics at a given point, x, depends only on the n values

$$(8.1) \qquad P[m_n(x) = 1], P[m_n(x) = 2], \ldots, P[m_n(x) = n]$$

or, as an equivalence alternative, on the n values

$$(8.2) \qquad S_{1,n}(x), S_{2,n}(x), \ldots, S_{n,n}(x)$$

and this is true for the most general case. In other words, complete knowledge of the joint cdf of the random variables $(X_1, X_2, \ldots, X_n)$ is not needed in order to derive the statistical behavior of its order statistics, but of only n functions $(S_{i,n}(x), i = 1, 2, \ldots, n)$. For this reason, many different joint cdf's for the sample lead to the same distribution for the order statistics. One interesting approach is that of exchangeable variables which are defined as follows.

Definition 8.1. (Exchangeable variables). The random variables $X_1, X_2, \ldots, X_n$ are said to be exchangeable if the distribution of the vector $(X_{i_1}, X_{i_2}, \ldots, X_{i_n})$ is the same for all permutations of the subscripts $(i_1, i_2, \ldots, i_n)$. ■

Example 8.1. (System of n identical elements). Let us consider a system of n equal elements such that all of them fall under the same working conditions. Then, their lifetimes $X_1, X_2, \ldots, X_n$ are exchangeable random variables. ■

Definition 8.2. (Exchangeable events). The events $C_1, C_2, \ldots, C_n$ are said to be exchangeable if the probability $P[C_{i_1}, C_{i_2}, \ldots, C_{i_k}]$ is independent of the permutations of the subscripts $1 \le i_1 < i_2 < \ldots < i_k \le n$. ■

It can be demonstrated (see Galambos (1978)) that for any given set of events $A = \{A_1, A_2, \ldots, A_n\}$ a set of exchangeable events $C = \{C_1, C_2, \ldots, C_n\}$ can be found such that

(8.3) $$P[m_n(A) = t] = P[m_n(C) = t]$$

where $m_n(A)$ and $m_n(C)$ represent the number of events of the sets A and C, respectively, which occur.

The main implication of all the above is that a set of events can be replaced by an exchangeable set for which the calculus of probabilities becomes easier. In fact, for exchangeable events we have

(8.4) $$S_{k,n}(C) = \binom{n}{k} P[C_1, C_2, \ldots, C_k] = \binom{n}{k} \alpha_k$$

where the meaning of α_k is obvious.

If the set of random variables can be extended to a larger set of N exchangeable variables, we have the following theorem (see Galambos (1978))

Theorem 8.1.   *Let* $X_1, X_2, \ldots, X_n$ *be such that there exist additional random variables* $X_{n+1}, X_{n+2}, \ldots, X_N$ *with distributions which satisfy*

(8.5) $$S_{k,N}(x) = \frac{\binom{N}{k} S_{k,n}(x)}{\binom{n}{k}}$$

and

(8.6) $$\lim_{n \to \infty} \frac{N}{n} = \infty$$

Then there exist normalizing constants a_n *and* $b_n > 0$ *such that*

(8.7) $$\lim_{n \to \infty} P[X_{n-k+1:n} < a_n + b_n x] = E_k(x)$$

if and only if $P\left[m_N(a_n + b_n x) < \binom{N}{n} y \right]$ *converge in distribution to* $U(y) = U(y; x)$, *and* $E_k(x)$ *is given by*

(8.8) $$E_k(x) = \sum_{t=0}^{k-1} \int_0^\infty \frac{y^t \exp(-y)\, dU(y; x)}{t!}$$

Example 8.2. (*Mardia's multivariate distribution*).   Assume a sequence $\{X_n\}$ of random variables such that the joint distribution of the first n terms in the sequence follows a multivariate Mardia's distribution with hazard function

$$G(x_1, x_2, \ldots, x_n) = \left[\sum_{j=1}^{n} \exp\left(\frac{x_j}{e_n}\right) - n + 1 \right]^{-e_n}; \qquad n = 1, 2, \ldots$$

and such that

$$\lim_{n \to \infty} e_n = \infty$$

Note that this distribution has unit exponential marginals and that it is a sequence of exchangeable variables.

The minimum $W_n = \min(X_1, X_2, \ldots, X_n)$ has cdf (see example 2.47)

$$L_n(x) = 1 - \left[n \exp\left(\frac{x}{e_n}\right) - n + 1 \right]^{-e_n} = 1 - \left\{ 1 - n \left[1 - \exp\left(\frac{x}{e_n}\right) \right] \right\}^{-e_n}$$

Thus, we can write

$$\lim_{n \to \infty} L_n(c_n + d_n x) = \lim_{n \to \infty} 1 - \left\{ 1 - n \left[1 - \exp\left(\frac{c_n + d_n x}{e_n}\right) \right] \right\}^{-e_n}$$

From this, it is clear that we get a non-degenerated limit distribution if

$$\lim_{n \to \infty} \frac{n(c_n + d_n x)}{e_n} = 0$$

Hence, we have

$$\lim_{n \to \infty} L_n(c_n + d_n x) = \lim_{n \to \infty} \left[1 - \left(1 + \frac{n(c_n + d_n x)}{e_n} \right)^{-e_n} \right]$$

$$= \lim_{n \to \infty} \{ 1 - \exp[-n(c_n + d_n x)] \}$$

from which we can choose

$$c_n = 0$$

$$d_n = \frac{1}{n}$$

to get

$$\lim_{n \to \infty} L_n(c_n + d_n x) = 1 - \exp(-x)$$

giving a non-degenerated limit.

Note that the limit coincides with the case of independence with the same marginals (see example 3.3). ∎

8.3. Dependence Conditions

We include here some dependence conditions which play a central role for a dependent sequence to have a similar limit behaviour to that of independent sequences.

Definition 8.3. (Strong mixing sequence). A sequence $\{X_n\}$ of random variables is said to satisfy the strong mixing condition if

$$(8.9) \qquad \phi(j) = |P(A \cap B) - P(A)P(B)|$$

where A is any event generated by $(X_1, X_2, \ldots, X_n)$ and B is any event generated by $(X_{n+j}, X_{n+j+1}, \ldots)$, goes to zero when $j - > \infty$, and this holds for any value of n. ∎

Condition (8.9) is difficult to check in application. However, in some particular cases it can be substituted with a simpler one. Chernick (1981) gives the following theorem and corollary for Markov process of order p, which we define previously.

Definition 8.4. (Markov sequence of order p). Let $\{X_n\}$ be a sequence of random variables. This sequence is said to be a Markov sequence of order p, if $(\ldots, X_{m-1}, X_m)$ is independent of $(X_{m+r}, X_{m+r+1}, \ldots)$, given $(X_{m+1}, X_{m+2}, \ldots, X_{m+p})$ for any $r > p$ and $m > 0$. ∎

This definition in practical terms means that the past $(\ldots, X_{m-1}, X_m)$ and the future $(X_{m+r}, X_{m+r+1}, \ldots)$ are independent, given the present $(X_{m+1}, X_{m+2}, \ldots, X_{m+p})$.

Theorem 8.2. (Strong mixing condition for Markov sequences). Let $\{X_n\}$ be a stationary Markov sequence of order p. If

$$\lim_{n \to \infty} \int_{-\infty}^{\infty} \int_{-\infty}^{\infty} \cdots \int_{-\infty}^{\infty} |dF(x_{m+1-p}, \ldots, x_m, x_{m+j+1}, \ldots, x_{m+j+p})$$

$$(8.10) \qquad - dF(x_{m+1-p}, \ldots, x_m) \, dF(x_{m+j+1}, \ldots, x_{m+j+p})| = 0$$

where the Fs are the joint cdf's. Then, the strong mixing condition is satisfied.

From this, the following corollary follows.

Corollary 8.3. If the joint density of $(X_1, \ldots, X_p, X_{p+j+1}, \ldots, X_{2p+j})$ converges point-wise to the product of the densities of $(X_1, \ldots, X_p)$ and $(X_{p+j+1}, \ldots, X_{2p+j})$, then the strong mixing condition is satisfied.

The following two examples are from Chernick (1981b).

Example 8.3. (AR(1) model with normal marginals).   Let us consider the model

$$X_n = \rho X_{n-1} + \varepsilon_n; \qquad n \ge 1; \quad -1 < \rho < 1$$

where ε_n are i.i.d. $N(0, 1 - \rho^2)$ random variables, X_0 is $N(0, 1)$ and X_{n-1} and ε_n are independent for $n > 0$. Then, X_n is $N(0, 1)$ and the sequence $\{X_n\}$ is Markov of order 1. The pair (X_i, X_{i+j}) has the joint density

$$f(x_1, x_2) = \exp - \frac{\dfrac{x_1^2 - 2\rho^j x_1 x_2 + x_2^2}{2(1 - \rho^{2j})}}{2\pi(1 - p^{2j})^{1/2}}$$

and the densities of X_i and X_{i+j} are

$$f(x_1) = \frac{\exp\left(-\dfrac{x_1^2}{2}\right)}{(2\pi)^{1/2}}$$

$$f(x_2) = \frac{\exp\left(-\dfrac{x_2^2}{2}\right)}{(2\pi)^{1/2}}$$

from which we find that

$$\lim_{j \to \infty} |f(x_1, x_2) - f(x_1)f(x_2)| = 0; \qquad \text{for any } x_1 \text{ and } x_2$$

Then, corollary 8.3 applies and the sequence $\{X_n\}$ satisfies the strong mixing condition.   ∎

This example can be easily extrapolated to any finite autoregressive process.

Example 8.4. (AR(1) model with Cauchy marginals).   Assume the same AR(1) model in example 8.3, but now with ε_n having a Cauchy $(0, \beta)$ pdf.

$$f(x) = (\beta\pi)^{-1}\left[1 + \left(\frac{x}{\beta}\right)^2\right]^{-1}; \qquad \beta = 1 - |\rho|; \quad -1 < \rho < 1$$

Then, X_n is Cauchy $(0, 1)$ and the joint pdf of (X_i, X_{i+j}) is

$$f(x_1, x_2) = [(1 - |\rho|^j)\pi]^{-1}\left[1 + \left(\frac{x_2 - \rho^j x_1}{1 - |\rho|^j}\right)^2\right]^{-1}[\pi(1 + x_1^2)]^{-1}$$

and the pdf's of X_i and X_{i+j} are

$$f(x_1) = \pi^{-1}[1 + x_1^2]^{-1}$$
$$f(x_2) = \pi^{-1}[1 + x_2^2]^{-1}$$

from which

$$\lim_{j \to \infty} |f(x_1, x_2) - f(x_1)f(x_2)| = 0; \qquad \text{for any } x_1 \text{ and } x_2$$

Thus, the sequence $\{X_n\}$ satisfies the strong mixing condition.   ∎

Definition 8.5. (Condition D). Let $X_1. X_2, \ldots$ be a sequence of random variables and let $F_{i_1, i_2, \ldots, i_r}(x_1, x_2, \ldots, x_r)$ be the joint cdf of members $X_{i_1}, X_{i_2}, \ldots, X_{i_r}$ in the sequence. The condition D is said to hold if for any set of integers $i_1 < i_2 < \ldots < i_p$ and $j_1 < j_2 < \ldots j_q$ such that $j_1 - i_p \geq s$, and any real number u we have

(8.11) $|F_{i_1, \ldots, i_p j_1, \ldots, j_q}(u, \ldots, u) - F_{i_1, \ldots, i_p}(u, \ldots, u)F_{j_1, \ldots, j_p}(u, \ldots, u)| \leq g(s)$

with

(8.12) $$\lim_{s \to \infty} g(s) = 0$$ ∎

Note that this condition involves only events defined by intersections of events of the type $C_i = \{X_i \leq u\}$, which are the only ones influencing the behavior of limit distributions of order statistics.

An equivalent definition can be formulated if the cdf's, F, are replaced by (8.11) by the hazard functions, G.

Note that the strong mixing condition implies condition D. It is enough to take $g(s) = \phi(s)$ and

$$A = \{X_{i_1} \leq u, \ldots, X_{i_p} \leq u\}$$
$$B = \{X_{j_1} \leq u, \ldots, X_{j_q} \leq u\}.$$

Definition 8.6. (Condition $D(u_n)$). The condition $D(u_n)$ is said to hold if for any integers satisfying the conditions above, we have

(8.13) $|F_{i_1, \ldots, i_p j_1, \ldots, j_q}(u_n) - F_{i_1, \ldots, i_p}(u_n)F_{j_1, \ldots, j_p}(u_n)| \leq \alpha_{n,s}$

where $\alpha_{n,s}$ is non-increasing in s and

(8.14) $$\lim_{n \to \infty} \alpha_{n,[n\delta]} = 0 \qquad \text{for each } \delta > 0$$ ∎

Note that condition D implies condition $D(u_n)$ for any sequence $\{u_n\}$.

The following theorem shows that condition $D(u_n)$ is satisfied for Markov sequences of order 1 (Chernick (1981b)).

Theorem 8.3. (Condition $D(u_n)$ for Markov sequences of order 1). *Let $\{X_n\}$ be a stationary Markov sequence of order 1. Let $F(x)$ be the cdf of X_n. Then, $D(u_n)$ is satisfied for any sequence $\{u_n\}$ such that*

$$\lim_{n \to \infty} F(u_n) = 1.$$

Definition 8.7. (Condition $D'(u_n)$). The condition $D'(u_n)$ is said to hold for the stationary sequence $\{X_n\}$ of random variables and the sequence $\{u_n\}$ of constants if

(8.15) $$\lim_{k \to \infty} \limsup_{n \to \infty} n \sum_{j=2}^{\left[\frac{n}{k}\right]} P[X_1 > u_n, X_j > u_n] = 0 \qquad \blacksquare$$

8.4. Limit Distributions of Maxima and Minima
8.4.1. Stationary sequences.

In this section we analyze the case of stationary sequences. We begin by defining stationary and m-dependent sequences of random variables.

Definition 8.8. (Stationary sequence). A sequence $X_1, X_2, \ldots$ of random variables is called stationary if

(8.16) $$F_{i_1, i_2, \ldots, i_k}(x_1, x_2, \ldots, x_k) = F_{i_1 + s, i_2 + s, \ldots, i_k + s}(x_1, x_2, \ldots, x_k)$$

for any integers k and s. $\blacksquare$

It is clear that exchangeable variables are stationary.

Theorem 8.4. (Sufficient conditions for i.i.d. limit distributions). *Let $\{X_n\}$ be a stationary sequence and let $\{a_n\}$ and $\{b_n\}$ be two sequences of real numbers such that*

(8.17) $$\lim_{n \to \infty} P[X_{n:n} \le a_n + b_n x] = G(x)$$

If the sequence $\{u_n = a_n + b_n x\}$ satisfies the condition $D(u_n)$ for each x, then $G(x)$ is one of the three limit distributions for the independent case. $\blacksquare$

Because condition D implies condition $D(u_n)$, the theorem remains true if condition D holds.

Example 8.5. (Marshall-Olkin model). Assume a longitudinal element (electric cord, railway, conductor rail, wire, chain, etc.), hypothetically sub-divided into n sections of unit length. Assume also that independent Poisson processes govern the occurrence of shocks destroying k consecutive sections starting at the j-th section ($j = 1, 2, \ldots, n - k + 1$; for $k = 1, 2, \ldots, n$). Assume also that the intensity of such processes is $\lambda^k (\lambda < 1)$. This means that we have n Poisson processes of intensity λ destroying one section, $(n - 1)$ processes of intensity λ^2 destroying two sections, and so on. Note that the intensity of the Poisson processes decrease as the damaged length of the element increases. Note also that, due to boundary effects, the extreme sections are the strongest and the central sections are the weakest. This is due to the fact that the extreme sections are affected by fewer Poisson processes than central sections.

The hazard function of the n sections is given by the Marshall-Olkin model

$$G(x_1, x_2, \ldots, x_n) = P(X_1 > x_1, X_2 > x_2, \ldots, X_n > x_n)$$

$$= \exp\left[-\left(\lambda \sum_{i=1}^{n} x_i + \lambda^2 \sum_{i-1}^{n-1} \max(x_i, x_{i+1}) + \cdots + \lambda^n \right. \right.$$

$$\left. \left. \cdot \sum_{i=1}^{1} \max(x_1, x_{i+1}, \ldots, x_n) \right) \right]$$

where $X_1, X_2, \ldots, X_n$ are the lifetimes of the sections.

Hence, we have

$$G(x, x, \ldots, x) = \exp\{ -[\lambda n + \lambda^2(n - 1) + \cdots + \lambda^n]x \}$$

$$= \exp\left[\left(\frac{-\lambda(n(1 - \lambda) - \lambda + \lambda^{n+1})}{(1 - \lambda)^2} \right) x \right]$$

If the element is assumed to fail as soon as one of the sections fails, the lifetime of the element is that of its weakest section. Thus, the cdf, $F(x)$, of its lifetime is given by

$$F(x) = 1 - G(x, x, \ldots, x) = 1 - \exp\left[\left(\frac{-\lambda(n(1 - \lambda) - \lambda + \lambda^{n+1})}{(1 - \lambda)^2} \right) x \right]$$

This proves that the choice of the normalizing sequences

$$c_n = 0$$

$$d_n = \frac{1 - \lambda}{\lambda n}$$

leads to

$$\lim_{n \to \infty} F(c_n + d_n x) = 1 - \exp(-x)$$

which is a Weibull type distribution.   ∎

Example 8.6. (Marshall-Olkin model). Assume now that we modify example 8.5 in such a way that all sections are affected by the same number of Poisson processes. To this aim we assume that there are $(n + k - 1)$ Poisson processes of intensity λ^k, the j-th process destroying k consecutive sections (real or hypothetical) starting at the (real or hypothetical) section $j - k + 1$, for $k = 1, 2, \ldots, n$. This means that any given section is affected by one Poisson process destroying that section alone, two Poisson processes destroying that section and a continguous one, three Poisoon processes destroying that section and two contiguous sections, and so on.

In this case the sequence $X_1, X_2, \ldots, X_n$ is stationary and we have

$$G(x_1, x_2, \ldots, x_n) = P(X_1 > x_1, X_2 > x_2, \ldots, X_n > x_n)$$

$$= \exp\left[-\left(\lambda \sum_{i=1}^{n} x_i + \lambda^2 \sum_{i=0}^{n} \max(x_i, x_{i+1}) + \cdots + \lambda^n \right. \right.$$

$$\left. \left. \cdot \sum_{i=2-n}^{n} \max(x_i, x_{i+1}, \ldots, X_{i+n-1}) \right) \right]^|$$

where $x_i = 0$ if $i < 1$ or $i > n$.

Condition $D(u_n)$ holds because we can write

$$|G_{i_1, \ldots, i_p j_1, \ldots, j_q}(u_n) - G_{i_1, \ldots, i_p}(u_n) G_{j_1, \ldots, j_p}(u_n)| \leq 1 - \exp[-\lambda^{s+1} p(\lambda) u_n]$$

where $p(\lambda)$ is a polynomial of order less that $p + q$ and s has the meaning given in definition 8.6.

If now we choose $u_n = c_n + d_n x$ with $c_n = 0$ and $d_n = 1/n$, as in the independent case (see example 3.3), we get

$$\alpha_{n,s} = 1 - \exp\left(-\lambda^{s+1} p(\lambda) \frac{x}{n} \right)$$

and

$$\lim_{n \to \infty} \alpha_{n, [n\delta]} = 1 - \exp\left(-\lambda^{n\delta+1} p(\lambda) \frac{x}{n} \right) = 0$$

This shows that condition $D(u_n)$ is satisfied. Thus, according to theorem 8.4, the limit distribution coincides with the independent case.

The same conclusion could have been obtained by noting that the cdf of the lifetime of the element is

$$F(x) = 1 - G(x, x, \ldots, x) = 1 - \exp\{-[\lambda n + \lambda^2(n + 1) + \cdots + \lambda^n(2n - 1)]x\}$$

$$= 1 - \exp\left[-\lambda\left(\frac{n(1 - 2\lambda^n)}{1 - \lambda} + \frac{\lambda(1 - \lambda^n)}{(1 - \lambda)^2}\right)x\right],$$

from which we get

$$\lim_{n\to\infty} F\left(\frac{(1 - \lambda)x}{\lambda n}\right) = 1 - \exp(-x)$$

which is the Weibull distribution. ∎

Theorem 8.5. (Sufficient conditions for i.i.d. associated limit distribution). *Let $\{X_n\}$ be a stationary sequence of random variables and let $\{a_n\}$ and $\{b_n\}$ be two sequences of real numbers. Assume that condition $D(u_n)$ and $D'(u_n)$ are satisfied with $u_n = a_n + b_n x$. Assume also that $\{\underline{X}_n\}$ is a sequence of i.i.d. random variables having the same cdf as each member of the sequence $\{X_n\}$. then*

(8.18) $$\lim_{n\to\infty} P[X_{n:n} \le a_n + b_n x] = G(x)$$

if, and only if,

(8.19) $$\lim_{n\to\infty} P[\underline{X}_{n:n} \le a_n + b_n x] = G(x)$$

Note that the sequences of constants $\{a_n\}$ and $\{b_n > 0\}$ coincide.

8.4.1.1. m-dependent sequences. *Definition 8.9. (m-dependent sequence).* A sequence $X_1, X_2, \ldots$ of random variables is called m-dependent if the random vectors $(X_{i_1}, X_{i_2}, \ldots, X_{i_k})$ and $(X_{j_1}, X_{j_2}, \ldots, X_{j_k})$ where

$$\min(j_1, j_2, \ldots, j_k) - \max(i_1, i_2, \ldots, i_k) \ge m$$

are independent. ∎

Note that X_i and X_j can be dependent if they are close ($|i - j| < m$) but they are independent if they are far apart.

Example 8.7 (Finite moving average stationary models). Any MA(q) model, i.e. any model of the form

$$X_t = \varepsilon_t + c_1 \varepsilon_{t-1} + c_2 \varepsilon_{t-2} + \cdots + c_q \varepsilon_{t-q}$$

where $\{\varepsilon_t\}$ is a sequence of i.i.d. random variables is obviously m-dependent. ∎

Any stationary m-dependent sequence obviously satisfies the D condition with $g(s)$ identically zero for $s \geq m$.

For an m-dependent stationary sequence we have the following result (Galambos (1978))

Theorem 8.6. (Asymptotic distribution of the maxima of m-dependent stationary sequences). *Let $X_1, X_2, \ldots$ be an m-dependent stationary sequence with common cdf $F(x)$ such that*

$$(8.20) \qquad \lim_{n \to \infty} n[1 - F(a_n + b_n x)] = u(x); \qquad 0 < u(x) < \infty$$

Then

$$(8.21) \qquad \lim_{n \to \infty} P[Z_n < a_n + b_n x] = \exp(-u(x))$$

if and only if

$$(8.22) \qquad \lim_{u \to \omega(F)} \frac{P[X_1 \geq u, X_i \geq u]}{1 - F(u)} = 0; \qquad 1 \leq i < m$$

For m-dependent sequences, condition (8.22) implies condition $D'(u_n)$ for any sequence $\{u_n\}$ such that $\lim u_n = \omega(F)$ when $n \to \infty$.

Example 8.8. (Marshall-Olkin model). Assume an m-dependent sequence $\{X_n\}$ of random variables such that the joint hazard function of any m consecutive elements, $\{X_i, X_{i+1}, \ldots, X_{i+m-1}\}$, in the sequence is the Marshall-Olkin function

$$G(x_1, x_2, \ldots, x_n) = \exp\left(-\lambda_1 \sum_{i=1}^{m} x_i - \lambda_2 \sum_{i<j} \max(x_i, x_j)\right)$$

Then, condition (8.22) becomes

$$\lim_{u \to \infty} \frac{P[X_1 \geq u, X_i \geq u]}{1 - F(u)} = \lim_{u \to \infty} \frac{\exp[-(2\lambda_1 + \lambda_2)u]}{\exp[-(\lambda_1 + \lambda_2)u]}$$

$$= \lim_{u \to \infty} \exp[-\lambda_1 u] = \begin{cases} 0 & \text{if } \lambda_1 > 0 \\ 1 & \text{if } \lambda_1 = 0 \end{cases}$$

Consequently, the limit distribution of the maximum of $\{X_n\}$ coincides with the independent case (Gumbel type) only if $\lambda_1 > 0$. ∎

8.4.1.2.   Moving-average models.   In this section we include one theorem which gives the asymptotic distribution of sequences following moving-average models (see Leadbetter et al. (1983)).

Theorem 8.7. (Asymptotic distribution of the maxima for the moving average model of stable distributions).   *Suppose the following moving average model*

$$(8.23) \qquad Y_t = \sum_{i=-\infty}^{\infty} C_i X_{t-i}; \qquad t \geq 1$$

where the X_t for $t > 1$ are independent and stable random variables, i.e. with characteristic function of the form

$$(8.24) \qquad \varphi(t) = \exp\left[-\gamma^\alpha |t|^\alpha \left(1 - \frac{i\beta h(t, \alpha) t}{|t|} \right) \right]$$

where

$$(8.25) \qquad h(t, \alpha) = \begin{cases} \dfrac{2 \log|t|}{\pi} & \text{if } \alpha = 1 \\[2ex] \tan\left(\dfrac{\pi\alpha}{2} \right) & \text{otherwise} \end{cases}$$

and

$$(8.26) \qquad \gamma \geq 0; \quad 0 < \alpha \leq 2; \quad |\beta| \leq 1$$

Assume also that the constants $C_i(-\infty < i < \infty)$ satisfy

$$(8.27) \qquad \sum_{i=-\infty}^{\infty} |C_i|^\alpha < \infty$$

$$(8.28) \qquad \sum_{i=-\infty}^{\infty} C_i \log|C_i| < \infty; \qquad \text{if } \alpha = 1, \quad \beta \neq 0$$

Then, we have

$$\lim_{n \to \infty} P[X_{n:n} \leq n^{1/\alpha} x] = \begin{cases} \exp\{ -K_\alpha (c_+^\alpha (1 + \beta) + c_-^\alpha (1 - \beta)] x^{-\alpha} \} & \text{if } x > 0 \\ 0 & \text{if } x \leq 0 \end{cases}$$

$$(8.29)$$

where

$$(8.30) \qquad c_+ = \max_{-\infty < i < \infty} \max(0, C_i)$$

$$(8.31) \qquad c_- = \max_{-\infty < i < \infty} \max(0, -C_i)$$

$$(8.32) \qquad K_\alpha = \frac{\Gamma(\alpha) \sin\left(\dfrac{2\pi}{2}\right)}{\pi}$$

Note that the normal distribution is a stable distribution ($\alpha = 2$). Note also that any ARMA model is a moving average model of the form (8.23) where the coefficients C_i can be obtained by inverting the original ARMA model.

Davis and Resnick (1985) have studied the limit distribution of extremes of moving average models of random variables belonging to a Frechet-type domain of attraction and give conditions under which the extremes belong to a Frechet-type domain of attraction. In particular, they derive the asymptotic joint distribution of maxima and minima and show that a necessary and sufficient condition for their asymptotic independence is that all coefficients C_i in (8.23) must have the same sign.

8.4.1.3. Gaussian sequences. An important case of stationary sequences is that of Gaussian sequences for which we have the following results (for the proof, the reader is referred to Galambos (1978)).

Theorem 8.8. (Asymptotic distribution of stationary normal sequences). *If* $X_1, X_2, \ldots$ *is a stationary sequence of standard normal* $N(0, 1)$ *random variables with correlation function*

$$(8.33) \qquad r_m = E[X_j X_{j+m}]$$

we have that

(i) *if*

$$(8.34) \qquad \lim_{m \to \infty} r_m \log m = 0$$

then

$$(8.35) \qquad \lim_{n \to \infty} P[Z_n < a_n + b_n x] = H_{3,0}(x);$$

(ii) *if*

$$(8.36) \qquad \lim_{m \to \infty} r_m \log m = \tau; \qquad 0 < \tau < \infty$$

then

$$(8.37) \qquad \lim_{n \to \infty} P[Z_n < a_n + b_n x] = H(x)$$

where $H(x)$ *is the convolution of* $H_{3,0}(x + \tau)$ *and* $\Phi[x(2\tau)^{-1/2}];$

(iii) *if*

(8.38)
$$\lim_{m \to \infty} r_m \log m = \infty$$

(8.39)
$$\lim_{m \to \infty} r_m (\log m)^{1/3} = 0$$

and r_m is decreasing, then

(8.40)
$$\lim_{n \to \infty} P[Z_n < (1 - r_n)^{1/2} a_n + x r_n^{1/2}] = \Phi(x)$$

where a_n and b_n are given by

(8.41)
$$a_n = \frac{1}{b_n} - \frac{b_n [\log \log n + \log(4\pi)]}{2}$$

(8.42)
$$b_n = (2 \log n)^{-1/2}.$$

Example 8.9. (Arma (p,q) Box-Jenkins model). Let us assume a sequence $\{X_n\}$ such that it follows an ARMA(p, q) Box-Jenkins autoregressive moving average model of orders p and q of the form

$$X_t = \phi_1 X_{t-1} + \cdots + \phi_p X_{t-p} + \varepsilon_t - \theta_1 \varepsilon_{t-1} - \cdots - \theta_q \varepsilon_{t-q}$$

where $\phi_1, \phi_2, \ldots, \phi_p$ and $\theta_1, \theta_2, \ldots, \theta_q$ are constants and $\{\varepsilon_t\}$ is a sequence of i.i.d. $N(0, \sigma_a^2)$ random variables. Assume that the model above is stationary and invertible, i.e. that the roots of the polynomials

$$P(x) = x^p - \phi_1 x^{p-1} - \cdots - \phi_p$$

and

$$Q(x) = x^q - \theta_1 x^{q-1} - \cdots - \theta_q$$

lie inside the unit circle.

The autocorrelation function satisfies the difference equation

$$r_k = \phi_1 r_{k-1} + \phi_2 r_{k-2} + \cdots + \phi_p r_{k-p}; \qquad k > q + 1$$

which has as solution

$$r_k = \sum_{s=1}^{r} \left\{ (m_s)^k \left[\left(\sum_{j=0}^{p_s - 1} c_{ij} k^j \right) \cos(k\alpha_s) + \left(\sum_{j=0}^{p_s - 1} d_{ij} k^j \right) \sin(k\alpha_s) \right] \right\}$$

where c_{ij} and d_{ij} are constants, m_s and α_s are the modulus and argument, respectively, of the r different roots of the polynomial $P(x)$, and p_s are their degrees of multiplicity, which satisfy the equation

$$\sum_{s=1}^{r} p_s = p$$

It is clear that

$$\lim_{m \to \infty} r_m \log m = 0$$

Then, if σ_a^2 is chosen such that X_t is $N(0, 1)$, theorem 8.8 guarantees that the limit distribution of $Z_n = \max(X_1, X_2, \ldots, X_n)$ is

$$\lim_{n \to \infty} P[Z_n < a_n + b_n x] = \exp[-\exp(-x)]$$

with a_n and b_n given by (8.41) and (8.42), respectively. ∎

Leadbetter et al. (1983) have shown for normal sequences that if

$$\lim_{m \to \infty} r_m \log m = 0$$

and $n[1 - \Phi(u_n)]$ is bounded, then conditions $D(u_n)$ and $D'(u_n)$ hold. This implies (see example 8.9) that ARMA (p, q) Box-Jenkins stationary and invertible models satisfy these conditions for $u_n = a_n + b_n x$.

Example 8.10. (SO$_2$ concentrations). Figure 8.1 shows the monthly and annual maxima of the hourly average concentrations (pphm) of SO_2 at Long Beach, California between 1956 and 1974, drawn on Gumbel probability paper for maxima, as given by Leadbetter et al. (1983).

Continuous lines in figure 8.1. show the Gumbel cdfs fitted by Roberts, by means of a variant of the least-squares method to the monthly and yearly data. As can be seen, the two lines are far from being parallel, as theory suggests. The reason for this fact is that the speed of convergence to the Gumbel limit distribution is small and one month is not enough to get a straight cdf on Gumbel probability paper. Note the curvature of the empirical cdf for the monthly data. On the contrary, the yearly data appears as a perfect linear trend.

One way to overcome this difficulty consists of fitting only the right tail data, as was indicated in chapter 4. Table 8.1. shows the estimated parameters by the method of Roberts and the maximum likelihood method using a complete sample and only the $2n^{1/2}$ largest order statistics. Note that the values of the shape parameter (slope of the straight line on the probability paper) range from 8.70 to 12.34 for the Roberts method, or from 7.87 to 10.65

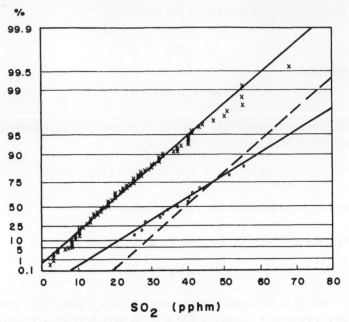

FIGURE 8.1.   Gumbel cdf fitted by Roberts by means of a variant of the least-squares method to the monthly and yearly data

		Estimation Method		
Data	Parameter	Maximum Likelihood		Roberts Method
		complete sample	right tail data	complete sample
Monthly Data	λ	14.63	15.76	14.5
	δ	7.87	8.27	8.70
Yearly Data	λ	31.91	34.52	31.5
	δ	10.65	8.91	12.34

TABLE 8.1.   Gumbel parameters estimated by different methods

for the maximum likelihood, when using monthly or yearly data, if the complete sample is used. On the contrary, if only right tail data is utilized the shape parameter goes from 8.27 to 8.91. In other words, by using right tail data only, one gets two nearly parallel straight lines.   ∎

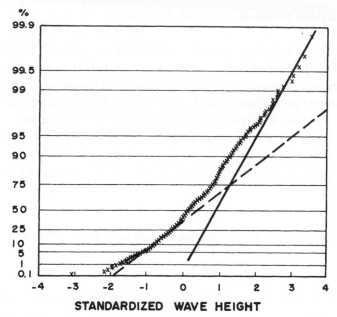

FIGURE 8.2.   Observed and theoretical cdfs of the standardized maximum wave heights over 12 minute periods on Gumbel probability paper for maxima

Example 8.11. (Maximum wave heights).   Figure 8.2. shows the observed and theoretical cdf of the standardized maximum wave heights over 12 minute periods on Gumbel probability paper for maxima, as given by Leadbetter et al. (1983).

The sea levels were observed by a floating buoy off South Uist in the Hebrides, the theoretical cdf (dotted line) was obtained by using a normal process and the parameters were estimated by spectral theory.

The considerable difference between the observed and the theoretical cdf's is not surprising because on one hand the estimation method used is only good for central values and not for extremes as is the case and on the other hand, the observation period of 12 minutes is too short (the convergence of maxima from a parent normal to Gumbel is very slow). The example is a warning to the reader about using assumptions based on a good fitting to central values. In this case the normal model is undoubtedly bad, even though it fits the overall values well. The correct procedure here is to fit a straight line, such as that shown in figure 8.2, to the right tail, which, on the other hand, shows an almost perfect linear trend.   ■

8.5. Asymptotic Distributions of k-th Extremes

In the case of i.i.d. variables, the distribution of the k-th extremes was deduced based on the distributions of maxima and minima. However, in the case of dependence new problems arise, because even when the maxima and minima can be properly normalized, in order to give a limiting distribution, it is not certain that the same is true for the k-th extremes. Nevertheless, the exact distribution of the k-th extremes, as shown in section 2.4 can be reduced to the distribution of $m_n(x)$, i.e. of the number of events $C_i(x)$ that occur (see section 2.4). The following theorem from (Galambos (1978)) gives an important result

Theorem 8.9. (Asymptotic distribution of k-th extremes). *Assume that*

$$(8.43) \qquad \lim_{n \to \infty} \underline{S}_{j,n}(a_n + b_n x) = u_j(x),$$

where $\underline{S}_{j,n}(x)$ is defined in section 2.4.1 exist and are finite in some interval (a, b).
 If the series

$$(8.44) \qquad U_t(x) = \sum_{k=0}^{\infty} (-1)^k \binom{k+t}{t} u_{k+t}(x); \qquad a < x < b$$

converges, then we have

$$(8.45) \qquad \lim_{n \to \infty} P[X_{n-k+1:n} < a_n + b_n x] = \sum_{t=0}^{k-1} U_t(x); \qquad a < x < b$$

An important corollary of this theorem is the following

Corollary 8.10. *If*

$$(8.46) \qquad u_j(x) = \frac{u^j(x)}{j!}; \qquad u(x) \geq 0$$

then the series $U_t(x)$ converges for all x for which $u(x)$ is finite and

$$(8.47) \qquad \lim_{n \to \infty} P[X_{n-k+1:n} < a_n + b_n x] = \exp\{-u(x)\} \sum_{t=0}^{k-1} \frac{u^t(x)}{t!}$$

Theorem 8.11. (Sufficient conditions for i.i.d. limit distributions). *Assume $\{X_n\}$ is a sequence of random variables and $\{a_n\}$, $\{b_n > 0\}$ are sequences of real numbers such that $D(u_n)$ and $D'(u_n)$ hold for $u_n = a_n + b_n x$ and every x.*

If

$$\text{(8.48)} \qquad\qquad \lim_{n\to\infty} P[X_{n:n} \le a_n + b_n x] = G(x)$$

then, for each r = 1, 2, ... we have

$$\text{(8.49)} \qquad \lim_{n\to\infty} P[X_{n-r+1:n} \le a_n + b_n x] = G(x) \sum_{i=0}^{r-1} \frac{[-\log G(x)]^i}{i!}$$

Note that the sequences of normalizing constants $\{a_n\}$ and $\{b_n\}$ coincide and that this theorem sets conditions under which theorem 7.1 remains true for dependent sequences.

Example 8.12. (Finite moving average stationary models). Because finite moving average stationary models satisfy condition D and $D'(u_n)$ for sequences $\{u_n\}$ such that $\lim u_n = \omega(F)$ when $n \to \infty$, if the normalized maximum has a limit distribution $G(x)$, the r-th largest order statistic has (8.49) as its limit distribution. ∎

Example 8.13. (ARMA(p,q) Box-Jenkins stationary and invertible models). Because these models satisfy conditions $D(u_n)$ and $D'(u_n)$ with $u_n = a_n + b_n x$, theorem 8.11 can be applied. ∎

Example 8.14. (Fatigue strength of a cable). Assume the fatigue strength model in example 7.2 with the only difference that now assumption (v) is replaced by (v'), see Castillo et al. (1983b).

(v') The logarithm of the number of cycles to failure, N_t, of the t-th section, when subjected to a fatigue test with constant $\Delta\sigma$ follows the Box-Jenkins stationary and invertible model

$$z_t = b_1 z_{t-1} + \cdots + b_p z_{t-p} + \varepsilon_t - c_1 \varepsilon_{t-1} - \cdots - c_q \varepsilon_{t-q}$$

where

$$z_t = N_t(\Delta\sigma) - E\{N_t(\Delta\sigma)\}$$

and $N_t(\Delta\sigma)$ is used to show that N_t depends on the $\Delta\sigma$ level. This implies that the fatigue strength of any given element is normal.

We study here the following two asymptotic cases:

(a) very long elements.
(b) very high number of elements.

(a) The cdf of the strength, Y_n, of a single element for $n \to \infty$ (very long element) satisfies (see example 8.9)

$$\lim_{n \to \infty} P\left(\frac{Y_n - \mu(\Delta\sigma)}{\sigma(\Delta\sigma)} \le c_n + d_n x\right) = 1 - \exp[-\exp(-x)]$$

where $\mu(\Delta\sigma)$ and $\sigma(\Delta\sigma)$ are the mean and standard deviation of N_t at the $\Delta\sigma$ level, and

$$d_n = (2 \log n)^{-1/2}$$

$$c_n = -\frac{1}{d_n} + \frac{d_n(\log \log n + \log 4\pi)}{2}$$

Then, the cdf of the k-th order statistics (k-th weakest element) is given by (see expression 2.2)

$$G_k(c_n + d_n V; \Delta\sigma) = \sum_{j=k}^{m} \binom{m}{j} \{1 - \exp[-\exp(x)]\}^i \exp[-(m-i)\exp(x)]$$

and

$$V = \frac{N - \mu(\Delta\sigma)}{\sigma(\Delta\sigma)}$$

(b) In this case theorem 8.11 is applicable and the cdf of the k-th order statistic is (see (8.49) arranged for lower order statistics)

$$G_k(c_{nm} + d_{nm} V) = 1 - \exp\{-\exp(x)\} \sum_{i=0}^{k-1} \frac{\exp(ix)}{i!} \qquad \blacksquare$$

For more results on central or moderately upper or lower order statistics we refer the reader to Watts et al. (1982) and Cheng (1985) and their references.

V   Multivariate Case

Chapter 9

Multivariate and Regression Models Related to Extremes

9.1. Introduction

As was mentioned in previous chapters, slight errors in the parent cdf can lead to very important errors for extremes. This fact makes it crucial to avoid subjective selection of models. In this context, the characterization of models becomes important.

The statistical theory of extremes discussed in previous chapters is the basis for many models in engineering and scientific applications. In fact, the Gumbel, Weibull and Frechet models are justified in many cases because of their limit properties. This justification is valid not only for random i.i.d. samples but for many cases of dependence, as shown in chapter 8. Similarly, when r-th order statistics play the fundamental role in the problem under study, other models possess very strong theoretical bases, as shown in chapter 7. In this way, extreme value theory permits some objectivity in the model selection procedure, which is unavoidable in applications.

In this chapter we derive some multivariate and regression models which are fully characterized by physical and extreme value properties. This means that only the consideration of the physical problem and its extreme value character leads to these models without any additional assumption.

We begin the chapter by deriving some regression models which can be applied to a large variety of lifetime problems. The models are based on the equality of two conditional cdf's. These, when forced by extreme value considerations to be Weibull or Gumbel distributions, give two functional equations which fully characterize the models.

The second part of the chapter is devoted to the characterization of three bivariate distributions with Weibullian conditionals. In the so-called Weibull-Weibull model both conditional families are Weibull and in the other two, one is normal or gamma. In fact, when a bivariate distribution, $F(x, y)$, is forced to have parametric families of conditionals, with parameters depending on the values of x and y, a functional equation is obtained as will be shown. This functional equation is the basis for the characterization.

In some cases regression models can be treated as part of multivariate models. This is possible when the dependent and the regressor variables are all random. However, there are cases, as the example below, for which the regressor variables are not all random. This is the main reason for making a distinction here between the models to be described in the following sections.

The main goal of this chapter is to show how physical and statistical considerations of a given practical problem can lead to a complete characterization of the models to be applied. Though only a few models are studied here, we encourage the readers to follow a similar process in their respective areas of knowledge.

9.2. Regression Models

For reasons of clarity, we divide this section into two parts. In the first part we derive a basic equation founded on physical considerations of a lifetime problem. In the second part, we add some extreme value conditions which lead to two fundamental functional equations which are then solved. In this way, starting from a physical problem and by using physical and extreme value considerations, we get as the only possibilities some parametric families, with no more subjective assumptions than those implied by the physical model itself.

9.2.1. Derivation of the basic equation.

Consider a manufacturing process in which identical elements are produced. By identical here we mean elements manufactured under the same conditions (material, environmental, etc.) but possibly random. Assume that its lifetime V

is dependent on a regressor variable U in such a way that there exists for all of them a common function $P(u, v)$ and, for a given element of production, a constant p such that the function

$$(9.1) \qquad\qquad p = P(U, V)$$

gives the lifetime V of the element as a function of the values of the regressor variable U and the regressor variable U as a function of V.

This implies that for any given p, i.e. for any given element, U can be written as a function of V, and V can be written as a function of U:

$$(9.2) \qquad\qquad V = V(U, p)$$

$$(9.3) \qquad\qquad U = U(V, p)$$

All the above permits us to consider (9.2) and (9.3) as stochastic processes where p gives these two functions their random character. The physical interpretation of this process is clear. Any given element of production has a lifetime V which depends on the value of the regressor variable U and this dependence is given by the function (9.2) (see figure 9.1.). A similar statement can be formulated for U and the function (9.3).

In order to be more specific, and with the aim of facilitating the comprehension of the following, let, for example, the elements of production be pieces of wire of a given length and let U be the stress amplitude in a fatigue test. It is very well known that the fatigue lifetime, V, measured as the number of cycles to failure or a function of it (normally its logarithm), is dependent on the stress amplitude and that this dependence is of the type shown in figure 9.1., showing a decreasing lifetime for increasing stress amplitude.

If the elements are selected at random from the total lot, the probability $P[V \leq v/U = u]$ of having a lifetime less than or equal to V when the stress amplitude is held constant at a level $U = u$, can be viewed as the percentage of elements of the total lot that fail before v under such a circumstances. However, we can think of the necessary stress amplitude, U, for an element of having a lifetime $V = v$. In this way, U can also be viewed as a random variable and then the probability $P[U \leq u/V = v]$ becomes the percentage of elements of the total lot that need a stress amplitude value smaller than u to fail at $V = v$. For brevity, we shall call these two functions the conditional distributions of V and U given the other, though they are not real conditional distributions because the pair (U, V) here is not a bivariate random variable.

One important fact to be noted is that the curves (9.1) cannot intersect. This is due to the fact that the lifetime, V, of a given element of wire for a given stress amplitude $U = u$ is a monotone function of the size of the maximum crack,

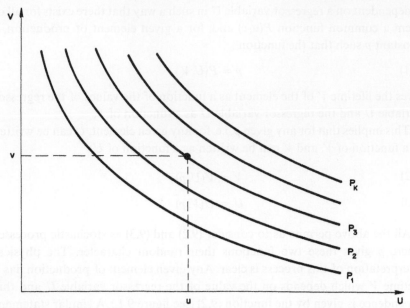

FIGURE 9.1.   Illustration of lifetime V depending on a regressor variable U

and this implies that if element A is weaker than B for a given stress amplitude $U = u_0$, A must also be weaker than B for any other stress amplitude. But this implies (see figures 9.1. and 9.2.) that

$$(9.4) \qquad P(V \le v/U = u) = P(U \le u/V = v)$$

Note that in figures 9.1. and 9.2. both terms in expression (9.4) are the percentages of curves such that the point (u, v) lies above the line. Note also that in the derivation process it is unimportant whether or not the curves intersect.

Although we have used the wire example, many other practical applications have the same properties as the one described above. Think, for example, of the life, V, of a breakwater when subject to a wave height, U, or of the lifetime, V, of electrical insulator when subject to a voltage, U. The functions (9.1) in these cases are of the same form as those shown in figure 9.1., i.e. lifetime decreases for increasing values of the regression variables wave height and voltage, respectively. The equality (9.4) also holds.

In order to have a physically feasible model, we shall introduce some natural assumptions on the function $P(u, v)$. As in the previous example, we permit that U and V be determined as random variables only if the other one remains

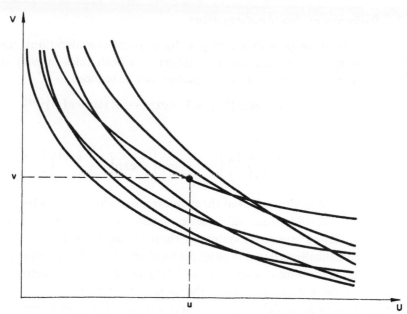

FIGURE 9.2. Illustration of intersecting curves for different elements

fixed. We also permit (U, V) to be a proper bivariate random vector. Finally, we allow the curves (9.1) to intersect. However, we impose the feasibility condition

$$(9.5) \qquad P(U_1, V_1) = P(U_2, V_2) = p \quad \Leftrightarrow \quad U_1 \geq U_2 \text{ if, and only if, } V_1 \leq V_2$$

Under these conditions we shall prove that (9.4) continues to hold. For this, we define A as the set of those elements for which $U \leq u$, given $V = v$, and B as the set of elements such that $V \leq v$ for $U = u$. The proof of identity (9.4) is equivalent to the proof of $A = B$. Let $a \in A$, then, there exist values U_a and p_a such that

$$(9.6) \qquad P(U_a, v) = p_a \quad \text{and} \quad U_a \leq u$$

Let v_a be the value of V such that

$$(9.7) \qquad P(u, v_a) = p_a$$

Assumption (9.5) and equations (9.6) and (9.7), for element a, yield $v_a \leq v$ if $U = u$. That is $a \in B$, and $A \subset B$. By an identical argument, it can be shown that $B \subset A$, and thus $A = B$, establishing our basic equation (9.4) for the general model.

9.2.2.   Derivation of functional equations.

If now some result of probability theory (by physical considerations, past experience, data, etc.) justifies the assumption that both sides in (9.4) are Weibull or Gumbel distributions, this equation can be reduced to

$$(9.8) \qquad 1 - \exp\{-[a(x)y + b(x)]^{c(x)}\} = 1 - \exp\{-[d(y)x + e(y)]^{f(y)}\}$$

or

$$(9.9) \qquad 1 - \exp\left[-\exp\left(\frac{y - c(x)}{d(x)}\right)\right] = 1 - \exp\left[-\exp\left(\frac{x - a(y)}{b(y)}\right)\right]$$

This justification will be achieved through extreme value theory when the elements are assembled into a so-called series system. In our previous example a wire can be viewed as a series system of short pieces. Note that it can be assumed hypothetically to be cut in short pieces of equal length (subelements). Note also that if the wire is under stress of amplitude U, by continuity, all subelements of the wire must be under the same stress amplitude U.

So, in building our model, we can assume a series of a large number of elements, n such that their lives V_i, $1 \le i \le n$, are identically distributed. We further assume that the value of the regressor variable applied to the i-th component is independent of i. The weakest link principle guarantees that the life V_0 of the system is the minimum of V_i, $1 \le i \le n$. A certainly surprising result will be that the common value U of the regressor variable applied to each element can also be expressed as the minimum of n random variables. Namely, assume the j-th element to be the weakest, and thus $V_j = V_0$. If u is the value of the regressor variable which leads to a lifetime V_0 for the i-th element, then $U_j = u$, and

$$p_i = P(u, V_i) = P(u, V_0)$$

Since $V_0 \le V_i$ for all i, (9.5) yields $u \le U_i$ for all i, and because $u = U_j$, we have $u = \min(U_j)$.

Consequently, since both U given V, and V given U, can be expressed as the minimum of a large number of identically distributed random variables, we appeal to extreme value theory to say that only Weibull, Gumbel or Frechet distributions are feasible limits. Out of these three laws we can select the Weibull distribution by its positive character (only the Weibull law represents a positive random variable) or the Gumbel distribution because it can also be the limit distribution of non-negative random variables (see chapter 3). Note that the Frechet distribution is excluded because $\alpha(F) \ne -\infty$ (see theorem 3.4).

9.2.3.   General solution of the functional equations.

Because of its simplicity, we start with equation (9.9), which can be written as

$$(9.10) \qquad \frac{x - a(y)}{b(y)} = \frac{y - c(x)}{d(x)}$$

and upon rearrangement of terms we get

$$(9.11) \qquad xd(x) - a(y)d(x) - yb(y) + c(x)b(y) = 0$$

which is a functional equation of the form

$$(9.12) \qquad \prod_{i=1}^{n} f_i(x)g_i(y) = 0$$

where $a(y)$, $b(y)$, $c(x)$ and $d(x)$ are the unknown functions to be determined.

According to Aczél (1966, page 161), all solutions of the equation (9.12) can be written in the form

$$(9.13) \qquad f_k(x) = \sum_{i=1}^{r} p_{ki}\phi_i(x); \qquad k = 1, 2, \ldots, n$$

$$(9.14) \qquad g_k(y) = \sum_{j=r+1}^{n} q_{kj}\psi_j(y); \qquad k = 1, 2, \ldots, n$$

where $0 \le r \le n$ and $\phi_1, \phi_2, \ldots, \phi_r$, on one hand, and $\psi_{r+1}, \psi_{r+2}, \ldots, \psi_n$, on the other, are arbitrary systems, of mutually linearly independent functions and

$$(9.15) \qquad \sum_{k=1}^{n} p_{ki}q_{kj} = 0; \qquad i = 1, 2, \ldots, r; \quad j = r+1, \ldots, n$$

Here we have $n = 4$ and $r \neq 0, 1, 3, 4$ because $xd(x)$ and $d(x)$, on one hand, and $yb(y)$ and $b(y)$, on the other hand, are linearly independent. Then equation (9.13), on account of (9.11) and (9.12), can be written

$$(9.16) \qquad \begin{pmatrix} f_1(x) \\ f_2(x) \\ f_3(x) \\ f_4(x) \end{pmatrix} = \begin{pmatrix} 1 & 0 \\ 0 & -1 \\ a_{31} & a_{32} \\ a_{41} & a_{42} \end{pmatrix} \begin{pmatrix} xd(x) \\ d(x) \end{pmatrix} = \begin{pmatrix} xd(x) \\ -d(x) \\ -1 \\ c(x) \end{pmatrix}$$

Similarly, equation (9.14) can be written

$$(9.17) \qquad \begin{pmatrix} g_1(y) \\ g_2(y) \\ g_3(y) \\ g_4(y) \end{pmatrix} = \begin{pmatrix} b_{13} & b_{14} \\ b_{23} & b_{24} \\ 1 & 0 \\ 0 & 1 \end{pmatrix} \begin{pmatrix} yb(y) \\ b(y) \end{pmatrix} = \begin{pmatrix} 1 \\ a(y) \\ yb(y) \\ b(y) \end{pmatrix}$$

where, by equation (9.15) and a and b constants must satisfy

$$
(9.18) \qquad \begin{pmatrix} 1 & 0 & a_{31} & a_{41} \\ 0 & -1 & a_{32} & a_{42} \end{pmatrix} \begin{pmatrix} b_{13} & b_{14} \\ b_{23} & b_{24} \\ 1 & 0 \\ 0 & 1 \end{pmatrix} = 0
$$

or equivalently

$$
(9.19) \qquad \left. \begin{array}{l} b_{13} = -a_{31} = A \\ b_{14} = -a_{41} = -B \\ b_{23} = a_{32} = C \\ b_{24} = a_{42} = -D \end{array} \right\}
$$

where A, B, C and D are arbitrary constants.

Substitution of these values in (9.16) and (9.17) leads to

$$
(9.20) \qquad b(y) = \frac{1}{Ay - B}
$$

$$
(9.21) \qquad a(y) = \frac{Cy - D}{Ay - B}
$$

$$
(9.22) \qquad d(x) = \frac{1}{Ax - C}
$$

$$
(9.23) \qquad c(x) = \frac{Bx - D}{Ax - C}
$$

which is the most general solution of equation (9.9) or its equivalent (9.11).
Finally, substitution in (9.9) yields the following conditional distribution

$$
F(y/x) = F(x/y) = 1 - \exp\{-\exp(Axy - Bx - Cy + D)\};
$$
$$
(9.24) \qquad\qquad -\infty < x < \infty, \quad -\infty < y < \infty
$$

which for $A \neq 0$ can also be written

$$
(9.25) \qquad F(y/x) = F(x/y) = 1 - \exp\{-\exp[A^*(x - C^*)(y - B^*) + D^*]\}
$$

where

(9.26)

$$A^* = A$$
$$B^* = \frac{B}{A}$$
$$C^* = \frac{C}{A}$$
$$D^* = D - BCA^3$$

Expression (9.25) cannot be a valid model because if it is an increasing function of y for $x > C^*$, it becomes decreasing for $x < C^*$. Consequently, A must be zero and the model (9.24) becomes

$$F(y/x) = F(x/y) = 1 - \exp\{-\exp(-Bx - Cy + D)\};$$

(9.27)
$$-\infty < x < \infty, \quad -\infty < y < \infty$$

where $B < 0$ and $C < 0$. This model is illustrated in figure 9.3.

The model (9.27), which is a 3-parameter model is the most general model satisfying the above required conditions. It will be called the Gumbel-Gumbel

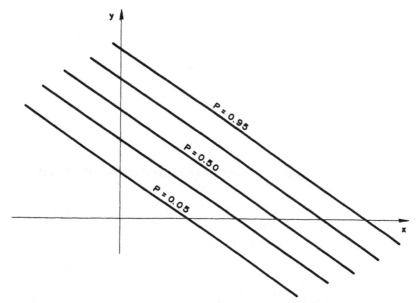

FIGURE 9.3.   Percentiles of the only feasible Gumbel-Gumbel model

regression model because of the first equality in (9.27) and its Gumbel character.

We now turn to equation (9.8) which can be written

$$(9.28) \qquad [a(x)y + b(x)]^{c(x)} = [d(y)x + e(y)]^{f(y)}$$

and upon taken logarithms we get

$$(9.29) \qquad r(x) + c(x)\log[y + s(x)] - t(y) - f(y)\log[x + u(y)] = 0$$

where

$$(9.30) \qquad r(x) = c(x)\log[a(x)]$$

$$(9.31) \qquad s(x) = \frac{b(x)}{a(x)}$$

$$(9.32) \qquad t(y) = f(y)\log[d(y)]$$

$$(9.33) \qquad u(y) = \frac{e(y)}{d(y)}$$

Castillo and Galambos (1987) have shown that either

$$(9.34) \qquad s(x) = -F; \quad u(y) = -E$$

or

$$(9.35) \qquad c(x) = f(y) = E$$

where E and F are constants.

If (9.34) holds, equation (9.29) becomes

$$(9.36) \qquad r(x) + c(x)\log(y - F) - t(y) - f(y) - f(y)\log(x - E) = 0$$

which is a functional equation of the form (9.12), and can be solved by the same technique as before.

Now we also have $n = 4$ and $r \neq 0, 1, 3, 4$, because the functions 1 and $\log(x - E)$, on one hand, and 1 and $\log(Y - F)$, on the other, are linearly independent. Equation (9.13) with (9.12) and (9.36) becomes

$$(9.37) \qquad \begin{pmatrix} f_1(x) \\ f_2(x) \\ f_3(x) \\ f_4(x) \end{pmatrix} = \begin{pmatrix} a_{11} & a_{12} \\ a_{21} & a_{22} \\ -1 & 0 \\ 0 & 1 \end{pmatrix} \begin{pmatrix} 1 \\ \log(x - E) \end{pmatrix} = \begin{pmatrix} r(x) \\ c(x) \\ -1 \\ -\log(x - E) \end{pmatrix}$$

and equation (9.14)

$$(9.38) \quad \begin{pmatrix} g_1(y) \\ g_2(y) \\ g_3(y) \\ g_4(y) \end{pmatrix} = \begin{pmatrix} 1 & 0 \\ 0 & 1 \\ b_{33} & b_{34} \\ b_{43} & b_{44} \end{pmatrix} \begin{pmatrix} 1 \\ \log(y-F) \end{pmatrix} = \begin{pmatrix} 1 \\ \log(y-F) \\ t(y) \\ f(y) \end{pmatrix}$$

where (see equation (9.15))

$$(9.39) \quad \begin{matrix} a_{11} = b_{33} = A^* \\ a_{12} = b_{43} = B \\ a_{21} = b_{34} = C \\ a_{22} = b_{44} = D \end{matrix} \Bigg\}$$

where A^*, B, C and D are arbitrary constants.
   Substitution of these values in (9.37) and (9.38) leads to

$$(9.40) \qquad r(x) = A^* + B \log(x - E)$$

$$(9.41) \qquad c(x) = C + D \log(x - E)$$

$$(9.42) \qquad t(y) = A^* + C \log(y - F)$$

$$(9.43) \qquad f(y) = B + D \log(y - F)$$

which is the general solution of (9.36).
   Consequently, the conditional distribution (9.8) can be written as

$$F(x/y) = F(y/x) = 1 - \exp\{-A(x-E)^B(y-F)^C \exp[D \log(x-E)\log(y-F)]\}$$

$$(9.44) \hspace{6cm} x \geq E, \quad y \geq F$$

If (9.35), instead of (9.34), holds, equation (9.28) becomes

$$(9.45) \qquad a(x)y + b(x) - d(y)x - e(y) = 0$$

which, once more, is of the form (12), with $n = 4$ and $r = 2$.
   Equations (9.13) and (9.14) with (9.12) and (9.45) give

$$(9.46) \quad \begin{pmatrix} f_1(x) \\ f_2(x) \\ f_3(x) \\ f_4(x) \end{pmatrix} = \begin{pmatrix} a_{11} & a_{12} \\ a_{21} & a_{22} \\ 0 & -1 \\ -1 & 0 \end{pmatrix} \begin{pmatrix} 1 \\ x \end{pmatrix} = \begin{pmatrix} a(x) \\ b(x) \\ -x \\ -1 \end{pmatrix}$$

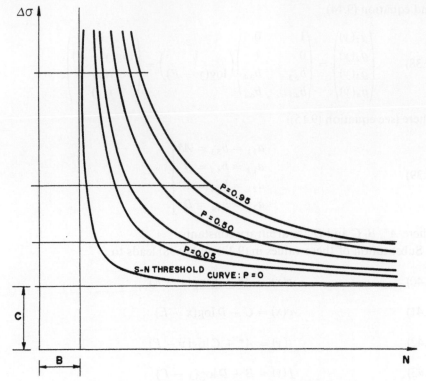

FIGURE 9.4.   Illustration of model in expression (9.53)

and

(9.47)
$$\begin{pmatrix} g_1(y) \\ g_2(y) \\ g_3(y) \\ g_4(y) \end{pmatrix} = \begin{pmatrix} 0 & 1 \\ 1 & 0 \\ b_{33} & b_{34} \\ b_{43} & b_{44} \end{pmatrix} \begin{pmatrix} 1 \\ y \end{pmatrix} = \begin{pmatrix} y \\ 1 \\ d(y) \\ e(y) \end{pmatrix}$$

where, according to (9.15),

(9.48)
$$\left. \begin{aligned} a_{21} = b_{43} = A \\ a_{11} = b_{44} = C \\ a_{22} = b_{33} = B \\ a_{12} = b_{34} = D \end{aligned} \right\}$$

where A, B, C and D are arbitrary constants.

Substitution of these values in (9.46) and (9.47) gives

(9.49) $$a(x) = C + Dx$$

(9.50) $b(x) = A + Bx$

(9.51) $d(y) = B + Dy$

(9.52) $e(y) = A + Cy$

which is the general solution of (9.45).

Finally, substitution of these in (9.8) leads to

$$F(x/y) = F(y/x) = 1 - \exp\{-[A + Bx + Cy + Dxy]^E\};$$

(9.53) $A + Bx + Cy + Dxy \geq 0$

which for $D > 0$ is illustrated in figure 9.4.

We have shown then that for the Weibull assumption we get models (9.44) and (9.53).

9.3.   Bivariate Models with Weibull Conditionals

In this section we characterize three bivariate models which are useful when conditional distributions are the minima of given sequences of random variables. This fact will justify the use of Weibullian conditional distributions which form the basis for the characterization. One of the conditional families is then assumed to be Weibullian and the other is assumed to belong to the Weibull, normal and gamma families. This will lead to three functional equations which are solved in the following sections. For a more general treatment of the problem see Castillo and Galambos (1986a).

9.3.1.   The Weibull-Weibull model.

We assume a two-dimensional random variable (X, Y) such that all conditional distributions can be defined as the minima of given sequences of random variables. We also assume that, for large number of terms in the sequences, they converge to the Weibull family.

The assumption of the limit distribution to be Weibull is not very restrictive, because if it is a Gumbel or Frechet distribution, known simple transformations can change them to Weibull.

Any joint probability density function, $f(x, y)$, of (X, Y) can be written

(9.54) $f(x, y) = f(y/x)g(x) = f(x/y)h(y)$

where $f(x/y)$, $f(y/x)$, $g(x)$ and $h(y)$ are the conditional and marginal probability density functions, respectively. If now we assume that $f(x/y)$ and $f(y/x)$ are Weibull, Equation (9.54) becomes

$$m(x)\exp\{-[a(x)(y-K)]^{c(x)}\}[a(x)(y-K)]^{c(x)-1}$$

$$= n(y)\exp\{-[d(y)(x-L)]^{f(y)}\}[d(y)(x-L)]^{f(y)-1} \qquad x > L; \quad y > K$$

(9.55)

where K and L are the location parameters and the functions $a(x)$, $c(x)$, $d(y)$, and $f(y)$ are related to the scale and shape factors of the Weibullian laws, and they must satisfy the following restrictions

(9.56) $$a(x) > 0; \quad c(x) > 0; \quad d(y) > 0; \quad f(y) > 0$$

and

(9.57) $$m(x) = a(x)c(x)g(x)$$

(9.58) $$n(y) = d(y)f(y)h(y)$$

Equation (9.55), where $a(x)$, $c(x)$, $m(x)$, $d(y)$, $f(y)$ and $n(y)$ are unknown functions, is the functional equation to be solved.

A procedure to solve this functional equation is given by Castillo and Galambos (1985). However, no explicit general solution has been given yet.

9.3.1.1.   A particular solution of practical importance.   In this section we study the particular case of equation (9.55) for which one explicit solution can be obtained. This is the case where the shape parameters are constant, i.e. we assume, from now on, that

(9.59) $$f(y) = A$$

(9.60) $$c(x) = B$$

where A and B are constants.

Upon taking logarithms in (9.55) and substitution of (9.59) and (9.60) we get

$$\log[n(y)] - d(y)^A(x-L)^A + (A-1)\log[d(y)] + (A-1)\log(x-L)$$

$$= \log[m(x)] - a(x)^B(y-K)^B + (B-1)\log[a(x)] + (B-1)\log(y-K)$$

(9.61)

which is a functional equation of the form

(9.62) $$\sum_{i=1}^{n} f_k(x)g_k(y) = 0$$

and its solution, which can be obtained by the same method above, is given by

(9.63) $a(x) = [E + F(x - L)^A]^{1/B}$

(9.64) $c(x) = B$

(9.65) $d(y) = [G + F(y - K)^B]^{1/A}$

(9.66) $f(y) = A$

(9.67) $m(x) = (x - L)^{A-1} \dfrac{[E + F(x - L)^A]^{(1-B)}}{B \exp[-H - G(x - L)^A]}$

(9.68) $n(y) = (y - K)^{B-1} \dfrac{[G + F(y - K)^B]^{(1-A)}}{A \exp[-H - E(y - K)^B]}$

(9.69) $g(x) = \dfrac{(x - L)^{A-1} \exp[-H - G(x - L)^A]}{B[E + F(x - L)^A]}$

(9.70) $h(y) = \dfrac{(y - K)^{B-1} \exp[-H - E(y - K)^B]}{A[G + F(y - K)^B]}$

$$f(x, y) = (x - L)^{A-1}(y - K)^{B-1} \exp\{-[H + G(x - L)^A$$
(9.71)
$$+ E(y - K)^B + F(x - L)^A(y - K)^B]\}$$

where E, F, G and H are constants.

The normalizing constant H is given by

(9.72) $$\exp(-H) = \dfrac{ABF \exp\left(\dfrac{-GE}{F}\right)}{V\left(\dfrac{GE}{F}\right) - \gamma}$$

where

(9.73) $$V(u) = \sum_{i=1}^{\infty} \dfrac{(-1)^{i+1} u^i}{i!i} - \log u$$

and γ is the Euler constant.

Conditions (9.56) are equivalent to

(9.74) $A > 0; \quad B > 0; \quad E > 0; \quad F \geq 0; \quad G > 0$

Equation (9.71) implies that X and Y are independent if and only if $F = 0$. So a test of $F = 0$ is equivalent to a test of independence under this model.

One of the most important applications of this model is based on the fact that the joint distribution of (X, Y) can be estimated from observations of its marginals, i.e. estimation can be done, for example, by maximizing, with respect to the parameters, the likelihood function

(9.75)
$$L = \prod_{i=1}^{n_1} \prod_{j=1}^{n_2} g(x_i)h(y_j)$$

where x_i $(i = 1, 2, \ldots, n_1)$ and y_j $(j = 1, 2, \ldots, n_2)$ are independent samples from the marginals.

This possibility becomes very important in some cases in which it is not possible to obtain samples of the two-dimensional random variable (X, Y). This is the case of competing risks where items can be tested at a single risk.

Another important conclusion that can be drawn from this model is that if one of the variables is Weibullian, X say, then the other, Y, must be Weibullian too. In effect, in order (9.69) to be a Weibull distribution we must have

(9.76)
$$F = 0$$
$$\exp[-H] = AGBE$$

but this implies that (9.70) is also Weibull and that they are independent as well.

The bivariate model with exponential conditionals is a particular case $(A = B = 1)$ of (9.71).

9.3.2.   The Weibull-gamma model.

If $f(x/y)$ is now Gamma, Equation (9.55) transforms into

(9.77)
$$m(x)\exp\{-[a(x)(y - K)]^{c(x)}\}[a(x)(y - K)]^{c(x)-1}$$
$$= h(y)d(y)x^{e(y)}\exp[-xf(y)]; \qquad x \geq 0; \quad y \geq K$$

where

(9.78)
$$d(y) = \frac{f(y)^{e(y)+1}}{\Gamma[e(y) + 1]}$$

and

(9.79)
$$a(x) > 0; \qquad c(x) > 0; \quad x \geq 0$$
$$e(y) > -1; \quad f(y) > 0; \quad y \geq K$$

Upon taking logarithms, (9.77) transforms into

(9.80)
$$u(x) - a(x)^{c(x)}(y - K)^{c(x)} + \log[a(x)^{c(x)-1}(y - K)^{c(x)-1}]$$
$$= v(y) + e(y)\log x - xf(y)$$

where

(9.81) $u(x) = \log[m(x)]$

(9.82) $v(y) = \log[h(y)d(y)]$

Equation (9.77) is the functional equation to be solved.
Castillo and Galambos (1987) show that its general solution is given by

(9.83) $a(x) = (A + Bx + G \log x)^{1/C}$

(9.84) $c(x) = C$

(9.85) $e(y) = F - G(y - K)^C$

(9.86) $f(y) = -E + B(y - K)^C$

(9.87) $g(x) = \dfrac{\exp\{D + Ex + F \log x\}}{C(A + Bx + G \log x)}$

(9.88) $h(y) = \dfrac{\Gamma(1 + e(y)) \exp\{(C - 1)\log(y - K) + D - A(y - K)^C\}}{[-E + B(y - K)^C]^{F+1-G(y-K)^C}}$

$$f(x, y) = (y - K)^{C-1} \exp[(D + Ex + F \log x)$$

(9.89) $$- (A + Bx + G \log x)(y - K)^C]$$

and inequalities (9.79) transform into

(9.90)
$$\begin{aligned}
a(x) &= (A + Bx + G \log x)^{1/C} > 0; & x &\geq 0 \\
c(x) &= C > 0 & & \\
e(y) &= F - G(y - K)^C > -1; & y &\geq K \\
f(y) &= -E + B(y - K)^C > 0; & y &\geq K
\end{aligned}$$

which leads to the models in Table 9.1.

$B > 0$	$B = 0$
$A > G\left[1 - \log\left(-\dfrac{G}{B}\right)\right]$	$A > 0$
$C > 0$	$C > 0$
$E < 0$	$E < 0$
$F > -1$	$F > -1$
$G < 0$	$G = 0$

TABLE 9.1.   Feasible Weibull-Gamma models

In this case independence holds if and only if

(9.91) $$B = G = 0$$

9.3.3.   The Weibull-normal model.

Finally, if $f(x/y)$ is normal, Equation (9.54) becomes

(9.92)
$$m(x)\exp\{-[a(x)(y-K)]^{c(x)}\}[a(x)(y-K)]^{c(x)-1}$$
$$= \frac{h(y)\exp\left[-\frac{1}{2}\left(\frac{x-d(y)}{e(y)}\right)^2\right]}{e(y)(2\pi)^{1/2}} \qquad -\infty < x < \infty; \quad y > K$$

and upon taking logarithms

(9.93)
$$u(x) - a(x)^{c(x)}(y-K)^{c(x)} + [c(x) - 1]\log[a(x)(y-K)]$$
$$= v(y) - \frac{\left(\dfrac{x-d(y)}{e(y)}\right)^2}{2}$$

where

(9.94) $$u(x) = \log\{m(x)\}$$

(9.95) $$v(x) = \log\left(\frac{h(y)}{e(y)(2\pi)^{1/2}}\right)$$

and

(9.96)
$$a(x) > 0; \qquad c(x) > 0; \quad -\infty < x < \infty$$
$$e(y) > 0; \qquad y > K$$

The solution of the functional equation (9.93) (see Castillo and Galambos (1985)) is

(9.97) $$a(x) = (A + Bx + Gx^2)^{1/C}$$

(9.98) $$c(x) = C$$

(9.99) $$d(y) = \frac{E - B(y-K)^C}{2G(y-K)^C - 2F}$$

(9.100) $$e^2(y) = \frac{1}{2G(y-K)^C - 2F}$$

(9.101) $$g(x) = \frac{\exp\{D + Ex + Fx^2\}}{C(A + Bx + Gx^2)}$$

$$h(y) = \frac{\pi^{1/2}}{[G(y - K)^C - F]^{1/2}} \exp\bigg((C - 1)\log(y - K)$$

(9.102) $$+ D - A(y - K)^C + \frac{[B(y - K)^C - E]^2}{4[G(y - K)^C - F]}\bigg)$$

$$f(x, y) = (y - K)^{C-1} \exp\{(D + Ex + Fx^2)$$

(9.103) $$- (A + Bx + Gx^2)(y - K)^C\}$$

Inequalities (9.96) now become

(9.104)
$$\left.\begin{array}{l} a(x) = (A + Bx + Gx^2)^{1/C} > 0; \quad -\infty < x < \infty \\ c(x) = C > 0 \\ e^2(y) = \dfrac{1}{2G(y - K)^C - 2F} > 0; \quad y > K \end{array}\right\}$$

which are equivalent to

(9.105) $$G \geq 0; \quad 4GA \geq B^2; \quad C > 0; \quad F < 0$$

Chapter 10 — Multivariate Extremes

10.1. Introduction

In the last chapters we analyzed the distributions of order statistics of a one-dimensional sample of size n. In this chapter we extend some of the previous results to the case of m-dimensional samples. More precisely, $(\mathbf{X}_1, \mathbf{X}_2, \ldots, \mathbf{X}_n)$ is now a sample of size n coming from a given population with cdf $\mathbf{F}(x)$ where $X_1, X_2, \ldots, X_n$ are m-dimensional vectors. The m-th component of X_i $(i = 1, 2, \ldots, n)$ will be denoted by X_{ij} $(j = 1, 2, \ldots, m)$. From now on, boldfaced letters will be used for vectors and distributions of m-dimensional vectors.

The aim of this chapter is to analyze the existence of vector sequences $\mathbf{a}_n, \mathbf{b}_n$, $\mathbf{c}_n$ and $\mathbf{d}_n$, such that the vectors of maxima and minima, $\mathbf{Z}_n$ and $\mathbf{W}_n$, respectively, where $Z_1, Z_2, \ldots, Z_n$ and $W_1, W_2, \ldots, W_n$ are the maxima and minima of the respective components in the sample, satisfy

$$(10.1) \qquad \lim_{n \to \infty} P[\mathbf{Z}_n < \mathbf{a}_n + \mathbf{b}_n \mathbf{x}] = \mathbf{H}(x)$$

$$(10.2) \qquad \lim_{n \to \infty} P[\mathbf{W}_n < \mathbf{c}_n + \mathbf{d}_n \mathbf{x}] = \mathbf{L}(x)$$

where $\mathbf{H}(x)$ and $\mathbf{L}(x)$ are non-degenerated distributions and the product of vectors $\mathbf{xy}$ must be interpreted as component-wise, i.e. $\mathbf{xy} = (x_1 y_1, x_2 y_2, \ldots, x_n y_n)$. If (10.1) is satisfied we say that $\mathbf{F}(x)$ belongs to the domain of attraction of $\mathbf{H}(x)$ in the right tail. Similarly, if (10.2) holds, we say that $\mathbf{F}(x)$ belongs to the domain of attraction of $\mathbf{L}(x)$ in the left tail. Finally, if $H(x)$ and $L(x)$ satisfy (10.1) or (10.2) they are called limit distributions.

Because in the case of multidimensional random variables, the survival function $\mathbf{G}(x_1, x_2, \ldots, x_n)$, defined as

(10.3) $\mathbf{G}(x_1, x_2, \ldots, x_n) = P[X_1 > x_1, X_2 > x_2, \ldots, X_n > x_n]$

is very frequently used, instead of the cdf $\mathbf{F}(x_1, x_2, \ldots, x_n)$ we give the relation between both functions, which is based on expression (2.137) with

(10.4) $C_i = \{X_i \leq x_i\}$

leading to

(10.5) $\mathbf{F}(x_1, x_2, \ldots, x_n) = \sum_{i=0}^{n} (-1)^i \underline{S}_{i,n}; \qquad n \geq 2$

But taking into account that

(10.6) $\mathbf{G}(x_1, x_2, \ldots, x_n) = \underline{S}_{n,n}$

we finally get

(10.7) $\mathbf{F}(x_1, x_2, \ldots, x_n) = \sum_{i=0}^{n-1} (-1)^i \underline{S}_{i,n} + (-1)^n \mathbf{G}(x_1, x_2, \ldots, x_n)$

which for the case of two-dimensional cdfs becomes

$\mathbf{F}(x_1, x_2) = 1 - [1 - F_{X_1}(x_1)] - [1 - F_{X_2}(x_2)] + \mathbf{G}(x_1, x_2)$

(10.8) $= \mathbf{G}(x_1, x_2) - 1 + F_{X_1}(x_1) + F_{X_2}(x_2)$

where $F_{X_1}(x_1)$ and $F_{X_2}(x_2)$ are the marginal cdfs.

10.2.   Dependence Functions

In this section we introduce the concept of dependence function which has been shown to be extremely useful in the analysis of the problem of extremes for multidimensional variables.

Definition 10.1. (Dependence function).   Let $\mathbf{F}(x)$ be the cdf of an m-dimensional random variable with univariate marginal $F_i(x)$ $(i = 1, 2, \ldots, m)$. We define the dependence function $\mathbf{D}_F(y_1, y_2, \ldots, y_m)$ of $\mathbf{F}(x)$ by means of

(10.9) $\mathbf{D}_F(F_1(x_1), F_2(x_2), \ldots, F_m(x_m)) = \mathbf{F}(x_1, x_2, \ldots, x_m)$

which for the particular case of increasing $F_i(x_i)$ $(i = 1, 2, \ldots, m)$ can be written

(10.10) $\mathbf{D}_F(y_1, y_2, \ldots, y_m) = \mathbf{F}(F_1^{-1}(y_1), F_2^{-1}(y_2), \ldots, F_m^{-1}(y_m))$ ∎

Distribution	Mardia's
$F(x, y)$	$[\exp(x) + \exp(y) - 1]^{-1} + 1 - \exp(-x) - \exp(-y)$
$G(x, y)$	$[\exp(x) + \exp(y) - 1]^{-1}$
$H(x, y)$	$H_{3,0}(x)H_{3,0}(y)\exp\{[(\exp(x) + \exp(y)]^{-1}\}$
$D(x, y)$	$\left(\dfrac{1}{1 - x} + \dfrac{1}{1 - y} - 1\right)^{-1} - 1 + x + y$
Marginals	$F_X(x) = 1 - \exp(-x);\quad F_Y(y) = 1 - \exp(-y)$

Distribution	Morgenstern's
$F(x, y)$	$\exp(-x - y)\{1 + \alpha[1 - \exp(-x)][1 - \exp(-y)]\}$ $+ 1 - \exp(-x) - \exp(-y)$
$G(x, y)$	$\exp(-x - y)\{1 + \alpha[1 - \exp(-x)][1 - \exp(-y)]\}$
$H(x, y)$	$H_{3,0}(x)H_{3,0}(y)$
$D(x, y)$	$y_1 y_2[1 + \alpha(1 - x)(1 - y)]$
Marginals	$F_X(x) = 1 - \exp(-x);\quad F_Y(y) = 1 - \exp(-y)$

Distribution	Gumbel's Type I
$F(x, y)$	$\exp(-x - y + \theta xy) + 1 - \exp(-x) - \exp(-y)$
$G(x, y)$	$\exp(-x - y + \theta xy)$
$H(x, y)$	$H_{3,0}(x)H_{3,0}(y)$
$D(x, y)$	$(1 - x)(1 - y)\exp[\theta \log(1 - x)\log(1 - y)] - 1 + x + y$
Marginals	$F_X(x) = 1 - \exp(-x);\quad F_Y(y) = 1 - \exp(-y)$

Distribution	Gumbel's Type II
$F(x, y)$	$\exp\{-(x^m + y^m)^{1/m}\} + 1 - \exp(-x) - \exp(-y)$
$G(x, y)$	$\exp\{-(x^m + y^m)^{1/m}\}$
$H(x, y)$	$H_{3,0}(x)H_{3,0}(y)$
$D(x, y)$	$\exp\{-[[-\log(1 - x)]^m + [-\log(1 - y)]^m]^{1/m}\} - 1 + x + y$
Marginals	$F_X(x) = 1 - \exp(-x);\quad F_Y(y) = 1 - \exp(-y)$

Distribution	Marshall-Olkin
$F(x, y)$	$\exp\{-x - y - \lambda \max(x, y)\} + 1 - \exp\{-(1 + \lambda)x\} - \exp\{-(1 + \lambda)y\}$
$G(x, y)$	$\exp\{-x - y - \lambda \max(x, y)\}$
$H(x, y)$	$H_{3,0}(x)H_{3,0}(y)$
$D(x, y)$	$\{(1 - x)(1 - y)[1 - \max(x, y)]^\lambda\}^{1/(1 + \lambda)} - 1 + x + y$
Marginals	$F_X(x) = 1 - \exp(-(1 + \lambda)x);\quad F_Y(y) = 1 - \exp - (1 + \lambda)y)$

TABLE 10.1.   Important functions of some bivariate exponentials

Note that this function is defined on the unit hypercube $0 \leq y_i \leq 1$, $(i = 1, 2, \ldots, m)$.

In order to clarify some concepts in this chapter we shall use several examples of bivariate exponential distributions, which are included in table 10.1.

Example 10.1. (Mardia's distribution). The dependence function of Mardia's distribution is given by

$$\mathbf{D}(y_1, y_2) = \mathbf{F}(F_{X_1}^{-1}(y_1), F_{X_2}^{-1}(y_2)) = \mathbf{F}(-\log(1 - y_1), -\log(1 - y_2))$$

$$= \left(\frac{1}{1 - y_1} + \frac{1}{1 - y_2} - 1\right)^{-1} + 1 - (1 - y_1) - (1 - y_2)$$

$$= \left(\frac{1}{1 - y_1} + \frac{1}{1 - y_2} - 1\right)^{-1} - 1 + y_1 + y_2 \qquad \blacksquare$$

Example 10.2. (Morgenstern's distribution). The dependence function of Morgenstern's distribution is given by

$$\mathbf{D}(y_1, y_2) = \mathbf{F}(-\log(1 - y_1), -\log(1 - y_2))$$

$$= (1 - y_1)(1 - y_2)[1 + \alpha y_1 y_2] - 1 + y_1 + y_2$$

$$= y_1 y_2 [1 + \alpha(1 - y_1)(1 - y_2)] \qquad \blacksquare$$

Example 10.3. (Gumbel's type I distribution). The dependence function of Gumbel's type I distribution is given by

$$\mathbf{D}(y_1, y_2) = \mathbf{F}(-\log(1 - y_1), -\log(1 - y_2))$$

$$= \exp[\log(1 - y_1) + \log(1 - y_2) + \theta \log(1 - y_1)\log(1 - y_2)]$$

$$+ 1 - (1 - y_1) - (1 - y_2)$$

$$= (1 - y_1)(1 - y_2)\exp[\theta \log(1 - y_1)\log(1 - y_2)] - 1 + y_1 + y_2$$

$$\blacksquare$$

Example 10.4. (Gumbel's type II distribution). The dependence function of Gumbel's type II distribution is given by

$$\mathbf{D}(y_1, y_2) = \mathbf{F}(-\log(1 - y_1), -\log(1 - y_2))$$

$$= \exp\{-[[-\log(1 - y_1)]^m + [-\log(1 - y_2)]^m]^{1/m}\} - 1 + y_1 + y_2$$

$$\blacksquare$$

Example 10.5. (Marshall-Olkin distribution). The dependence function of the Marshall-Olkin distribution is

$$\mathbf{D}(y_1, y_2) = \mathbf{F}\left(\frac{-\log(1 - y_1)}{1 + \lambda}, \frac{-\log(1 - y_2)}{1 + \lambda}\right)$$

$$= \{(1 - y_1)(1 - y_2)[1\text{-max}(y_1, y_2)]^\lambda\}^{1/(1 + \lambda)} - 1 + y_1 + y_2 \quad \blacksquare$$

Example 10.6. (Independent bivariate exponential distribution). The dependence function of the independent exponential bivariate distribution is given by

$$\mathbf{D}(y_1, y_2) = \mathbf{F}(-\log(1 - y_1), -\log(1 - y_2))$$

$$= \{1 - \exp[\log(1 - y_1)]\}\{1 - \exp[\log(1 - y_2)]\} = y_1 y_2 \quad \blacksquare$$

10.3.   Limit Distributions

A very easy way of obtaining the limit distribution of a multivariate parent is by means of the following theorem (see Galambos (1978)).

Theorem 10.1. (Convergence of marginal distributions). *Let* $\mathbf{F}_n(\mathbf{x})$ *be a sequence of m-dimensional cdf's with univariate marginals* $F_{t_n}(xt)$. *If* $\mathbf{F}_n(\mathbf{x})$ *converges in distribution (weakly) to a nondegenerate continuous cdf* $\mathbf{F}(\mathbf{x})$, *then* $F_{t_n}(x_t)$ *converges in distribution to the t-th marginal* $F_t(x_t)$ *of* $\mathbf{F}(\mathbf{x})$ *for* $1 \leq t \leq m$.

This theorem states that if the limit distribution for the m-dimensional case exists, the same sequence of constants leads to the convergence of marginal sequences. In other words, one possible method of finding the m-dimensional limit distribution is to analyze the convergence of the marginals by the techniques given in Chapter 4, obtain the sequences of constants a_{t_n}, b_{t_n}, c_{t_n} and d_{t_n} and then calculate the limit with these constants as m-dimensional sequences. This technique is illustrated in the following example by its application to the bivariate exponential distributions in Table 10.1.

Example 10.7. (Mardia's distribution). For the Mardia's distribution, because the marginals are exponentials we can choose (see example 3.5)

$$\mathbf{a}_n = (\log n, \log n); \quad \mathbf{b}_n = (1, 1)$$

Thus

$$\lim_{n \to \infty} \mathbf{F}^n(\mathbf{a}_n + \mathbf{b}_n\mathbf{x}) = \lim_{n \to \infty} \mathbf{F}^n(\log n + x_1, \log n + x_2)$$

$$= \lim_{n \to \infty} \left([n\exp(x_1) + n\exp(x_2) - 1]^{-1} + 1 \right.$$

$$\left. - \frac{[\exp(-x_1) + \exp(-x_2)]}{n} \right)^n$$

$$= \exp\left(-\exp(-x_1) - \exp(-x_2) + \frac{1}{\exp(x_1) + \exp(x_2)} \right)$$

$$= H_{3,0}(x_1)H_{3,0}(x_2)\exp\{[\exp(x_1) + \exp(x_2)]^{-1}\} \qquad \blacksquare$$

Example 10.8. *(Morgenstern's distribution).*   For the Morgenstern distribution we have

$$\lim_{n \to \infty} \mathbf{F}^n(\mathbf{a}_n + \mathbf{b}_n\mathbf{x})$$

$$= \lim_{n \to \infty} \mathbf{F}^n(\log n + x_1, \log n + x_2)$$

$$= \lim_{n \to \infty} \mathbf{F}^n(\log[n\exp(x_1)], \log[n\exp(x_2)])$$

$$= \lim_{n \to \infty} \left[\exp(-x_1 - x_2)\frac{1 + \alpha\left(1 - \dfrac{\exp(-x_1)}{n}\right)\left(1 - \dfrac{\exp(-x_1)}{n}\right)}{n^2} \right.$$

$$\left. + 1 - \frac{\exp(-x_1) + \exp(-x_2)}{n} \right]^n$$

$$= \exp\{-\exp(-x_1) - \exp(-x_2)\} = H_{3,0}(x_1)H_{3,0}(x_2) \qquad \blacksquare$$

Example 10.9. *(Gumbel's type I distribution).*   For the Gumbel's type I distribution we have

$$\lim_{n \to \infty} \mathbf{F}^n(\mathbf{a}_n + \mathbf{b}_n\mathbf{x}) = \lim_{n \to \infty} \mathbf{F}^n(\log n + x_1, \log n + x_2)$$

$$= \lim_{n \to \infty} \left(\frac{\exp[-x_1 - x_2 + \theta(\log n + x_1)(\log n + x_2)]}{n^2} \right.$$

$$\left. + 1 - \frac{\exp(-x_1) + \exp(-x_2)}{n} \right)^n$$

$$= \exp[-\exp(-x_1) - \exp(-x_2)] = H_{3,0}(x_1)H_{3,0}(x_2)$$

where it has been taken into account that $\theta < 0$. $\blacksquare$

Example 10.10. (Gumbel's type II distribution). For the Gumbel's type II distribution we have

$$\lim_{n \to \infty} F^n(a_n + b_n x) = \lim_{n \to \infty} F^n(\log n + x_1, \log n + x_2)$$

$$= \lim_{n \to \infty} \left(\exp[-[(\log n + x_1)^m + (\log n + x_2)^m]^{1/m}] \right.$$

$$\left. + 1 - \frac{\exp(-x_1) + \exp(-x_2)}{n} \right)^n$$

$$= \lim_{n \to \infty} \left(\exp(-2^{1/m} \log n) + 1 - \frac{\exp(-x_1) + \exp(-x_2)}{n} \right)^n$$

$$= \exp\{-\exp(-x_1) - \exp(-x_2)\} = H_{3,0}(x_1) H_{3,0}(x_2) \quad \blacksquare$$

Example 10.11. (Marshall-Olkin distribution). For the Marshall-Olkin distribution

$$\lim_{n \to \infty} F^n(a_n + b_n x)$$

$$= \lim_{n \to \infty} F^n\left(\frac{\log n + x_1}{1 + \lambda}, \frac{\log n + x_2}{1 + \lambda} \right)$$

$$= \lim_{n \to \infty} \left\{ \exp\left[-\left(\log n + x_1 + \log n + x_2 + \frac{\lambda[\log n + \max(x_1, x_2)]}{1 + \lambda} \right) \right. \right.$$

$$\left. \left. + 1 - \frac{\exp(-x_1) + \exp(-x_2)}{n} \right] \right\}^n$$

$$= \lim_{n \to \infty} \left[\frac{\exp\left(-\frac{x_1 + x_2 + \lambda \max(x_1, x_2)}{1 + \lambda} \right)}{n^{(2 + \lambda)/(1 + \lambda)}} \right.$$

$$\left. + 1 - \frac{\exp(-x_1) + \exp(-x_2)}{n} \right]^n$$

$$= \exp[-\exp(-x_1) - \exp(-x_2)] = H_{3,0}(x_1) H_{3,0}(x_2) \quad \blacksquare$$

The following theorem (see Galambos (1978)) gives necessary and sufficient conditions for the maximum of a sequence of i.i.d. variables to have a limit distribution. Extensions of theorems 3.3 and 3.4 to the m-dimensional case (see Marshall-Olkin (1983)) can be used for the same purpose.

Theorem 10.2. (Necessary and sufficient conditions for the existence of a limit distribution for Z_n). *Let $X_1, X_2, \ldots, X_n$ be i.i.d. m-dimensional vectors with common cdf $F(x)$. Then, there are vectors a_n and $b_n > 0$ such that $(Z_n - a_n)/b_n$ converges in distribution to a non-degenerate distribution $H(x)$ if, and only if, each of its marginals belongs to the domain of attraction of some $H_c(x)$ in the right tail and if*

$$(10.11) \qquad \lim_{n \to \infty} D_F^n(y_1^{1/n}, y_2^{1/n}, \ldots, y_m^{1/n}) = D_H(y_1, y_2, \ldots, y_m)$$

This theorem gives one way to determine if $F(x)$ is in the domain of attraction of $H(x)$ and allows determining whether or not the maximum of a sequence of random variables has a limit distribution. If this is so, it also allows one to obtain the limit distribution through its dependence function. The following examples illustrate this technique.

Example 10.12. (Mardia's distribution). For the Mardia distribution we have

$$D_H(y_1, y_2) = \lim_{n \to \infty} D_F^n(y_1^{1/n}, y_2^{1/n})$$

$$= \lim_{n \to \infty} \left[y_1^{1/n} + y_2^{1/n} - 1 + \left(\frac{1}{1 - y_1^{1/n}} + \frac{1}{1 - y_2^{1/n}} - 1 \right)^{-1} \right]^n$$

$$= \lim_{n \to \infty} \left[y_1^{1/n} + y_2^{1/n} - 1 + \frac{(1 - y_1^{1/n})(1 - y_2^{1/n})}{1 - (y_1 y_2)^{1/n}} \right]^n$$

$$= \lim_{n \to \infty} \left[y_1^{1/n} y_2^{1/n} \left(1 + \frac{(1 - y_1^{1/n})(1 - y_2^{1/n})}{1 - (y_1 y_2)^{1/n}} \right) \right]^n$$

$$= y_1 y_2 \lim_{n \to \infty} \left[1 - \frac{\log y_1 \log y_2}{n \log(y_1 y_2)} \right]^n$$

$$= y_1 y_2 \exp \left[\frac{-\log y_1 \log y_2}{\log(y_1 y_2)} \right]$$

$$= y_1 y_2 \exp \left[-\left(\frac{1}{\log y_1} + \frac{1}{\log y_2} \right)^{-1} \right]$$

In consequence, the cdf of the limit distribution is

$$H(x_1, x_2) = D_H(H_{3,0}(x_1), H_{3,0}(x_2))$$

$$= H_{3,0}(x_1), H_{3,0}(x_2) \exp\{[\exp(x_1) + \exp(x_2)]^{-1}\} \qquad \blacksquare$$

Example 10.13. (Morgenstern's distribution).   For the Morgenstern distribution we have

$$\mathbf{D_H}(y_1, y_2) = \lim_{n \to \infty} \mathbf{D_F^n}(y_1^{1/n}, y_2^{1/n})$$

$$= y_1 y_2 \lim_{n \to \infty} [1 + \alpha(1 - y_1^{1/n})(1 - y_2^{1/n})]^n$$

$$= y_1 y_2 \lim_{n \to \infty} [1 + n\alpha(1 - y_1^{1/n})(1 - y_2^{1/n})]$$

$$= y_1 y_2 \lim_{n \to \infty} \left(1 + \frac{n\alpha[\log y_1 \log y_2]}{n^2}\right) = y_1 y_2$$

where it has been taken into account that

$$\text{If } u_n \quad \to \quad 0 \quad \Rightarrow \quad (1 + u_n)^n \approx n u_n$$

$$\text{If } u_n \quad \to \quad 1 \quad \Rightarrow \quad 1 - u_n \approx -\log u_n$$

Consequently, the cdf of the limit distribution is (see expression (10.10))

$$\mathbf{H}(x_1, x_2) = \mathbf{D_H}(H_{3,0}(x_1), H_{3,0}(x_2)) = H_{3,0}(x_1) H_{3,0}(x_2) \qquad \blacksquare$$

Example 10.14. (Gumbel's type I distribution).   For the Gumbel type I distribution we have

$$\mathbf{D_H}(y_1, y_2) = \lim_{n \to \infty} \mathbf{D_F^n}(y_1^{1/n}, y_2^{1/n})$$

$$= \lim_{n \to \infty} \{(1 - y_1^{1/n})(1 - y_2^{1/n}) \exp[\theta \log(1 - y_1^{1/n}) \log(1 - y_2^{1/n})] - 1$$

$$+ y_1^{1/n} + y_2^{1/n}\}^n$$

$$= \lim_{n \to \infty} \left(\frac{\log y_1 \log y_2 \exp[\theta \log(1 - y_1^{1/n}) \log(1 - y_2^{1/n})]}{n^2}\right.$$

$$\left. + 1 - 2 + y_1^{1/n} + y_2^{1/n}\right)^n$$

$$= \lim_{n \to \infty} \exp[n(y_1^{1/n} + y_2^{1/n} - 2)] = \lim_{n \to \infty} \exp\left[\frac{y_1^{1/n} + y_2^{1/n} - 2}{\left(\frac{1}{n}\right)}\right]$$

$$= \lim_{n \to \infty} \exp[y_1^{1/n} \log y_1 + y_2^{1/n} \log y_2]$$

$$= \exp[\log(y_1 y_2)] = y_1 y_2 \qquad \blacksquare$$

Example 10.15. (Gumbel's type II distribution).   For the Gumbel type II distribution we get

$$\mathbf{D_H}(y_1, y_2) = \lim_{n \to \infty} \mathbf{D_F^n}(y_1^{1/n}, y_2^{1/n})$$

$$= \lim_{n \to \infty} \{\exp\{-[[-\log(1 - y_1^{1/n})]^m + [-\log(1 - y_2^{1/n})]^m]^{1/m}\} - 1$$

$$+ y_1^{1/n} + y_2^{1/n}\}^n$$

$$= \lim_{n \to \infty} \left\{\exp\left\{-[-\log(1 - y_2^{1/n})]\left[1 + \left(\frac{\log(1 - y_1^{1/n})}{\log(1 - y_2^{1/n})}\right)^m\right]^{1/m}\right\} - 1\right.$$

$$\left. + y_1^{1/n} + y_2^{1/n}\right\}^n$$

$$= \lim_{n \to \infty} \exp[n(y_1^{1/n} + y_2^{1/n} - 2)] = y_1 y_2 \qquad \blacksquare$$

Example 10.16. (Marshall-Olkin distribution).   For the Marshall-Olkin distribution we get

$$\mathbf{D_H}(y_1, y_2) = \lim_{n \to \infty} \mathbf{D_F^n}(y_1^{1/n}, y_2^{1/n})$$

$$= \lim_{n \to \infty} \{\{(1 - y_1^{1/n})(1 - y_2^{1/n})[1 - \max(y_1^{1/n}, y_2^{1/n})]^\lambda\}^{1/(1 + \lambda)}$$

$$- 1 + y_1 + y_2\}^n$$

$$= \lim_{n \to \infty} \left[\log y_1 \log y_2 \left(\frac{[1 - \max(y_1^{1/n}, y_2^{1/n})]^\lambda}{n^2}\right)^{1/(1 + \lambda)} - 1 + y_1 + y_2\right]^n$$

$$= \lim_{n \to \infty} \left[\left(\frac{\log y_1 \log y_2 \{\log[\max(y_1, y_2)]\}^\lambda}{n^3}\right)^{1/(1 + \lambda)} - 1 + y_1 + y_2\right]^n$$

$$= \lim \exp\{n(y_1^{1/n} + y_2^{1/n} - 2)\} = y_1 y_2 \qquad \blacksquare$$

We include, finally, a theorem which enables one to decide whether or not a given m-dimensional cdf can appear as a limit distribution of some sequence (see Galambos (1978)).

Theorem 10.3. (Necessary and sufficient conditions for a feasible limit distribution of maxima).   *An m-dimensional cdf $\mathbf{H}(x)$ is a limit distribution of maxima if, and only if, its univariate marginals belong to the domain of attraction of $H_c(x)$ in the right tail and if its dependence function $\mathbf{D_H}(y_1, y_2, \ldots, y_m)$ satisfies the* *condition*

(10.12) $D_H^k(y_1^{1/k}, y_2^{1/k}, \ldots, y_m^{1/k}) = D_H(y_1, y_2, \ldots, y_m)$

where $k \geq 1$ is arbitrary.

This theorem gives a functional equation that characterizes the dependence functions of all limit distributions of maxima.

By introducing the function

(10.13) $d_H(y_1, y_2, \ldots, y_m) = -\log D_H(\exp(-y_1), \exp(-y_2), \ldots, \exp(-y_m))$

the functional equation (10.12) can also be written as

(10.14) $k\, d_H(y_1, y_2, \ldots, y_m) = d_H(ky_1, ky_2, \ldots, ky_m)$

which defines $d_H(y_1, y_2, \ldots, y_m)$ as an Euler's homogeneous function of order one. The general solution of this equation leads to a general solution of $H(x_1, x_2, \ldots, x_m)$ of the form

$$H(x_1, x_2, \ldots, x_m) = [H_{3,0}(x_1), H_{3,0}(x_2), \ldots, H_{3,0}(x_m)]^{v(x_2 - x_1, x_3 - x_1, \ldots, x_m - x_1)}$$

(10.15)

where $v(x_2 - x_1, x_3 - x_1, \ldots, x_m - x_1)$ is a function that must satisfy some complicated extra conditions in order to make H a feasible cdf.

For $m = 2$, these conditions are simpler and some results can be obtained. The interested reader is referred to Geffroy (1958,1959), Tiago de Oliveira (1958,1962,1963), Sibuya (1960), Pickands III (1977) and Galambos (1978).

Example 10.17. (Mardia's distribution).   The cdf in example 10.12 can be a limit distribution because

(i) its marginals belong to the domain of attraction of $H_{3,0}(x)$ in the right tail.

(ii) its dependence function satisfies (10.12). In fact we have

$$D_H^k(y_1^{1/k}, y_2^{1/k}, \ldots, y_m^{1/k}) = \left[y_1^{1/k} y_2^{1/k} \exp\left(-\frac{\log y_1^{1/k} \log y_2^{1/k}}{\log(y_1^{1/k} y_2^{1/k})} \right) \right]^k$$

$$= y_1 y_2 \exp\left(-\frac{\log y_1 \log y_2}{\log(y_1 y_2)} \right) = D_H(y_1, y_2, \ldots, y_m)$$

To conclude this section we just add that everything said above for maxima can be easily translated for minima if the sequence $X_1, X_2, \ldots$ is changed into

$-\mathbf{X}_1, -\mathbf{X}_2, \ldots$ Then the corresponding versions of theorems 10.1, 10.2 and 10.3 are obtained.   ∎

10.4.   Concomitants of Order Statistics

On some occasions in daily practice, engineers and scientists are interested in the values of certain random variables associated with extremes of other variables. This is, for example, the case of the period of sea waves having the greatest height. It is well known that waves of height larger than certain threshold values produce damage to breakwaters which is dependent on their period. In this case, the design of the breakwater is not only a function of the maximum wave height but of the pair, maximum height-period.

In the design of snow emergency routes the engineer is interested in the snowfall intensities associated with maximum traffic intensities. A structural designer is interested in the temperatures associated with maximum snowfall or wind intensities in order to analyze some thermical effects on the structure under study. It is not likely to have high temperatures with large snow loads and this fact must be adequately introduced into the analysis. All of the above and many other cases are clear examples of the so-called concomitants of order statistics. In the following section we define them more precisely.

Definition 10.2. (Concomitants of order statistics). Let (X_i, Y_i) $(i = 1, 2, \ldots, n)$ be an i.i.d. two-dimensional sample coming from a given population. Let us consider the order statistics of the first component $X_{1:n}, X_{2:n}, \ldots, X_{n:n}$ and let us denote $Y_{[1:n]}, Y_{[2:n]}, \ldots, Y_{[n:n]}$ the associated values of the second component. Then, $Y_{[r:n]}$ is called the concomitant of the r-th order statistic of X.   ∎

In the three examples above the X values are the wave height, the traffic intensities and the snowfall or wind intensities, and the Y values are the periods, the snowfall intensities and the temperatures, respectively.

In the following paragraphs we give the exact distribution of concomitants.

Assume (X_i, Y_i) are i.i.d., have common cdf $F(x, y)$ and pdf $f(x, y)$. Hence, we have

(10.16) $f_{X_{r:n}Y_{[r:n]}}(x, y) = f(y/x) f_{r:n}(x)$

where $f(y/x)$ is the conditional pdf of $Y_{[r:n]}$ given $X_{r:n} = x$, and $f_{r:n}(x)$ is the pdf of $X_{r:n}$. Thus,

(10.17) $$f_{Y_{[r:n]}} = \int_{-\infty}^{\infty} f(y/x) f_{r:n}(x)$$

This expression can be immediately generalized to the joint pdf of k concomitants, as follows

$$f_{Y_{[r_1:n]},\ldots,Y_{[r_k:n]}}(y_1, y_2, \ldots, y_k)$$

(10.18) $$= \int_{-\infty}^{\infty} \int_{-\infty}^{x_k} \cdots \int_{-\infty}^{x_2} \prod_{i=1}^{k} f(y_i/x_i) f_{r_1,\ldots,r_k:n}(x_1,\ldots,x_k) \, dx_i \ldots dx_k$$

From (10.17) and (10.18) we get (see Yang (1977))

(10.19) $$E[Y_{[r:n]}] = E\{E[Y_1/X_1 = X_{r:n}]\}$$

(10.20) $$\mathrm{Var}[Y_{[r:n]}] = E\{\mathrm{Var}[Y_1/X_1 = X_{r:n}]\} + \mathrm{Var}\{E[Y_1/X_1 = X_{r:n}]\}$$

(10.21) $$\mathrm{Cov}[Y_{[r:n]}, Y_{[s:n]}] = \mathrm{Cov}\{E[Y_1/X_1 = X_{r:n}], E[Y_1/X_1 = X_{s:n}]\};$$

$$r \neq s$$

(10.22) $$\mathrm{Cov}[Y_{[r:n]}, X_{[s:n]}] = \mathrm{Cov}\{X_{s:n}, E[Y_1/X_1 = X_{r:n}]\}$$

The asymptotic joint distribution of concomitants of central order statistics are given by the following theorem (Yang (1977)).

Theorem 10.4. (Asymptotic joint distribution of concomitants of order statistics). *Assume that $f(x, y)$ is continuous and let $1 \leq r_1 < r_2 < \ldots < r_k \leq n$ be a sequence of integers such that*

(10.23) $$\lim_{n \to \infty} \frac{r_i}{n} = \lambda_i; \qquad 0 < \lambda_i < 1: \quad i = 1, 2, \ldots, k$$

Then

(10.24) $$\lim_{n \to \infty} P[Y_{[r_1:n]} \leq y_1, \ldots, Y_{[r_k:n]} \leq y_k] = \prod_{i=1}^{k} P[Y_i \leq y_i/X_i = \lambda_i]$$

The following theorem (see Galambos (1978)) gives a powerful tool to derive the asymptotic distribution of concomitants of the maximum of X.

Theorem 10.5. (Limit distribution of concomitants of maxima). *Let (X_i, Y_i) $(i = 1, 2, \ldots, n)$ be an i.i.d. sample from a population with absolutely continuous cdf $F(x, y)$. Let $F_1(x)$ be the marginal cdf of X such that $\omega(F_1) = \infty$, $F_1''(x)$ exists for large x, and $F_1'(x) = f_1(x) \neq 0$.*

Furthermore, let

$$(10.25) \qquad \lim_{x \to \infty} \frac{d}{dx}\left(\frac{1 - F_1(x)}{f_1(x)}\right) = 0$$

If $a_n, b_n > 0$ and A_n and $B_n > 0$ are sequences such that

$$(10.26) \qquad \lim_{n \to \infty} F_1^n(a_n + b_n x) = H_{3,0}(x)$$

and

$$(10.27) \qquad \lim_{n \to \infty} P(Y_1 < A_n + B_n u / X_1 = a_n + b_n z) = T(u, z)$$

where $T(u, z)$ is a nondegenerate distribution function, then

$$(10.28) \qquad \lim_{n \to \infty} P[Y_{[n:n]} < A_n + B_n u] = \int_{-\infty}^{\infty} T(u, z) H_{3,0}(z) \exp(-z)\, dz$$

Note that X is required to belong to a Gumbel type domain of attraction and that an important role is played by the cdf of Y conditional to given X.

Example 10.18. (*Limit distribution of periods associated with largest wave heights*).   Longuet-Higgins (1975) derives the following joint pdf for height, X, and period, Y, of sea waves.

$$f(x, y) = \frac{x^2 \exp\left(\dfrac{-x_2(1 + y^2)}{2}\right)}{\sqrt{2\pi}}$$

which has Rayleigh and normal marginals for X and Y, respectively. The X marginal is

$$F_1(x) = 1 - \exp\left\{\frac{-x^2}{a^2}\right\}; \qquad x > 0; \quad a = \sqrt{2}$$

from which

$$\lim_{x \to \infty} \frac{d}{dx}\left(\frac{1 - F_1(x)}{f_1(x)}\right) = \lim_{x \to \infty} \frac{d}{dx}\left(\frac{a^2}{2x}\right)$$

$$= \lim_{x \to \infty}\left(\frac{-a^2}{2x^2}\right) = 0$$

Thus, it satisfies all conditions in theorem 10.5. From example 3.4 we know that

$$a_n = a\sqrt{\log n}$$

$$b_n = \frac{a}{2\sqrt{\log n}}$$

The conditional pdf of Y given $X = x$ is

$$f(y/x) = \frac{x\exp\left[-\dfrac{y^2}{\left(\dfrac{2}{x^2}\right)}\right]}{\sqrt{2\pi}}$$

i.e. it is a $N(0, 1/x^2)$.

Thus expression (10.27) if A_n and B_n are taken as

$$A_n = B_n = \frac{1}{\sqrt{\log n}}$$

becomes

$$\lim_{n\to\infty} P[Y_1 < A_n + B_n u/X_1 = a_n + b_n z] = \lim_{n\to\infty} \Phi[(A_n + B_n u)(a_n + b_n z)]$$

$$= T(u, z)$$

which leads to

$$T(u) = \int_{-\infty}^{\infty} \Phi[a(1 + u)]H_{3,0}(z)\exp(-z)\,dz$$

$$= \Phi[a(1 + u)]\int_{-\infty}^{\infty} H_{3,0}(z)\exp(-z)\,dz = \Phi[a(1 + u)]$$

and then

$$\lim_{n\to\infty} P\left(Y_{[n:n]} < \frac{1 + u}{\sqrt{\log n}}\right) = \Phi[a(1 + u)]$$

which is equivalent to

$$\lim_{n\to\infty} P\left(Y_{[n:n]} < \frac{y}{\sqrt{\log n}}\right) = \Phi[ay] \qquad \blacksquare$$

Example 10.19. (*Concomitants of maxima in Gumbel's type I distribution*). Let (X, Y) be a two-dimensional random variable which follows a Gumbel's type I distribution. The marginals are unit exponentials and

(10.25) becomes

$$\lim_{x \to \infty} \frac{d}{dx}\left(\frac{1 - F_1(x)}{f_1(x)}\right) = \lim_{x \to \infty} \frac{d}{dx}[\exp(-x + x)] = 0$$

Example 3.5 shows that the x-marginal belongs to the domain of attraction of the Gumbel distribution in the right tail and that

$$a_n = \log n; \quad b_n = 1$$

The conditional cdf of Y given $X = x$ is

$$F(y/x) = \frac{\dfrac{\partial F(x, y)}{\partial x}}{f_X(x)} = \frac{\exp(-x - y + \theta xy)(\theta y - 1) + \exp(-x)}{\exp(-x)}$$

$$= 1 + \exp[y(\theta x - 1)][\theta y - 1]$$

Thus, expression (10.27) with

$$A_n = B_n = \frac{1}{\log n}$$

gives

$$\lim_{n \to \infty} P[Y_1 < A_n + B_n u / X_1 = a_n + b_n z]$$

$$= \lim_{n \to \infty} F(A_n + B_n u / X_1 = a_n + b_n z)$$

$$= \lim_{n \to \infty} \{1 + \exp\{(A_n + B_n u)[\theta(a_n + b_n z) - 1]\}[\theta(A_n + B_n u) - 1]\}$$

$$= 1 - \exp\{\theta(1 + u)\} = T(u, z)$$

and then

$$T(u) = 1 - \exp[\theta(1 + u)]$$

Consequently, we get

$$\lim_{n \to \infty} P\left[Y_{[n:n]} \le \frac{y}{\log n}\right] = 1 - \exp(\theta y)$$

i.e. the random variable $Y_{[n:n]} \log n$ converges in distribution to an exponential, $E(-\theta)$, variable. ∎

If $F(x, y)$ is bivariate normal, the reader is referred to David and Galambos (1974). For the distribution of the rank of a concomitant of an order statistic see also David et al. (1977).

Appendix A The Orthogonal Inverse Expansion Method

In this appendix we derive several of the approximating formulas for moments of order statistics by the orthogonal expansion method. They are a direct generalization of the method given by Sugiura (1962, 1964).

As the starting point for this paragraph we shall use the well known Schwarz inequality

$$(A.1) \qquad \left| \int_0^1 f(u)g(u)\,du \right| \leq \left(\int_0^1 f^2(u)\,du \int_0^1 g^2(u)\,du \right)^{1/2}$$

which becomes an equality if and only if $f(u)$ and $g(u)$ are linearly dependent, i.e. If

$$(A.2) \qquad\qquad f(u) = \lambda g(u)$$

where λ is a constant.

A particular case of application of this inequality, which we shall use frequently in this paragraph is as follows. Let $\{\psi_k(u)\}$ for $k = 0, 1, \ldots$ be an orthonormal system of functions defined on the interval $(0, 1)$, i.e. that satisfies the following relations

$$(A.3) \quad \psi_0(u) = 1; \qquad \int_0^1 \psi_k(u)\,du = 0; \qquad \int_0^1 \psi_k^2(u)\,du = 1; \qquad k = 1, 2, \ldots$$

(A.4) $\displaystyle\int_0^1 \psi_k(u)\psi_{k'}(u)\,du = 0; \qquad k \ne k'$

(A.5) $\displaystyle a_k = \int_0^1 f(u)\psi_k(u)\,du; \quad b_k = \int_0^1 g(u)\psi_k(u)\,du; \qquad k = 0, 1, \ldots$

Then, applying the Schwarz inequality to $[f(u) - \Sigma a_k\psi_k(u)]$ and $[g(u) - \Sigma b_k\psi_k(u)]$ we get

$$\left| \int_0^1 \left(f(u) - \sum_{k=0}^m a_k\psi_k(u) \right)\left(g(u) - \sum_{k=0}^m b_k\psi_k(u) \right) du \right|$$

(A.6) $\displaystyle \le \left[\int_0^1 \left(f(u) - \sum_{k=0}^m a_k\psi_k(u) \right)^2 du \int_0^1 \left(g(u) - \sum_{k=0}^m b_k\psi_k(u) \right)^2 du \right]^{1/2}$

which taking into account (A.3) to (A.5) becomes

$$\left| \int_0^1 f(u)g(u)\,du - \sum_{k=0}^m a_k b_k \right|$$

(A.7) $\displaystyle \le \left(\int_0^1 f^2(u)\,du - \sum_{k=0}^m a_k^2 \right)^{1/2}\left(\int_0^1 g^2(u)\,du - \sum_{k=0}^m b_k^2 \right)^{1/2}$

This will be our basic inequality for the approximate formulas and distribution free bounds

A.1.   Case of a General Parent

From (2.3), the α-th moment of the r-th order statistic of a sample of size n is given by

(A.8) $\displaystyle \mu_{r:n}^{(\alpha)} = \int_{-\infty}^{\infty} \frac{x^\alpha F^{r-1}(x)[1 - F(x)]^{n-r} f(x)\,dx}{B(r, n-r+1)}$

which, by means of the transformation $u = F(x)$, becomes

(A.9) $\displaystyle \mu_{r:n}^{(\alpha)} = \int_0^1 \frac{U^\alpha(u)u^{r-1}(1 - u)^{n-r}\,du}{B(r, n-r+1)}$

where $U(u)$ stands for $F^{-1}(u)$, for simplicity.

This expression shows that $\mu_{r:n}^{(\alpha)}$ is a functional which for α, r and n given, depends only on the function $U(u)$. If there exists a function $U^*(u)$ such that it maximizes (minimizes) this functional, then its associated moment becomes a distribution free upper (lower) bound.

In order to get useful bounds for the functional (A.9), functions $U(u)$ will be assumed to have associated finite moments $\mu^{(\alpha)}$ and $\mu^{(2\alpha)}$ of order α and 2α, respectively. This for $\alpha = 1$ means finite mean, μ, and variance, σ^2. In other words, its domain of definition will be the domain of all $U(u)$ admissible functions such that

$$(A.10) \qquad \int_{-\infty}^{\infty} x^\alpha \, dF(x) = \int_0^1 U^\alpha(u) \, du = \mu^{(\alpha)}$$

$$(A.11) \qquad \int_{-\infty}^{\infty} x^{2\alpha} \, dF(x) = \int_0^1 U^{2\alpha}(u) \, du = \mu^{(2\alpha)}$$

Substitution in Schwarz inequality (A.7) of

$$(A.12) \qquad f(u) = U^\alpha(u)$$

$$(A.13) \qquad g(u) = \frac{u^{r-1}(1-u)^{n-r}}{B(r, n-r+1)}$$

leads to

$$\left| \mu_{r:n}^{(\alpha)} - \sum_{k=0}^m a_k b_k \right| \le \left(\int_0^1 U^{2\alpha}(u) \, du - \sum_{k=0}^m a_k^2 \right)^{1/2} \left(\int_0^1 g^2(u) \, du - \sum_{k=0}^m b_k^2 \right)^{1/2}$$
$$(A.14)$$

and taking into account that

$$(A.15) \qquad a_0 = \int_0^1 U^\alpha(u)\psi_0(u) \, du = \int_0^1 U^\alpha(u) \, du = \mu^{(\alpha)}$$

$$(A.16) \qquad b_0 = \int_0^1 g(u)\psi_0(u) \, du = \int_0^1 g(u) \, du = 1$$

$$\int_0^1 g^2(u) \, du = \int_0^1 \frac{u^{2r-2}(1-u)^{2n-2r} \, du}{B^2(r, n-r+1)}$$

$$(A.17) \qquad = \frac{B(2r-1, 2n-2r+1)}{B^2(r, n-r+1)}$$

(A.14) becomes

$$\left| \mu_{r:n}^{(\alpha)} - \sum_{k=0}^m a_k b_k \right| \le \left(\mu^{(2\alpha)} - \sum_{k=0}^m a_k^2 \right)^{1/2} \left(\frac{B(2r-1, 2n-2r+1)}{B^2(r, n-r+1)} - \sum_{k=0}^m b_k^2 \right)^{1/2}$$
$$(A.18)$$

This expression, if $U(u)$ is known, allows us to obtain approximations for $\mu_{r:n}^{(\alpha)}$ to a desired precision by taking m large enough, because if $\{\psi_k(u)\}$ is a complete system, the right hand side of (A.18) can be made zero. Note that the two factors on the right hand side of (A.18) decrease with increasing m.

If $m = 0$, (A.18) gives

$$(A.19) \qquad \left| \mu_{r:n}^{(\alpha)} - \mu^{(\alpha)} \right| \leq [\mu^{(2\alpha)} - (\mu^{(\alpha)})^2]^{1/2} \left(\frac{B(2r - 1, 2n - 2r + 1)}{B^2(r, n - r + 1)} - 1 \right)^{1/2}$$

which is a distribution free bound, i.e. valid for any distribution.

These last two expressions are especially interesting because they give bounds for the mean and second moments for $\alpha = 1$ and $\alpha = 2$, respectively.

The bound in (A.19), according to (A.2), (A.12) and (A.13) is attained when

$$(A.20) \qquad f(u) - \sum_{k=0}^{m} a_k \psi_k(u) = \lambda \left(g(u) - \sum_{k=0}^{m} b_k \psi_k(u) \right)$$

which now becomes

$$(A.21) \qquad U^{\alpha}(u) - \mu^{(\alpha)} = \lambda \left(\frac{u^{r-1}(1-u)^{n-r}}{B(r, n - r + 1)} - 1 \right)$$

However, (A.21) leads to a distribution function only when $r = 1$ or $r = n$, because in other cases, (A.21) fails to be nondecreasing in the interval $(0, 1)$. In other words, for $r \neq 1$ or $r \neq n$ the function $U(u)$ leading to the bound in (A.19) is not the inverse of a distribution function. This implies that the bounds (A.19) are sharp (attainable) for $r = 1$ or $r = n$ and can be improved for other cases.

In the case of the difference of two order statistics we have

$$(A.22) \qquad E[X_{s:n}^{(\alpha)} - X_{r:n}^{(\alpha)}] = \int_0^1 U^{\alpha}(u) \left(\frac{u^{s-1}(1-u)^{n-s}}{B(s, n - s + 1)} - \frac{u^{r-1}(1-u)^{n-r}}{B(r, n - r + 1)} \right) du$$

and setting

$$(A.23) \qquad f(u) = U^{\alpha}(u)$$

$$(A.24) \qquad g(u) = \frac{u^{s-1}(1-u)^{n-s}}{B(s, n - s + 1)} - \frac{u^{r-1}(1-u)^{n-r}}{B(r, n - r + 1)}$$

in the Schwarz inequality we get

$$\left| E[X_{s:n}^{(\alpha)} - X_{r:n}^{(\alpha)}] - \sum_{k=0}^{m} a_k b_k \right|$$

$$(A.25) \qquad \leq \left(\int_0^1 f^2(u)\, du - \sum_{k=0}^{m} a_k^2 \right)^{1/2} \left(\int_0^1 g^2(u)\, du - \sum_{k=0}^{m} b_k^2 \right)^{1/2}$$

Now, taking into account that

(A.26) $\qquad b_0 = \int_0^1 g(u)\psi_0(u)\,du = \int_0^1 g(u)\,du = 0$

$$\int_0^1 g^2(u)\,du = \int_0^1 \frac{u^{2s-2}(1-u)^{2n-2s}}{B^2(s,\,n-s+1)}\,du + \int_0^1 \frac{u^{2r-2}(1-u)^{2n-2r}}{B^2(r,\,n-r+1)}\,du$$

$$- 2\int_0^1 \frac{u^{r+s-2}(1-u)^{2n-s-r}}{B(s,\,n-s+1)B(r,\,n-r+1)}\,du$$

$$= \frac{B(2s-1,\,2n-2s+1)}{B^2(s,\,n-s+1)} + \frac{B(2r-1,\,2n-2r+1)}{B^2(r,\,n-r+1)}$$

(A.27) $\qquad\qquad\qquad - 2\,\dfrac{B(r+s-1,\,2n-s-r+1)}{B(s,\,n-s+1)B(r,\,n-r+1)}$

we get

$$\left| E[X_{s:n}^{(\alpha)} - X_{r:n}^{(\alpha)}] - \sum_{k=0}^m a_k b_k \right|$$

$$\leq \left(\mu^{(2\alpha)} - [\mu^{(\alpha)}]^2 - \sum_{k=1}^m a_k^2 \right)^{1/2} \left(\frac{B(2s-1,\,2n-2s+1)}{B^2(s,\,n-s+1)} \right.$$

$$\left. + \frac{B(2r-1,\,2n-2r+1)}{B^2(r,\,n-r+1)} - 2\,\frac{B(r+s-1,\,2n-s-r+1)}{B(s,\,n-s+1)B(r,\,n-r+1)} - \sum_{k=1}^m b_k^2 \right)^{1/2}$$

(A.28)

which for $m = 0$ becomes

$$|E[X_{s:n}^{(\alpha)} - X_{r:n}^{(\alpha)}]|$$

$$\leq \{\mu^{(2\alpha)} - [\mu^{(\alpha)}]^2\}^{1/2} \left(\frac{B(2s-1,\,2n-2s+1)}{B^2(s,\,n-s+1)} + \frac{B(2r-1,\,2n-2r+1)}{B^2(r,\,n-r+1)} \right.$$

$$\left. - 2\,\frac{B(r+s-1,\,2n-s-r+1)}{B(s,\,n-s+1)B(r,\,n-r+1)} \right)^{1/2}$$

(A.29)

which is a distribution free bound.

The range is a special case for $s = n,\ r = 1$.

The bound (A.29) is attained when (A.20) holds, i.e.

(A.30) $\qquad U^\alpha(u) - \mu^{(\alpha)} = \lambda \left(\dfrac{u^{s-1}(1-u)^{n-s}}{B(s,\,n-s+1)} - \dfrac{u^{r-1}(1-u)^{n-r}}{B(r,\,n-r+1)} \right)$

which leads to a distribution function only for $r = 1$ and $s = n$. In this case, the distribution is symmetric.

By a similar argument (see Sugiura (1964)) we have

$$\left| E[X_{r:n}X_{s:n}] - \frac{1}{2}\sum_{\lambda,\,v=0}^{k} a_\lambda a_v (b_{\lambda,v} + b_{v,\lambda}) \right|$$

$$\leq \left(\sigma^4 - \sum_{\lambda,\,v=1}^{k} a_\lambda^2 a_v^2 \right)^{1/2} \left(\frac{B(2r-1, 2s-2r-1, 2n-2s+1)}{2B^2(r, s-r, n-s+1)} \right.$$

$$- \frac{B(2r-1, 2n-2r+1)}{2B^2(r, n-r+1)} - \frac{B(r+s-1, 2n-r-s+1)}{B(r, n-r+1)B(s, n-s+1)}$$

$$\left. - \frac{B(2s-1, 2n-2s+1)}{2B^2(s, n-s+1)} + 1 - \frac{1}{4}\sum_{\lambda,\,v=1}^{k} (b_{\lambda,v} + b_{v,\lambda}) \right)^{1/2}; \quad 1 \leq r < s \leq n$$

(A.31)

where

$$\text{(A.32)} \qquad\qquad\qquad B(p,q,r) = \frac{\Gamma(p)\Gamma(q)\Gamma(r)}{\Gamma(p+q+r)}$$

and

$$b_{\lambda,v} = \frac{1}{B(r, s-r, n-s+1)} \iint\limits_{0<u<v<1} u^{r-1}(v-u)^{s-r-1}(1-v)^{n-s}\psi_\lambda(u)\psi_v(v)\,du\,dv$$

(A.33)

A.2. Case of a Symmetric Parent

We now turn to the case of symmetric parent population, for which the above bounds can be improved.

For symmetric parents $U(u) = -U(1-u)$, the moment (A.9) can be written as

$$\text{(A.34)} \qquad\qquad \mu_{r:n}^{(\alpha)} = \int_0^1 \frac{U^\alpha(u)[h_r(u) + (-1)^\alpha h_r^*(u)]}{2}\,du$$

where

$$\text{(A.35)} \qquad\qquad h_r(u) = \frac{u^{r-1}(1-u)^{n-r}}{B(r, n-r+1)}$$

$$\text{(A.36)} \qquad\qquad h_r^*(u) = \frac{u^{n-r}(1-u)^{r-1}}{B(r, n-r+1)} = h_{n-r+1}(u)$$

Setting now

(A.37) $$f(u) = U^\alpha(u)$$

(A.38) $$g(u) = \frac{h_r(u) + (-1)^\alpha h_r^*(u)}{2}$$

in Schwarz inequality (A.7), we get

$$\left| \mu_{r:n}^{(\alpha)} - \sum_{k=0}^{m} a_k b_k \right| \leq \left(\int_0^1 U^{2\alpha}(u)\, du - \sum_{k=0}^{m} a_k^2 \right)^{1/2} \left(\int_0^1 g^2(u)\, du - \sum_{k=0}^{m} b_k^2 \right)^{1/2}$$
(A.39)

Thus, upon taking into account that

(A.40) $$a_0 = \int_0^1 U^\alpha(u)\psi_0(u)\, du = \int_0^1 U^\alpha(u)\, du = \mu^{(\alpha)}$$

(A.41) $$b_0 = \int_0^1 g(u)\psi_0(u)\, du = \int_0^1 g(u)\, du = 0$$

$$\int_0^1 g^2(u)\, du = \int_0^1 \left(\frac{h_r(u) + (-1)^\alpha h_r^*(u)}{2} \right)^2 du$$

$$= \frac{1}{4} \int_0^1 [h_r^2(u) + h_r^{*2}(u) + 2(-1)^\alpha h_r(u) + h_r^*(u)]\, du$$

(A.42) $$= \frac{\left[\dfrac{B(2r - 1, 2n - 2r + 1) + (-1)^\alpha B(n, n)}{B^2(r, n - r + 1)} \right]}{2}$$

we get from (A.11)

$$\left| \mu_{r:n}^{(\alpha)} - \sum_{k=0}^{m} a_k b_k \right|$$

$$\leq \left(\mu^{(2\alpha)} - \sum_{k=0}^{m} a_k^2 \right)^{1/2} \left(\frac{B(2r - 1, 2n - 2r + 1) + (-1)^\alpha B(n, n)}{2B^2(r, n - r + 1)} - \sum_{k=0}^{m} b_k^2 \right)^{1/2}$$
(A.43)

which is the equivalent to (A.18) for symmetric distributions.

If $m = 0$ we get

(A.44) $$|\mu_{r:n}^{(\alpha)}| \leq \{\mu^{(2\alpha)} - [\mu^{(\alpha)}]^2\}^{1/2} \left(\frac{B(2r - 1, 2n - 2r + 1) + (-1)^\alpha B(n, n)}{2B^2(r, n - r + 1)} \right)^{1/2}$$

which is a bound valid for any symmetric distribution.

The bound in (A.44) is attainable only if (A.20) holds, i.e. if

(A.45) $$U^{\alpha}(u) = \lambda[u^{r-1}(1-u)^{n-r} - u^{n-r}(1-u)^{r-1}]$$

which, as for the general case, leads to a distribution function only when $r = 1$ or $r = n$.

Finally, for the difference of two order statistics we have

(A.46) $$E[X_{s:n}^{(\alpha)} - X_{r:n}^{(\alpha)}] = \int_0^1 U^{\alpha}(u)\left(\frac{h_s(u) - h_s^*(u) - h_r(u) + h_r^*(u)}{2}\right)du$$

and we can set

(A.47) $$f(u) = U^{\alpha}(u)$$

(A.48) $$g(u) = \frac{h_s(u) - h_s^*(u) - h_r(u) + h_r^*(u)}{2}$$

in the Schwartz inequality to get

$$\left| E[X_{s:n}^{(\alpha)} - X_{r:n}^{(\alpha)}] - \sum_{k=0}^m a_k b_k \right|$$

(A.49) $$\leq \left(\int_0^1 U^2(u)\,du - \sum_{k=0}^m a_k^2\right)^{1/2}\left(\int_0^1 g^2(u)\,du - \sum_{k=0}^m b_k^2\right)^{1/2}$$

and taking into account that

(A.50) $$a_0 = \mu^{(\alpha)}$$

(A.51) $$b_0 = 0$$

$$\int_0^1 g^2(u)\,du = \frac{1}{4}\int_0^1 \{[h_s(u) - h_s^*(u)]^2 + [h_r(u) - h_r^*(u)]^2$$

$$- 2[h_s(u) - h_s^*(u)][h_r(u) - h_r^*(u)]\}\,du$$

$$= \frac{1}{4}\left(\frac{2B(2s-1, 2n-2s+1) - B(n,n)}{B^2(s, n-s+1)}\right.$$

$$+ \frac{2B(2r-1, 2n-2r+1) - B(n,n)}{B^2(r, n-r+1)}$$

(A.52) $$\left. - 2\frac{2B(s+r-1, 2n-r-s+1) - 2B(n+s-r, n-s+r)}{B(s, n-s+1)B(r, n-r+1)}\right)$$

we get

$$\left| E[X_{s:n}^{(\alpha)} - X_{r:n}^{(\alpha)}] - \sum_{k=1}^{m} a_k b_k \right|$$

$$\leq \left(\mu^{(2\alpha)} - [\mu^{(\alpha)}]^2 - \sum_{k=1}^{m} a_k^2 \right)^{1/2} \left(\frac{B(2s-1, 2n-2s+1) - B(n, n)}{2B^2(s, n-s+1)} \right.$$

$$+ \frac{B(2r-1, 2n-2r+1) - B(n, n)}{2B^2(r, n-r+1)}$$

$$\left. - \frac{B(s+r-1, 2n-r-s+1) - B(n+s-r, n-s+r)}{B(r, n-r+1)B(s, n-s+1)} - \sum_{k=1}^{m} b_k^2 \right)^{1/2}$$

(A.53)

which for $m = 0$ gives the following distribution free bound for symmetric distributions

$$|E[X_{s:n}^{(\alpha)} - X_{r:n}^{(\alpha)}]$$

$$\leq \{\mu^{(2\alpha)} - [\mu^{(\alpha)}]^2\}^{1/2} \left(\frac{B(2s-1, 2n-2s+1) - B(n, n)}{2B^2(s, n-s+1)} \right.$$

$$+ \frac{B(2r-1, 2n-2r+1) - B(n, n)}{2B^2(r, n-r+1)}$$

(A.54)
$$\left. - \frac{B(s+r-1, 2n-r-s+1) - B(n+s-r, n-s+r)}{B(r, n-r+1)B(s, n-s+1)} \right)^{1/2}.$$

Appendix B   Computer Codes

This appendix includes computer codes in order that some of the techniques described in the book may be easily applied.

The codes are in the language BASIC, and have been prepared to be run on an IBM computer. With very small modifications the codes can also be run on any other personal or large computer supporting BASIC. Translation to FORTRAN or PASCAL languages is also simple and straightforward.

We warn the user about some precision problems when running the estimation program in computers with reduced precision variables. Some warning messages have been included in the program itself to avoid errors, though a warning message does not necessarily mean an erroneous solution.

The notation and data required for the running of the programs are explained in the listings of the programs as remarks.

Finally, testing of the programs can be done by means of the data in appendix C and the examples throughout the book.

B.1. Simulation Program

```
0001 REM ***********************************************************************
0002 REM PROGRAM SIMUL
0003 REM ***********************************************************************
0004 REM FOR SIMULATING SAMPLES FROM GIVEN PARENT DISTRIBUTIONS
0005 REM ***********************************************************************
0010 DIM X(500)
0020 CLS:INPUT "GIVE NAME OF THE OUTPUT FILE ";RES$
0030 OPEN "O",#1,RES$
0040 CLS:INPUT "GIVE NAME OF THE DATA TO BE SIMULATED, IF NO MORE TYPE END
";TITLE$
0041 IF TITLE$="END" THEN CLS:PRINT "SIMULATED DATA ARE IN FILE
";RES$:CLOSE:STOP
0050 CLS:INPUT "NUMBER OF DATA TO BE SIMULATED ";N
0060 CLS:PRINT "LIST OF POSSIBLE DISTRIBUTIONS "
0065 PRINT "-----------------------------"
0066 PRINT
0070 PRINT "GUMBEL (MAXIMA) ......................... 1"
0080 PRINT "GUMBEL (MINIMA) ......................... 2"
0090 PRINT "WEIBULL (MAXIMA) ........................ 3"
0100 PRINT "WEIBULL (MINIMA) ........................ 4"
0110 PRINT "FRECHET (MAXIMA) ........................ 5"
0120 PRINT "FRECHET (MINIMA) ........................ 6"
0130 PRINT "EXPONENTIAL ............................. 7"
0140 PRINT "UNIFORM ................................. 8"
0150 PRINT "CAUCHY .................................. 9"
0160 PRINT "RAYLEIGH ................................10"
0170 PRINT "PARETO ..................................11"
0175 PRINT "F(X)=EXP(-1/(X*X))......................12"
0180 PRINT
0190 INPUT "CHOOSE NUMBER OF DISTRIBUTION TO BE SIMULATED";J
0195 GOSUB 7500
0200 GOSUB 8000
0210 FOR I=1 TO N:GOSUB 6000:NEXT I
0220 GOSUB 2000
0230 PRINT "PUSH ANY KEY TO CONTINUE ":X$=INPUT$(1)
0400 PRINT#1,TITLE$
0430 FOR I=1 TO N:PRINT#1,X(I):NEXT I
0440 PRINT #1,9.99E+20
0500 GOTO 40
1990 REM ***********************************************************************
2000 REM SUBROUTINE TO PRINT DATA
2001 REM ***********************************************************************
2205 L=0:CLS
2206 FOR I=1 TO 4:PRINT "   I         X(I)      ";: NEXT I : PRINT
2207 FOR I=1 TO 5:PRINT "----------------";: NEXT I : PRINT:PRINT
2210 FOR I=1 TO N
2220 PRINT USING "###   ######.###      ";I,X(I);:L=L+1:IF L=4 THEN L=0:PRINT
2230 NEXT I
2240 PRINT :FOR I=1 TO 5:PRINT "----------------";: NEXT I : PRINT
2250 RETURN
5990 REM ***********************************************************************
6000 REM SUBROUTINE FOR TRANSFORMING UNIFORM TO SIMULATED DISTRIBUTION
6001 REM ***********************************************************************
6010 ON J GOTO 6100,6200,6300,6400,6500,6600,6700,6800,6900,7000,7100,7105
6100 X(I)=L-D*LOG(-LOG(X(I))):GOTO 7200
6200 X(I)=L+D*LOG(-LOG(1-X(I))):GOTO 7200
6300 X(I)=L-D*(-LOG(X(I)))^(1/B):GOTO 7200
6400 X(I)=L+D*(-LOG(1-X(I)))^(1/B):GOTO 7200
6500 X(I)=L+D*(-LOG(X(I)))^(-1/B):GOTO 7200
6600 X(I)=L-D*(-LOG(1-X(I)))^(-1/B):GOTO 7200
6700 X(I)=-LOG(1-X(I)):GOTO 7200
6800 X(I)=A+(B-A)*X(I):GOTO 7200
6900 X(I)=TAN(3.1416*(X(I)-.5)):GOTO 7200
7000 X(I)=SQR(-A*A*LOG(1-X(I))):GOTO 7200
```

```
7100 X(I)=(1-X(I))^(-1/B):GOTO 7200
7105 X(I)=SQR(-1/LOG(X(I))):GOTO 7200
7200 RETURN
7500 ON J GOTO 7510,7510,7520,7520,7530,7530,7540,7550,7700,7560,7570,7700
7510 INPUT "GIVE THE GUMBEL PARAMETERS LAMBDA AND DELTA ";L,D:GOTO 7700
7520 INPUT "GIVE THE WEIBULL PARAMETERS LAMBDA, DELTA AND BETA ";L,D,B:GOTO
7700
7530 INPUT "GIVE THE FRECHET PARAMETERS LAMBDA, DELTA AND BETA ";L,D,B:GOTO
7700
7540 INPUT "GIVE THE EXPONENTIAL PARAMETER LAMBDA ";L :GOTO 7700
7550 INPUT "GIVE THE UNIFORM PARAMETERS A AND B ";A,B:GOTO 7700
7560 INPUT "GIVE THE RAYLEIGH PARAMETER A ";A:GOTO 7700
7570 INPUT "GIVE THE PARETO PARAMETER BETA ";B:GOTO 7700
7700 RETURN
7999 STOP
8000 H=1
8010 FOR I=N TO 1 STEP -1
8020 V=RND
8030 H=H*V^(1/I)
8040 X(I)=H
8050 NEXT I
8060 RETURN
```

B.2. Drawing Program

```
0000 REM ********************************************************************
0001 REM PROGRAM DRAW
0002 REM ********************************************************************
0003 REM FOR DRAWING SAMPLES ON GUMBEL, WEIBULL OR FRECHET PROBABILITY PAPER
0004 REM ********************************************************************
0010 DIM X(500),Y(500),Z(30)
0011 CLS
0012 INPUT "GIVE NAME OF THE INPUT DATA FILE ";FILE$
0014 CLS
0015 CLOSE
0020 OPEN "I",#1,FILE$
0030 INPUT "GIVE NUMBER OF THE GROUP OF DATA TO BE USED, IF END GIVE 0 ",K:IF
K=0 THEN STOP
0031 FOR J=1 TO K-1:GOSUB 5500:NEXT J
0034 CLS
0035 GOSUB 5500
0038 INPUT "IF FITTING THE LEFT TAIL (MINIMA) 1, IF FITTING THE RIGHT TAIL
(MAXIMA) 2 ";ITAIL
0046 ON ITAIL GOTO 48,47
0047 PRINT "RIGHT TAIL FIT (MAXIMA)":GOTO 49
0048 PRINT "LEFT TAIL FIT (MINIMA)"
0049 PRINT
0063 PRINT "PROBABILISTIC PAPERS"
0064 PRINT "----------------------" :PRINT
0065 PRINT "1 ............. GUMBEL"
0066 PRINT "2 ............. WEIBULL"
0067 PRINT "3 ............. FRECHET":PRINT
0068 INPUT "TYPE NUMBER OF THE SELECTED PAPER ";TP
0069 CLS
0073 PRINT "PLOTTING POSITION FORMULAS"
0074 PRINT "--------------------------":PRINT
0075 PRINT "1 .................. P(I)=I/(N+1)"
0076 PRINT "2 .................P(I)=(I-0.5)/N"
0077 PRINT "3 .................P(I)=(I-3/8)/(N+.25)"
0078 PRINT "4 .................P(I)=(I-.44)/(N+.12)":PRINT
0079 INPUT "GIVE NUMBER OF THE SELECTED PLOTTING POSITION FORMULA.......",NP
0080 GOSUB 5000
0090 FOR I=1 TO N
```

```
0100 ON NP GOTO 110,120,130,140
0110 Y(I)=I/(N+1):GOTO 150
0120 Y(I)=(I-.5)/N:GOTO 150
0130 Y(I)=(I-3/8)/(N+.25):GOTO 150
0140 Y(I)=(I-.44)/(N+.12):GOTO 150
0150 NEXT I
0154 CLS
0155 FOR I=1 TO 3:PRINT          " I    P(I)      X(I) ";:NEXT I:PRINT
0156 PRINT  "------------------------------------------------------------"
0157 L=0:FOR I=1 TO N
0158 PRINT USING "###  #.###  ######.## ";I,Y(I),X(I);:L=L+1:IF L=3 THEN
L=0:PRINT
0159 NEXT I
0185 PRINT  "------------------------------------------------------------"
0200 INPUT"PUSH ANY KEY TO CONTINUE ",AA
0240 XMAX=X(N)
0250 XMIN=X(1)
0260 YMAX=Y(N)
0270 YMIN=Y(1)
0272 IF YMAX<.995 THEN YMAX=.995
0273 IF YMIN>.005 THEN YMIN=.005
0275 GOSUB 8500
0278 REM PRINT "XMAX,XMIN,YMAX,YMIN";XMAX,XMIN,YMAX,YMIN
0279 REM INPUT "PUSH ANY KEY TO CONTINUE ",AA:IF AA<>0 THEN STOP
0280 FOR I=1 TO N
0282 Q=Y(I):R=X(I):GOSUB 8000:GOSUB 8230:Y(I)=Q:X(I)=R
0283 NEXT I
0284 Q=YMAX:R=XMAX:GOSUB 8000:GOSUB 8230:YMAX=Q:XMAX=R
0285 Q=YMIN:R=XMIN:GOSUB 8000:GOSUB 8230:YMIN=Q:XMIN=R
0288 XLEN=70
0290 YLEN=20
0300 XD=XLEN/(XMAX-XMIN)
0310 YD=YLEN/(YMAX-YMIN)
0315 CLS
0318 LOCATE 3,10:PRINT TITLE$
0320 FOR I=1 TO N
0330 X(I)=(X(I)-XMIN)*XD+10
0340 Y(I)=(YMAX-Y(I))*YD+1
0350 LOCATE Y(I),X(I):PRINT "*"
0360 NEXT I
0365 RESTORE
0370 FOR I=1 TO 17
0380 READ T:T1=T
0392 Q=T:GOSUB 8000:T=Q
0400 T=(YMAX-T)*YD+1
0404 LOCATE T,1:PRINT T1
0410 NEXT I
0420 LOCATE 22,1
0440 GOTO 15
5000 M=0
5010 NT=N
5020 FOR I=2 TO NT
5030 IF X(I)>=X(I-1) GOTO 5090
5040 A=X(I)
5050 X(I)=X(I-1)
5060 X(I-1)=A
5070 M=1
5080 NT=I
5090 NEXT I
5100 IF M=1 GOTO 5000
5110 RETURN
5490 REM ***********************************************************************
5500 REM SUBROUTINE FOR READING DATA
5505 REM ***********************************************************************
5510 INPUT#1,TITLE$
5520 I=1
5530 INPUT#1,X(I)
5540 IF X(I)=9.99E+20 THEN 5580
```

```
5550 I=I+1
5560 GOTO 5530
5580 N=I-1
5590 RETURN
7990 REM ***********************************************************************
8000 REM SUBROUTINE FOR CHANGING COORDINATES (GUMBEL, WEIBULL OR FRECHET)
8010 REM ***********************************************************************
8100 ON ITAIL GOTO 8220,8210
8210 Q=-LOG(-LOG(Q)):RETURN
8220 Q=-LOG(-LOG(1-Q)):RETURN
8230 ON ITAIL GOTO 8270,8232
8232 ON TP GOTO 8310,8240,8250
8240 R=-LOG(L-R):GOTO 8310
8250 R=LOG(R-L):GOTO 8310
8270 ON TP GOTO 8310,8290,8280
8280 R=-LOG(L-R):GOTO 8310
8290 R-LOG(R-L)
8310 RETURN
8500 CLS
8501 ON TP GOTO 8800,8510,8600
8510 ON ITAIL GOTO 8530,8520
8520 PRINT "THRESHOLD PARAMETER (MINIMUM VALUE =";XMAX;")";:INPUT L:IF L<XMAX
THEN 8520
8525 GOTO 8800
8530 PRINT "THRESHOLD PARAMETER (MAXIMUM VALUE =";XMIN;")";:INPUT L:IF L>XMIN
THEN 8530
8535 GOTO 8800
8600 ON ITAIL GOTO 8520,8530
8800 RETURN
9000 DATA
0.005,0.01,0.02,0.05,0.1,0.2,0.3,0.4,0.5,0.6,0.7,0.8,0.9,0.95,0.98,0.99,0.995
```

B.3. Estimation Program

```
0000 REM ***********************************************************************
0001 REM * PROGRAM ESTIM
0002 REM ***********************************************************************
0010 REM ***********************************************************************
0020 REM   PROGRAM FOR ESTIMATING THE PARAMETERS OF GUMBEL,WEIBULL OR FRECHET
0030 REM ***********************************************************************
0040 DEFDBL A-H,O-Z
0050 OPTION BASE 1
0051 CLS:INPUT "GIVE NAME OF DATA FILE ";FILE$
0053 INPUT "GIVE NAME OF OUTPUT FILE ";RES$
0054 OPEN "O",#1,RES$
0070 DIM XS(500),W(500),P(500),DX(5),XO(5),COV(3),COVIN(3),X(3),Y(5)
0080 DIM T$(3),M$(7),P$(3)
0090 DIM EPS(5),Y8(5),DER1(5),DEX(5),XOO(5),X1(5),X2(5),X3(5)
0110 REM ***********************************************************************
0120 REM *                              NOTATION
0130 REM ***********************************************************************
0140 REM X( )    = VALUES OF THE PARAMETERS TO BE ESTIMATED
0150 REM XS( )   = ORDER STATISTICS
0160 REM P( )    = PLOTTING POSITIONS (CDF VALUES)
0170 REM W( )    = WEIGHTS
0180 REM NV      = NUMBER OF PARAMETERS
0190 REM IN      = TYPE OF DISTRIBUTION TO BE FITTED
0200 REM                 IN = 1 ..........GUMBEL
0210 REM                 IN = 2 ..........WEIBULL
0220 REM                 IN = 3 ..........FRECHET
0230 REM IP      = CODE OF PLOTTING POSITION FORMULA TO BE USED
0240 REM                 IP = 1 .......... P(I)=I/(N+1)
0250 REM                 IP = 2 .......... P(I)=(I-0.375)/(N+0.25)
0260 REM                 IP = 3 .......... P(I)=(I-0.5)/N
```

```
0270 REM N      = SAMPLE SIZE
0280 REM K1     = LOWER ORDER STATISTIC TO BE INCLUDED IN THE PROCESS
0290 REM K2     = UPPER ORDER STATISTIC TO BE INCLUDED IN THE PROCESS
0300 REM IT     = METHOD TO BE USED IN THE FITTING PROCEDURE
0310 REM           IT = 1 ........ LEAST-SQUARES PROBABILITY ABSOLUTE ERROR
0320 REM           IT = 2 ........ LEAST-SQUARES RETURN PERIOD RELATIVE ERROR
0330 REM                           OR TAIL EQUIVALENCE
0340 REM           IT = 3 ........ STANDARD WEIGHTED LEAST-SQUARES METHOD
0350 REM           IT = 4 ........ GIVEN WEIGHTS
0360 REM           IT = 5 ........ MAXIMUM LIKELIHOOD
0370 REM           IT = 6 ........ PERCENTILE METHOD
0380 REM           IT = 7 ........ METHOD OF MOMENTS (GUMBEL)
0390 REM ITAIL  = 1 ......... LEFT TAIL (DISTRIBUTION FOR MINIMA IS FITTED)
0400 REM ITAIL  = 2 ......... RIGHT TAIL (DISTRIBUTION FOR MAXIMA IS FITTED)
0410 REM AL     = THRESHOLD PARAMETER OF WEIBULL OR FRECHET LAW
0420 REM **************************************************************************
0430 REM *           DATA IS READ
0440 REM **************************************************************************
0450 REM THE PROGRAM ASSUMES THE EXISTENCE OF A FILE NAMED "DATA" WHERE ONE
0460 REM OR MORE DATA SETS WITH THE FOLLOWING INFORMATION APPEARS IN EACH
0470 REM      (A)     TITLE
0480 REM      (B)     SAMPLE SIZE
0490 REM      (C)     ITAIL
0500 REM      (D)     SAMPLE (IF DATA UNKNOWN, FICTITIUOUS VALUES MUST BE ADDED)
0501 REM **************************************************************************
0510 CLOSE#8:OPEN "I",#8,FILE$:CLS:INPUT "NUMBER OF SET OF DATA TO BE USED, IF
NO MORE 0 ";NSK :IF NSK=0 THEN PRINT "OUTPUT IS IN FILE ";RES$:CLOSE : STOP
0520 NV=2 : AL=0 : IP=3 : CLS
0530 T$(1)="GUMBEL TYPE"
0540 T$(2)="WEIBULL TYPE"
0550 T$(3)="FRECHET TYPE"
0560 FOR I=1 TO 3 : PRINT I;".......";T$(I) : NEXT I : PRINT
0570 INPUT "GIVE TYPE OF DISTRIBUTION TO FIT ACCORDING THE TABLE ABOVE ";IN
0580 M$(1)="LEAST-SQUARES PROBABILITY ABSOLUTE ERROR METHOD"
0590 M$(2)="LEAST-SQUARES RETURN PERIOD RELATIVE ERROR METHOD"
0600 M$(3)="STANDARD WEIGHTED LEAST-SQUARES METHOD"
0610 M$(4)="LEAST-SQUARES ERROR WITH GIVEN WEIGHTS METHOD"
0620 M$(5)="MAXIMUM LIKELIHOOD METHOD"
0630 M$(6)="PERCENTILE METHOD"
0640 M$(7)="METHOD OF MOMENTS (GUMBEL)"
0650 CLS : FOR I=1 TO 7 : PRINT I;".......";M$(I) : NEXT I : PRINT
0660 INPUT "GIVE ESTIMATION METHOD TO BE USED ACCORDING TABLE ABOVE ";IT
0670 P$(1)="P(I)=I/(N+1)"
0680 P$(2)="P(I)=(I-0.375)/(N+0.25)"
0690 P$(3)="P(I)=(I-0.5)/N"
0700 IF IT>4 GOTO 750
0710 CLS : FOR I=1 TO 3 : PRINT I;".............";P$(I) : NEXT I : PRINT
0720 INPUT "GIVE PLOTTING POSITION FORMULA ACCORDING THE TABLE ABOVE ";IP
0730 IF IT=4 THEN FOR I=1 TO N : PRINT "GIVE WEIGHT ";I:INPUT W(I):NEXT I
0750 FOR J=1 TO NSK:GOSUB 4250:NEXT J
0760 INPUT "FIT LEFT TAIL (MINIMA) 1, FIT RIGHT TAIL (MAXIMA) 2 ";ITAIL
0810 PRINT#1," ":PRINT#1,"TITLE :";TITLE$;"      SAMPLE SIZE=";N : PRINT#1," "
: PRINT#1," "
0820 IF ITAIL=1 THEN PRINT#1, T$(IN);" DISTRIBUTION FOR MINIMA IS FITTED"
0830 IF ITAIL=2 THEN PRINT#1, T$(IN);"DISTRIBUTION FOR MAXIMA IS FITTED"
0840 PRINT#1, "BY THE ";M$(IT)
0850 IF IT<5 THEN PRINT#1, "AND THE PLOTTING FORMULA ";P$(IP)
0860 K1=1 : K2=N : IF IT=7 GOTO 890
0870 INPUT "FIRST AND LAST ORDER STATISTICS TO BE USED ";K1,K2
0871 IF K1>=1 AND K1<K2-1 AND K2<=N AND N<=500 GOTO 880
0872 PRINT "ERROR, CONSTANTS K1,K2 AND N MUST BE IN ASCENDING ORDER"
0873 PRINT "WITH K2 LARGER THAN K1+1 AND N SMALLER THAN 200"
0874 GOTO 870
0880 PRINT#1, "ORDER STATISTICS FROM ";K1;" TO ";K2;" ARE USED"
0890 IF IN>1 THEN INPUT "THRESHOLD VALUE ";AL : PRINT#1, "THRESHOLD VALUE
=";AL
1000 REM **************************************************************************
1010 REM    SAMPLE IS REARRANGED IN INCREASING ORDER OF MAGNITUDE
```

```
1020 REM *********************************************************************
1030 GOSUB 4100
1040 GOSUB 5500
1050 GOSUB 5000
1090 REM *********************************************************************
1100 REM      IF RIGHT TAIL, DATA AND THRESHOLD PARAMETER ARE CHANGED SIGN
1110 REM *********************************************************************
1120 IF ITAIL=1 GOTO 1200
1130 FOR I=1 TO N :P(I)=-XS(I) : NEXT I
1140 FOR I=1 TO N : XS(I)=P(N-I+1) : NEXT I
1150 K3=K1 : K1=N-K2+1 : K2=N-K3+1
1160 AL=-AL
1170 REM *********************************************************************
1180 REM     TRANSFORM WEIBULL AND FRECHET TO GUMBEL POPULATIONS
1190 REM *********************************************************************
1200 ON IN GOTO 1320,1220,1260
1210 PRINT "INVALID VALUE OF IN=";IN : STOP
1220 FOR I=K1 TO K2
1230 IF AL>XS(I) THEN PRINT " ERROR THRESHOLD VALUE NOT CORRECT=";AL : STOP
1240 XS(I)=LOG(XS(I)-AL) : NEXT I
1250 GOTO 1320
1260 FOR I=K1 TO K2
1270 IF AL<=XS(I) THEN PRINT "ERROR THRESHOLD VALUE (FRECHET) NOT
CORRECT=";AL:STOP
1280 XS(I)=-LOG(AL-XS(I)) : NEXT I
1290 REM *********************************************************************
1300 REM                NORMALIZATION OF GUMBEL DATA
1310 REM *********************************************************************
1320 AME=0
1330 FOR I=K1 TO K2 : AME=AME+XS(I) : NEXT I
1340 AME=AME/(K2-K1+1)
1350 AS2=0
1360 FOR I=K1 TO K2 : AS2=AS2+(XS(I)-AME)^2 : NEXT I
1370 AS2=AS2/(K2-K1+1)
1380 AS2=SQR(AS2)
1390 FOR I=K1 TO K2 : XS(I)=(XS(I)-AME)/AS2 : NEXT I
1400 REM PRINT "NORMALIZED DATA" : FOR I=K1 TO K2 : PRINT I,XS(I) : NEXT I
1410 IF IT<>7 GOTO 1480
1420 XO(2)=SQR(6!)/3.1416
1430 XO(1)=.5772*XO(2)
1440 GOTO 2080
1450 REM *********************************************************************
1460 REM                INITIAL ESTIMATION
1470 REM *********************************************************************
1480 GOSUB 3670
1490 IF IT=6 GOTO 2080
1500 REM *********************************************************************
1510 REM      PLOTTING POSITIONS ARE CALCULATED
1520 REM *********************************************************************
1530 FOR I=K1 TO K2
1540 ON IP GOTO 1560, 1570, 1580
1550 PRINT " ERROR PLOTTING FORMULA NOT DEFINED" : STOP
1560 P(I)=I/(N+1) : GOTO 1590
1570 P(I)=(I-.375)/(N+.25) : GOTO 1590
1580 P(I)=(I-.5)/N
1590 NEXT I
1600 REM *********************************************************************
1610 REM                WEIGHTS ARE CALCULATED
1620 REM *********************************************************************
1630 FOR I=K1 TO K2
1640 ON IT GOTO 1660,1670,1680,1690,1690
1650 PRINT " ERROR WEIGHTING FUNCTION NOT DEFINED" : STOP
1660 W(I)=1 : GOTO 1690
1670 W(I)=1/(P(I)*P(I)) : GOTO 1690
1680 W(I)=1/(P(I)*(1-P(I))) : GOTO 1690
1690 NEXT I
1700 REM *********************************************************************
1710 REM                     ESTIMATION
```

```
1720 REM ***********************************************************************
1730 GOSUB 2440
1740 REM ***********************************************************************
1750 REM          IF MAXIMUM LIKELIHOOD COVARIANCE MATRIX IS CALCULATED
1760 REM ***********************************************************************
1770 IF IT<>5 GOTO 2080
1780 EP=0 : FOR I=1 TO NV : EP=EP+DX(I) : NEXT I : EP=EP/(2*NV) : PRINT
"EP=";EP
1790 K=0
1800 FOR I=1 TO NV
1810 FOR J=1 TO I
1820 FOR K91=1 TO NV : Y(K91)=X0(K91) : NEXT K91
1830 Y(I)=Y(I)+EP
1840 Y(J)=Y(J)+EP : U1=-F
1850 GOSUB 3440
1860 Y(J)=Y(J)-2*EP : GOSUB 3440 : U5=F : Y(I)=Y(I)-2*EP : GOSUB 3440 : U3=-F
1870 Y(J)=Y(J)+2*EP : GOSUB 3440 : U4=F : K=K+1
1880 COV(K)=(U1+U2+U3+U4)/(4*EP*EP)
1890 NEXT J
1900 NEXT I
1910 K91=1
1920 FOR I=1 TO NV
1930 FOR J=K91 TO K91+I-1 : PRINT COV(J); : NEXT J : PRINT
1940 K91=K91+I
1950 NEXT I
1960 IDGT=4
1970 D1=0 : D2=0
1980 GOSUB 3860
1990 K91=1
2000 FOR I=1 TO NV
2010 FOR J=K91 TO K91+I-1 : PRINT COVIN(J); : NEXT J : PRINT
2020 K91=K91+I
2030 NEXT I
2040 REM ***********************************************************************
2050 REM          CORRECTION FOR NORMALIZATION
2060 REM ***********************************************************************
2070 NV1=NV*(NV+1)/2
2080 X0(1)=AME+AS2*X0(1)
2090 X0(2)=AS2*X0(2)
2100 IF IT=5 THEN FOR I=1 TO NV1 : COVIN(I)=COVIN(I)*AS2*AS2 : NEXT I
2110 REM ***********************************************************************
2120 REM          CORRECTION FOR RIGHT TAIL
2130 REM ***********************************************************************
2140 IF ITAIL=2 THEN X0(1)=-X0(1)
2150 REM ***********************************************************************
2160 REM          CORRECTION FOR TRANSFORMATION FROM WEIBULL OR FRECHET
2170 REM ***********************************************************************
2180 ON IN GOTO 2230,2200,2220
2190 PRINT "ERROR INVALID IN=";IN : STOP
2200 X0(1)=EXP(X0(1)) : X0(2)=1/X0(2)
2210 GOTO 2230
2220 X0(1)=EXP(X0(1)) : X0(2)=1/X0(2)
2230 IF IN=1 GOTO 2300
2240 IF IT<>5 GOTO 2300
2250 IF IN=2 THEN A1=X0(1) : A2=-X0(2)*X0(2)
2260 IF IN=3 THEN A1=-X0(1) : A2=-X0(2)*X0(2)
2270 COVIN(1)=COVIN(1)*A1*A1
2280 COVIN(2)=COVIN(2)*A1*A2
2290 COVIN(3)=COVIN(3)*A2*A2
2300 PRINT#1," "
2310 IF IN=1 THEN PRINT#1, "PARAMETERS : LOCATION=";X0(1);" SCALE=";X0(2)
2320 IF IN>1 THEN PRINT#1, "PARAMETERS : SCALE=";X0(1);" SHAPE=";X0(2)
2330 IF IT<>5 GOTO 2370
2340 PRINT#1," " : PRINT#1, "COVARIANCE MATRIX" : PRINT#1," " : K91=1 : FOR
I=1 TO NV
2350 FOR J=K91 TO K91+I-1 : PRINT#1, COVIN(J); : NEXT J : PRINT#1," "
2360 K91=K91+I : NEXT I : PRINT#1," "
2370 PRINT#1,"-----------------------------------------------------------"
```

```
2380 GOTO 510
2390 REM ******************************************************************
2400 REM ********************* END OF MAIN PROGRAM    *********************
2410 REM ******************************************************************
2420 REM ******************** SUBROUTINE AMAX STARTS *********************
2430 REM ******************************************************************
2440 REM    SUBROUTINE AMAX
2450 REM ******************************************************************
2460 REM                               NOTATION
2470 REM ******************************************************************
2480 REM NV      = NUMBER OF INDEPENDENT VARIABLES IN FUNCTION TO MAXIMIZE
2490 REM RER     = RELATIVE ERROR USED IN SUBROUTINE
2500 REM AER     = ABSOLUTE ERROR USED IN SUBROUTINE
2510 REM X00(I)  = INITIAL VALUES OF THE INDEPENDENT VARIABLES
2520 REM FACT    = MODULUS OF INCREMENT VECTOR IN THE GRADIENT DIRECTION
2530 REM FANT    = PREVIOUS VALUE OF THE FUNCTION (PREVIOUS ITERATION)
2540 REM DX(I)   = INCREMENTS OF THE INDEPENDENT VARIABLES IN ORDER TO GET
2550 REM            THE DESIRED EXACT DIGIT IN THE DERIVATIVES
2560 REM DEX(I)  = INCREMENTS OF INDEPENDENT VARIABLES IN MAXIMIZATION
2570 REM NDIG1   = LOWER BOUND NUMBER OF DIGIT PRECISION (NORMALLY NDIG1=6)
2580 REM NDIG2   = UPPER BOUND NUMBER OF DIGIT PRECISION (NORMALLY NDIG2=4)
2590 REM NM      = NUMBER OF ITERATIONS IN THE PRECISION LOOP
2600 REM NM1     = NUMBER OF ITERATIONS IN THE MAXIMIZATION LOOP
2610 REM ******************************************************************
2620 ITERM=100
2630 ITERAM=100
2640 FOR I=1 TO NV:DX(I)=.0001:NEXT I
2650 NDIG1=5:NDIG2=4:EPS1=10^(-NDIG1):EPS2=10^(-NDIG2):RER=1E-09
2660 EDIS=1E-08:AER=0:EGR=.01:FACT=1:FANT=0#
2670 FOR I=1 TO NV:Y(I)=X0(I):NEXT I
2680 GOSUB 3440 :F0=F
2690 REM ******************************************************************
2700 REM *   CALCULUS OF FIRST DERIVATIVES
2710 REM * INCREMENTS OF VARIABLES ARE CALCULATED TO GET DESIRED PRECISION
2720 REM ******************************************************************
2730 ITERA=0
2740 FOR I=1 TO NV
2750 FOR J=1 TO NV:Y(J)=X0(J):NEXT J
2760 NM=0:NM1=0
2770 Y(I)=X0(I)+DX(I)
2780 GOSUB 3440:F1=F:Y(I)=X0(I)-DX(I):GOSUB 3440:F2=F:Y(I)=X0(I)
2790 R=ABS((F1-F2)/F0)
2800 IF R>EPS2 GOTO 2870
2810 IF R>EPS1 GOTO 2970
2820 IF MN<=30 GOTO 2840
2830 PRINT "CONSTANT FUNCTION OR INCORRECT EPS1,EPS2=";EPS1,EPS2:STOP
2840 NM=NM+1
2850 DX(I)=DX(I)*10
2860 GOTO 2770
2870 DX0=0
2880 DX2=(DX(I)+DX0)*.5
2890 IF NM1>100 THEN PRINT "ILL CONDITIONED FUNCTION":STOP
2900 NM1=NM1+1:Y(I)=X0(I)+DX2:GOSUB 3440:F1=F:Y(I)=X0(I)-DX2:GOSUB 3440:F2=F
2910 Y(I)=X0(I):R=ABS((F1-F2)/F0)
2920 IF R<=EPS1 GOTO 2950
2930 IF R<EPS2 GOTO 2960
2940 DX(I)=DX2:GOTO 2880
2950 DX0=DX2:GOTO 2880
2960 DX(I)=DX2
2970 DER1(I)=(F1-F2)/(2*DX(I))
2980 REM
2990 NEXT I
3000 F7=F0:ITER=0:INH=0:AM=0
3010 FOR I=1 TO NV:AM=AM+DER1(I)^2:NEXT I
3020 AMM=SQR(AM):AM=FACT/AMM
3030 FOR I=1 TO NV:DEX(I)=DER1(I)*AM:NEXT I
3040 FOR I=1 TO NV:Y(I)=X0(I)+DEX(I):NEXT I
3050 GOSUB 3440:F1=F
```

```
3060 IF ITER>ITERM-5 THEN PRINT "F0,F1,DEX(1),DEX(2)=";F0,F1,DEX(1),DEX(2)
3070 ITER=ITER+1
3080 IF ITER<ITERM GOTO 3100
3090 PRINT "ERROR LIMIT NUMBER OF ITERATIONS=";ITER," AMM=";AMM:GOTO 3200
3100 IF F1<F7 GOTO 3150
3110 FOR I=1 TO NV:DEX(I)=DEX(I)*2:NEXT I
3120 INH=1
3130 FOR I=1 TO NV:X1(I)=Y(I):NEXT I
3140 FACT=FACT*2:F7=F1:GOTO 3040
3150 IF INH=1 GOTO 3220
3160 FOR I=1 TO NV:DEX(I)=DEX(I)*.5:NEXT I
3170 FACT=FACT*.5
3180 IF FACT>=AER GOTO 3040
3190 PRINT "FACT,AER=";FACT,AER
3200 IF AMM>=EGR THEN PRINT#1, "WARNING : GRADIENT=";AMM;" LARGER THAN ";EGR
3210 RETURN
3220 FOR I=1 TO NV:Y8(I)=Y(I):NEXT I
3230 FOR I=1 TO NV:AIN=Y8(I)-X0(I):X2(I)=X0(I)+.4*AIN:X3(I)=X0(I)+.6*AIN
3240 NEXT I
3250 FOR I=1 TO NV:Y(I)=X2(I):NEXT I:GOSUB 3440:F2=F
3260 FOR I=1 TO NV:Y(I)=X3(I):NEXT I:GOSUB 3440:F3=F
3270 IF F2<=F3 THEN FOR I=1 TO NV:X0(I)=X2(I):NEXT I:F0=F2:GOTO 3290
3280 FOR I=1 TO NV:Y8(I)=X3(I):NEXT I:F1=F3
3290 DIST=0
3300 FOR I=1 TO NV:DIST=DIST+ABS(X0(I)-Y8(I)):NEXT I
3310 IF DIST>EDIS GOTO 3220
3320 IF ABS((FANT-F0)/F0)<=RER GOTO 3200
3330 FANT=F0:ITERA=ITERA+1
3340 IF ITERA=1 THEN CLS
3341 LOCATE 1,17:PRINT "ITERATIONS=";ITERA
3345 PRINT " GRADIENT=";AMM;" F=";F0
3350 IF ITERA<ITERAM GOTO 3380
3360 PRINT "LIMIT NUMBER OF ITERATIONS=";ITERA," AMM=";AMM:GOTO 3200
3370 REM
3380 FACT=FACT*.5:GOTO 2740
3390 REM ********************************************************************
3400 REM *   END OF SUBROUTINE AMAX *****************************************
3410 REM *
3420 REM *   STARTS SUBROUTINE FU  *****************************************
3430 REM ********************************************************************
3435 REM *   SUBROUTINE FU
3436 REM ********************************************************************
3440 IF Y(2)<=0 GOTO 3600
3450 F=0
3460 ON IT GOTO 3470,3470,3470,3470,3520
3470 FOR III=K1 TO K2:F19=EXP((XS(III)-Y(1))/Y(2)):F19=EXP(-F19)
3480 F18=1#:P1=F18-F19
3490 F=F+(P(III)-P1)^2*W(III):NEXT III
3500 F=-F
3510 RETURN
3520 XA=XS(K1):XAA=XS(K2)
3530 FOR III=K1 TO K2:F19=(Y(1)-XS(III))/Y(2):F=F+F19+EXP(-F19):NEXT III
3540 F18=(XA-Y(1))/Y(2):F18=EXP(F18):F18=EXP(-F18):F17=1#
3550 F18=F17-F18
3560 IF F18=0 GOTO 3600
3570 F16=Y(2):F16=LOG(F16):F17=(XAA-Y(1))/Y(2):F17=EXP(F17)
3580 F=-F+(K1-K2-1)*F16-(N-K2)*F17
3590 F=F+(K1-1)*LOG(F18):GOTO 3510
3600 F=-1.7E+30:GOTO 3510
3610 REM ********************************************************************
3620 REM * END OF SUBROUTINE FU   *****************************************
3630 REM ********************************************************************
3640 REM ********************************************************************
3650 REM *   STARTS SUBROUTINE ESTIM   ************************************
3660 REM ********************************************************************
3665 REM SUBROUTINE ESTIM
3668 REM ********************************************************************
3670 FOR I=1 TO 2
```

```
3680 ON I GOTO 3690,3700
3690 NS1=K1:NS2=INT((K1+K2)/2):GOTO 3710
3700 NS1=INT((K1+K2)/2+1):NS2=K2
3710 S1=0:S2=0
3720 FOR K=NS1 TO NS2:P9=1-(K-.5)/N:S1=S1+LOG(-LOG(P9)):S2=S2+XS(K)
3730 NEXT K
3740 ON I GOTO 3750,3760
3750 A11=NS2-NS1+1:A12=S1:C1=S2:GOTO 3770
3760 A21=NS2-NS1+1:A22=S1:C2=S2
3770 NEXT I
3780 DE=A11*A22-A21*A12:X0(1)=(C1*A22-C2*A12)/DE:X0(2)=(A11*C2-A21*C1)/DE
3790 RETURN
3800 REM *********************************************************************
3810 REM *   END OF SUBROUTINE ESTIM   ** ***********************************
3820 REM *********************************************************************
3830 REM *********************************************************************
3840 REM *   STARTS SUBROUTINE INVERSE    ** ********************************
3850 REM *********************************************************************
3860 REM SUBROUTINE INVERSE
3865 REM *********************************************************************
3870 IF NV=1 THEN COVIN(1)=1/COV(1):RETURN
3880 IF NV>2 GOTO 3910
3890 DE=COV(1)*COV(3)-COV(2)^2:COVIN(1)=COV(3)/DE:COVIN(2)=-COV(2)/DE
3900 COVIN(3)=COV(1)/DE:GOTO 4000
3910 IF NV>3 THEN PRINT "ERROR NV CANNOT BE LARGER THAN 3":STOP
3920 DE=COV(1)*COV(3)*COV(6)+2*COV(2)*COV(5)*COV(4)-COV(3)*COV(4)^2
3930 COVIN(1)=(COV(3)*COV(6)-COV(5)*COV(5))/DE
3940 COVIN(2)=(COV(4)*COV(5)-COV(2)*COV(6))/DE
3950 COVIN(3)=(COV(1)*COV(6)-COV(4)*COV(4))/DE
3960 COVIN(4)=(COV(2)*COV(5)-COV(4)*COV(3))/DE
3970 COVIN(5)=(COV(2)*COV(4)-COV(1)*COV(5))/DE
3980 COVIN(6)=(COV(1)*COV(3)-COV(2)*COV(2))/DE
3990 RETURN
4000 Z1=COV(1)*COVIN(1)+COV(2)*COVIN(2)
4010 Z2=COV(1)*COVIN(2)+COV(2)*COVIN(3)
4020 Z3=COV(2)*COVIN(2)+COV(3)*COVIN(3)
4030 RETURN
4040 REM *********************************************************************
4050 REM *   END OF SUBROUTINE INVERSE    ***********************************
4060 REM *********************************************************************

4070 REM *********************************************************************
4080 REM *   STARTS SUBROUTINE ORDER    *************************************
4090 REM *********************************************************************
4100 M=0
4110 NT=N
4120 FOR I=2 TO NT
4130 IF XS(I)>=XS(I-1) GOTO 4190
4140 A=XS(I)
4150 XS(I)=XS(I-1)
4160 XS(I-1)=A
4170 M=1
4180 NT=I
4190 NEXT I
4200 IF M=1 GOTO 4100
4210 RETURN
4220 REM *********************************************************************
4230 REM *   END OF SUBROUTINE ORDER    ************************************
4240 REM *********************************************************************
4250 REM *********************************************************************
4260 REM *   SUBROUTINE READ (FOR READING DATA)   **************************
4270 REM *********************************************************************
4280 INPUT#8,TITLE$
4290 I=1
4300 INPUT#8,XS(I)
4310 IF XS(I)=9.99D+20 THEN 4340
4320 I=I+1
4330 GOTO 4300
```

```
4340 N=I-1
4350 RETURN
4360 REM ******************************************************************
4370 REM *     END OF SUBROUTINE READ
4380 REM ******************************************************************
5000 REM ******************************************************************
5001 REM *    SUBROUTINE TO PRINT SAMPLE DATA
5002 REM ******************************************************************
5005 L=0 :PRINT#1," ":PRINT#1,"SAMPLE":PRINT#1,"------":PRINT#1," "
5010 FOR I=1 TO 4:PRINT#1,"   I         X(I)    ";:NEXT I:PRINT#1," "
5020 FOR I=1 TO 5:PRINT#1,"----------------";:NEXT I :PRINT#1," ":PRINT#1," "
5030 FOR I=1 TO N:PRINT#1,USING "###  ######.###    ";I,XS(I);:L=L+1:IF L=4
THEN L=0:PRINT#1," "
5031 NEXT I
5040 PRINT#1," " :FOR I=1 TO 5:PRINT#1,"----------------";:NEXT I:PRINT#1," "
5050 RETURN
5500 REM ******************************************************************
5501 REM *    SUBROUTINE TO PRINT SAMPLE DATA (PRINTER)
5502 REM ******************************************************************
5505 L=0 :CLS: PRINT:PRINT "SAMPLE":PRINT "------":PRINT
5510 FOR I=1 TO 4:PRINT "   I         X(I)    ";:NEXT I:PRINT
5520 FOR I=1 TO 5:PRINT "----------------";:NEXT I :PRINT :PRINT
5530 FOR I=1 TO N:PRINT USING "###  ######.###    ";I,XS(I);:L=L+1:IF L=4 THEN
L=0:PRINT
5531 NEXT I
5540 PRINT :FOR I=1 TO 5:PRINT "----------------";:NEXT I:PRINT
5550 RETURN
```

B.4. Selection of Domain of Attraction from Samples Program

```
0001 REM ******************************************************************
0002 REM *     PROGRAM SELEC
0003 REM ******************************************************************
0010 REM ******************************************************************
0020 REM  PROGRAM FOR DETERMINING THE DOMAIN OF ATTRACTION OF A
0030 REM     PARENT DISTRIBUTION FROM A SAMPLE
0040 REM ******************************************************************
0050 REM                   NOTATION
0060 REM ******************************************************************
0070 REM X( )      = SAMPLE VALUES IN ANY ORDER
0080 REM A$        = TITLE
0090 REM N         = SAMPLE SIZE
0100 REM ITAIL     = 1 ....... DOMAIN OF ATTRACTION FOR MINIMA IS DETERMINED
0110 REM           = 2 ....... DOMAIN OF ATTRACTION FOR MAXIMA IS DETERMINED
0120 REM K7        = 1 ....... THE CURVATURE METHOD IS USED
0130 REM           = 2 ....... THE PICKANDS METHOD IS USED
0140 REM ******************************************************************
0150 DIM X(500),Y(500),Z(500),A1$(4)
0155 CLS:INPUT "GIVE NAME OF DATA FILE ";FILE$
0156 OPEN "I",#1,FILE$
0157 INPUT "GIVE NAME OF OUTPUT FILE ";RES$
0158 OPEN "O",#2,RES$
0160 CLS
0170 A1$(4)="THIS PROGRAM DETERMINES THE DOMAIN OF ATTRACTION OF A PARENT"
0180 A1$(3)="POPULATION FROM A SAMPLE BY " : PRINT
0190 A1$(1)="THE CURVATURE METHOD"
0200 A1$(2)="THE PICKANDS METHOD"
0240 REM ******************************************************************
0250 REM      DATA IS READ
0260 REM ******************************************************************
0270 REM THE PROGRAM ASSUMES THE EXISTENCE OF A FILE (FILE$) WHERE ONE
0280 REM OR MORE DATA SETS WITH THE FOLLOWING INFORMATION APPEARS IN EACH
0290 REM (A)  TITLE
```

```
0320 REM (B)   SAMPLE (IF DATA IS UNKNOWN OR MISSING FICTITIOUS VALUES MUST
0330 REM       BE ADDED
0340 REM *************************************************************************
0350 B$="---------------------------------------------------------------"
0360 PRINT #2, B$
0390 CLS:INPUT "NUMBER OF SET DATA TO BE USED, IF NO MORE GIVE 0 ";NSK:IF
NSK=0 THEN CLS:PRINT "OUTPUT IS IN FILE ";RES$:CLOSE:STOP
0400 CLOSE#1:OPEN "I",#1,FILE$
0410 FOR J=1 TO NSK:GOSUB 2000:NEXT J
0412 CLS:PRINT "IF FITTING THE LEFT TAIL (MINIMA).........1"
0413 PRINT "IF FITTING THE RIGHT TAIL (MAXIMA)..........2"
0415 INPUT ITAIL
0420 CLS:PRINT A1$(4) : PRINT A1$(3) : PRINT
0430 PRINT A1$(1);".........";1 : PRINT A1$(2);".........";2 : PRINT
0440 INPUT "SELECT A NUMBER ACCORDING THE TABLE ABOVE ";K7
0442 PRINT #2,TITLE$:PRINT #2," "
0445 PRINT #2,A1$(4) : PRINT #2, A1$(3);A1$(K7) : PRINT #2, B$
0450 GOSUB 770
0460 IF K7=2 GOTO 520
0470 FOR I=1 TO N
0480 Y(I)=(I-.5)/N
0490 Y(I)=-LOG(-LOG(Y(I)))
0500 NEXT I
0510 GOTO 580
0520 FOR I=1 TO N
0530 Y(I)=X(I)
0540 NEXT I
0550 FOR I=1 TO N
0560 X(I)=Y(N-I+1)
0570 NEXT I
0580 IF ITAIL=2 GOTO 610
0590 FOR I=1 TO N : Z(I)=-X(I) : NEXT I
0600 FOR I=1 TO N : X(I)=Z(N-I+1) : NEXT I
0610 IF K7=2 GOTO 1110
0620 REM *************************************************************************
0630 REM                    THE CURVATURE METHOD
0640 REM *************************************************************************
0650 N11=INT(2*SQR(N)) : N22=INT(N11/2)
0660 N1=N-N11+1
0670 N2=N-N22+1
0680 GOSUB 940
0690 SLOPE1=SLOPE
0700 N1=N2
0710 N2=N
0720 GOSUB 940
0730 S=SLOPE1/SLOPE
0740 PRINT #2, "VALUE OF STATISTIC S=";S : PRINT #2, : PRINT #2, B$
0750 GOTO 390
0760 REM *************************************************************************
0770 REM       SAMPLE IS SORTED IN INCREASING ORDER OF MAGNITUDE
0780 REM *************************************************************************
0790 M=0
0800 NT=N
0810 FOR I=2 TO NT
0820 IF X(I)>=X(I-1) GOTO 880
0830 A=X(I)
0840 X(I)=X(I-1)
0850 X(I-1)=A
0860 M=1
0870 NT=I
0880 NEXT I
0890 IF M=1 GOTO 790
0900 RETURN
0910 REM *************************************************************************
0920 REM          A LEAST-SQUARES LINE IS FITTED
0930 REM *************************************************************************
0940 A1=0
0950 A2=0
```

```
0960 A01=0
0970 A11=0
0980 FOR I=N1 TO N2
0990 A1=A1+Y(I)
1000 A2=A2+Y(I)*Y(I)
1010 A01=A01+X(I)
1020 A11=A11+X(I)*Y(I)
1030 NEXT I
1040 N8=N2-N1+1
1050 SLOPE=(N8*A11-A1*A01)/(N8*A2-A1*A1)
1060 ORDOR=(A2*A01-A1*A11)/(N8*A2-A1*A1)
1070 RETURN
1080 REM ***********************************************************************
1090 REM              THE PICKANDS METHOD
1100 REM ***********************************************************************
1110 DM=9E+30
1120 DM1=9E+30
1130 N4=N/4
1140 FOR L=1 TO N4
1150 NT=4*L
1160 ANT=NT
1170 DL=0!
1180 Z24=X(2*L)-X(4*L)
1190 Z12=X(L)-X(2*L)
1200 C=LOG(Z12/Z24)/LOG(2!)
1210 A=C*Z24/(2!^C-1!)
1220 REM
1230 FOR I=1 TO NT
1240 AUX=1!+C*(X(I)-X(NT))/A
1250 IF AUX<0 GOTO 1400
1260 GL=1!-AUX^(-1!/C)
1270 FL=(NT-I+1)/ANT
1280 E=ABS(FL-GL)
1290 IF E>DL THEN DL=E
1300 FL=(NT-I)/ANT
1310 E=ABS(FL-GL)
1320 IF E>DL THEN DL=E
1330 NEXT I
1340 DL=DL
1350 IF DL>DM GOTO 1400
1360 DM=DL
1370 M=L
1380 A1=A
1390 C1=C
1400 NEXT L
1410 PRINT #2, "VALUE OF PARAMETER C=";C1
1420 PRINT #2, "VALUE OF PARAMETER A=";A1
1430 PRINT #2, : PRINT #2, B$
1440 CLS:PRINT "OUTPUT IS IN FILE ";RES$
1450 GOTO 160
2000 REM ***********************************************************************
2010 REM *        SUBROUTINE FOR READING DATA
2020 REM ***********************************************************************
2030 INPUT #1,TITLE$
2040 I=1
2050 INPUT#1,X(I)
2060 IF X(I)=9.99E+20 THEN 2090
2070 I=I+1
2080 GOTO 2050
2090 N=I-1
2100 RETURN
```

Appendix C   Data Examples

In order to illustrate the different methods to be described throughout the book, several sets of data are used sistematically. Some other sets appear only occasionally to show special facts. In this section we describe these sets of data paying special attention to their physical origin and the aim of the data analysis, because they both will play an important role in the model selection and parameter estimation procedures. This is clear because they decide whether upper, lower or central order statistics are of interest. A summary of the characteristics of these data is given in table C.15.

C.1. Wind Data

The yearly maximum wind speed, in miles per hour, registered at a given location during 50 years is included in table C.1. We assume here that this data will be used to determine a design wind speed for structural building purposes. Important facts to be taken into account for these data are its non-negative character and, perhaps, the existence of a non-clearly defined upper bound (the maximum conceivable wind speed is bounded).

22.64	22.80	23.75	24.01	24.04
24.24	24.74	25.45	25.55	25.66
25.99	26.63	26.69	26.88	26.89
27.12	27.43	27.69	27.71	28.12
28.58	28.88	29.12	29.45	29.48
30.18	31.31	31.55	31.57	32.54
32.98	33.83	33.86	34.64	35.21
36.82	37.23	38.09	38.26	38.82
38.96	38.90	42.99	43.66	44.61
45.24	47.91	54.75	69.40	98.16

TABLE C.1.   Wind data

C.2.   Flood Data

The yearly maximum flow discharge, in cubic meters per second, measured at a given location of a river during 60 years is shown in table C.2. The aim of the data analysis is supposed to be the design of a flood protection device at that location. Similar characteristics as those for the wind data appear here: a lower bound clearly defined (zero) and possibly an obscure upper bound.

24.21	26.46	29.48	30.32	31.60
32.88	33.03	33.63	35.14	35.23
35.59	35.89	35.95	36.07	36.49
36.50	37.13	37.48	38.01	38.21
38.53	38.91	39.26	39.45	40.32
40.36	40.49	40.69	41.03	41.05
41.54	42.62	42.82	42.91	43.05
43.31	43.34	43.42	43.65	43.87
44.71	45.04	45.58	46.00	48.29
48.76	49.28	49.43	50.17	50.45
50.73	51.90	52.54	52.94	54.01
57.84	60.10	61.95	67.76	75.70

TABLE C.2.   Flood data

C.3.   Wave Data

The yearly maximum wave heights, in feet, observed at a given location in 12 years are shown in table C.3. The data above, coming from shallow water, will be used for designing a breakwater. The wave height is, by definition, a non-negative random variable, which is bounded from above. In addition, we know that for shallow water an upper bound can be given, but for open sea water this bound becomes unclear.

2.91	3.74	4.09	5.88	6.42
6.93	7.21	7.92	8.26	8.79
9.17	9.50	9.62	10.00	10.14
10.28	10.45	10.77	11.65	11.65
11.82	12.27	12.68	13.28	13.46
13.88	13.98	14.32	14.38	14.46
14.86	15.03	15.30	16.07	16.23
17.36	18.68	18.72	19.44	20.09
21.06	21.13	21.53	21.80	23.15
24.75	25.45	28.13	29.95	37.19

TABLE C.3.   Wave Data

C.4.   Telephone Data

The times, in seconds, between 40 consecutive phone calls to a computerized center are shown in table C.4. The aim of the analysis is to determine the computer's ability to handle very close, consecutive calls because of a limited response time. A clear lower bound (zero) can be established from physical considerations.

0.000060	0.000074	0.000112	0.000200	0.000221
0.000236	0.000285	0.000298	0.000337	0.000374
0.000389	0.000416	0.000487	0.000559	0.000632
0.000645	0.000813	0.000960	0.001100	0.001130
0.001170	0.001280	0.001300	0.001350	0.001420
0.001560	0.001590	0.001720	0.002210	0.002380
0.002480	0.002640	0.003390	0.003480	0.005260

TABLE C.4.   Phone data

C.5.   Epicenter Data

The distances, in miles, to a nuclear power plant of the most recent 8 earthquakes of intensity larger than a given value are listed in table C.5. The

58.20	58.20	59.50	61.80	65.80
67.80	68.50	70.90	73.70	77.00
80.80	83.70	84.30	89.00	97.60
98.30	99.60	101.40	105.10	105.80
106.70	119.10	119.50	119.90	121.90
125.70	128.40	146.10	153.90	154.60
155.80	157.40	157.70	163.70	172.70
173.90	174.20	175.10	176.00	178.70
179.10	179.50	180.70	182.10	182.70
186.70	187.50	191.00	192.60	193.00
199.40	211.60	212.10	216.80	222.90
227.30	229.40	234.50	236.80	238.90

TABLE C.5. Epicenter data

data is needed in order to evaluate the risks associated with earthquakes occurring close to the central site. In addition, it is known that a fault is the main cause of earthquakes in the area, and the closest point of the fault is 50 miles.

C.6. Link Data

20 chain links have been tested for strength and the results are given in table C.6. The data is used for quality control and minimum strength characteristics are needed.

51.1	57.1	65.1	69.2	71.2
73.1	76.1	81.9	84.1	86.6
88.1	92.6	94.9	96.6	97.0
101.4	103.4	103.8	105.2	119.1

TABLE C.6. Link data

C.7. Electrical Insulation Data

The lifetimes of 30 electrical insulation elements are shown in table C.7. Quality control and minimum lifetime are of interest.

744	822	847	885	920
948	968	985	1010	1018
1019	1028	1029	1031	1040
1047	1071	1074	1097	1134
1147	1170	1174	1209	1251
1273	1320	1383	1388	1462

TABLE C.7.   Electrical insulation data

C.8.   Fatigue Data

40 specimens of wire were tested for fatigue strength to failure and the results are shown in table C.8. The aim of the study is to find a design fatigue stress.

39611	44132	44209	45898	50139
54625	58970	64703	64950	66508
70208	72098	75001	80393	81868
82202	82447	89268	90021	96136
96723	101610	101833	106055	112833
119154	122366	134511	135220	136395
138378	153790	184916	216370	240316

TABLE C.8.   Fatigue data

C.9.   Precipitation Data

The yearly total precipitation in Philadelphia for the last 40 years, measured in inches, is shown in table C.9. The aim of the study is related to drought risk determination.

29.34	29.88	32.20	33.03	33.27
34.04	34.95	35.15	35.45	36.77
37.78	38.35	38.37	39.14	39.52
40.00	40.47	40.48	41.05	41.15
41.75	42.62	43.05	43.36	44.46
44.82	44.85	45.4	45.84	45.95
46.00	46.06	46.62	47.79	47.87
48.13	49.06	49.42	49.63	52.13

TABLE C.9.   Precipitation data

C.10. Houmb Data

The yearly maximum significant wave height measured in Myken-Skomvaer (Norway) in the period 1949–1976 and published by Houmb et al (1978) is included in table C.10. We assume that this data will be used for the design of sea structures.

5.60	6.55	6.65	7.35	7.80
7.90	8.00	8.50	9.05	9.15
9.40	9.60	9.80	9.90	10.85
10.90	11.10	11.30	11.30	11.55
11.75	12.85	12.90	13.40	

TABLE C.10.   Houmb data

C.11. Ocmulgee River Data

The yearly maximum water discharge of the Ocmulgee river measured at two different locations, Macon and Hawkinsville, between 1910 and 1949, and published by Gumbel (1964), are given in tables C.11. and C.12. The aim of the analysis is assumed to be related to flood protection design.

4.8	7.3	7.9	8.5	10.7
14.2	14.3	16.9	19.0	19.1
19.6	21.0	22.7	24.0	25.4
28.3	28.3	28.8	31.0	31.0
32.6	33.3	33.9	37.0	40.4
44.8	44.8	47.1	47.8	50.2
51.0	57.6	64.4	65.3	66.2
72.5	73.4	73.4	78.6	84.0

TABLE C.11.   Ocmulgee River (Macon)

C.12. Oldest Ages at Death in Sweden Data

The oldest ages at death in Sweden during the period 1905 to 1958 for women and men, respectively, are given in tables C.13. and C.14. The analysis is needed in order to forecast oldest ages at death in the future.

5.9	5.9	6.9	7.6	12.2
13.3	13.5	14.3	15.2	16.2
17.4	18.8	19.3	19.9	20.1
25.8	26.2	27.0	28.2	30.0
30.3	33.0	34.8	35.4	37.9
40.0	40.4	41.6	42.4	44.0
44.4	45.2	46.8	50.0	52.0
57.0	61.0	68.0	70.5	79.0

TABLE C.12.   Ocmulgee River (Hawkinsville)

101.50	101.69	102.12	102.32	102.54
102.72	102.78	102.92	103.14	103.31
103.38	103.41	103.46	103.53	103.56
103.56	103.77	103.83	103.86	103.94
103.97	104.01	104.12	104.27	104.33
104.37	104.40	104.42	104.46	104.52
104.71	104.85	104.87	105.01	105.01
105.02	105.03	105.19	105.32	105.45
105.64	105.71	105.83	105.86	105.87
105.88	105.98	106.13	106.15	106.15
106.52	107.49	107.89	107.90	

TABLE C.13.   Oldest ages at death in Sweden (women)

100.08	100.49	100.82	100.88	100.90
101.17	101.26	101.41	101.63	101.66
101.67	101.70	101.76	102.41	102.52
102.54	102.55	102.57	102.57	102.61
102.63	102.69	102.78	102.88	102.94
103.00	103.06	103.15	103.17	103.24
103.25	103.36	103.40	103.43	103.47
103.57	103.80	103.98	104.01	104.22
104.65	104.88	104.92	105.00	105.12
105.12	105.12	105.48	105.55	105.72
105.83	106.09	106.48	106.50	

TABLE C.14.   Oldest ages at death in Sweden (men)

Data Set	Random Variable	Sample Size	Lower Bound	Upper Bound	Units	Aim of Analysis
Wind data	Maximum yearly wind speed	40	> 0	unknown	miles per hour	Sea structure design
Flood data	Maximum yearly flow discharges	50	> 0	unknown	cubic meters per second	Flood protection design
Wave data	Maximum yearly wave height	12	> 0	know for shallow waters	feet	Breakwater design
Phone data	Times between calls	40	0	unknown	seconds	Computer ability to handle calls
Epicenter data	Distance of epicenters to nuclear power plant	8	50	250	miles	earthquake risk assessment
Link data	Strength of links	20	≥ 0	unknown	Kgs	Chain design
Electrical Insulation data	Lifetime of electrical insulation elements	30	≥ 0	unknown	hours	Quality control
Fatigue data	Fatigue lifetime	40	> 0	unknown	cycles	Fatigue stress design
Precipitation data	Yearly total rainfall	40	> 0	unknown	inches	Drought risk assessment

TABLE C.15. Summary of characteristics of data sets

Bibliography

Aczél, J. (1966). *Lectures on Functional Equations and Their Applications.* Academic Press, New York.

Afanas'ev, N. N. (1940). "Statistical theory of the fatigue strength of metals." *J. Tech. Phys.*, **10**, 1553–1568.

Afanas'ev, N. N. (1953). *The Statistical Theory of Fatigue Resistance in Metals.* Izd-vo Akademii Nauk USSR, Kiev (AMR 12 #741).

Alexander, C. H. (1980). "Simultaneous confidence bounds for the tail of an inverse distribution function." *Ann. Statist.* **8**, 1391–1394.

American National Standard Building Requirements for Minimum Design Loads in Buildings and Other Structures A58.1. (1972). American National Standards Institute, New York.

Anderson, C. W. (1984). "Large deviations of extremes." In: *Statistical Extremes and Applications.* NATO ASI Series, D. Reidel Publishing Company.

Andrä, W. and Saul, R. (1974). "Versuche mit Bündeln aus parallelen Drähten und Litzen für die Nordbrücke Mannheim-Ludwigshafen und das Zeltdach in München." *Die Bautechnik* **9, 10** and **11,** 289–298, 332–340 and 371–373.

Andrä, W. and Saul, R. (1979). "Die Festigkeit insbesondere Dauerfestigkeit langer Paralleldrahtbündel." *Die Bautechnik* **4,** 128–130.

Ang, A. H. S. (1973). Structural risk analysis and reliability-based design. *J. Struct. Div., ASCE,* **99,** ST9, 1891–1910.

Antle, C. E. and Rademaker, F. (1972). An upper confidence limit on the maximum of m future observations from a type I extreme value distribution. *Biometrika* **59,** 475–477.

ARGON, A. S. (1972). *Fracture of composites*. Treatise on Materials Science and Technology (edited by Herman). Academic Press, New York.

ARGON, A. S. (1974). "Statistical aspects of fracture." Composite Materials, **5**: *Fracture and Fatigue* (edited by Lawrence J. Broutman), Academic Press, New York.

ARMENÀKAS, A. E., GARG, S. K., SCIAMMERELLA, C. A. and SVALBONAS, V. (1970). Statistical theories of strength of bundles and fiber-reinforced composites. AFML- TR-70-3, Air Force Materials Laboratory, Wright-Patterson AFB, Ohio. AD871745.

ARMENÀKAS, A. and SCIAMMERELLA, C. A. (1973). "Experimental investigation of the failure mechanism of fiber-reinforced composites subjected to uniaxial tension." *Experimental Mechanics,* **13,** 49–58.

ARNOLD, B. C. (1986). Bivariate distributions with Pareto conditionals. To appear.

ASTM (1963). A guide for fatigue testing and the statistical analysis of fatigue data. *ASTM Special Technical Publication,* N. 91 A.

ASTM (1980). "Standard practice for statistical analysis of linear and linearized stress-life (S-N) and strain-life (E-N) fatigue data." *ASTM,* E 739–780.

AVEN, T. (1985). "Upper (lower) bounds on the mean of the maximum (minimum) of a number of random variables." *J. Appl. Prob.* **22,** 723–728.

AZIZ, P. M. (1956). "Application of the statistical theory of extreme values to the analysis of maximum pit depth for aluminium." *Corrosion,* **12,** 35–46.

BAIN, L. J. (1972). "Inferences based on censored sampling from the Weibull or extreme-value distribution." *Technometrics,* **14,** 693–702.

BAIN, L. J. and ANTLE, C. E. (1967). "Estimation of parameters in the Weibull distribution." *Technometrics,* **9,** 621–627.

BAIN, L. J., ANTLE, C. E. and BILLMAN, B. R. (1971). Statistical analyses for the Weibull distribution with emphasis on censored sampling. ARL 71-0242. Aerospace Research Laboratories, Wright-Patterson AFB, Ohio.

BALKEMA, A. A. (1973). *Monotone transformations and limit laws.* Mathematical Centre Tracts, Amsterdam.

BALKEMA, A. A. and HAAN, L. de (1972). "On R. von-Mises' condition for the domain of attraction of $\exp[-\exp(-x)]$." *Ann. Math. Statist.* **43,** 1352–1354.

BALKEMA, A. A. and RESNICK, S. I. (1977). "Max-infinite divisibility." *J. Appl. Probability* **14,** 309–319.

BALKEMA, A. A. and HAAN, L. de (1978a). "Limit distributions for order statistics, I." *Theory Probab. Appl.* **23,** 77–92.

BALKEMA, A. A. and HANN, L. de (1978b). "Limit distributions for order statistics II." *Theory Probab. Appl.* **23,** 341–358.

BAR-DAVID, I. and EIN-GAL, M. (1975). "Passages and maxima for a particular Gaussian process." *Ann. Probab.* **3,** 549–556.

BARLOW, R. E., GUPTA, S. S. and PANCHAPAKESAN, S. (1969). "On the distribution of the maximum and minimum of ratios of order statistics." ANN. MATH. STATIST. **40,** 918–934.

. BARLOW, R. E. (1971). "Averaging time and maxima air pollution concentrations." Proc. 38th Session ISI, Washington D.C., 663–676.

BARLOW, R. E. and SINGPURWALLA, N. D. (1974). Averaging time and maxima for dependent observations. *Proc. Symposium on Statistical Aspects of Air Quality Data.*

BARLOW, R. E. and PROSCHAN, F. (1975). *Statistical Theory of Reliability and Life Testing: Probability Models.* Holt, Rinehart and Winston, New York.

BARNDORFF-NIELSEN, O. (1963). "On the limit behavior of extreme order statistics." *Ann. Math. Statist.* **34,** 992–1002.

BARNDORFF-NIELSEN, O. (1964). "On the limit distribution of the maximum of a random number of independent random variables." *Acta Math. Acad. Sci. Hungar.* **15,** 399–403.

BARROIS, W. G. (1970). *Manual for Fatigue of Structures. Fundamental and Physical Aspects.* AGARDMAN-8-70. NATO Advisory Group for Aerospace Research and Development, Paris.

BARTENEV, G. M. and SIDOROV, A. B. (1966). "Statistical theory of the strength of glass fibers." *Polymer Mechanics* **2,** 52–56.

BATDORF, S. B. (1982). "Tensile strength of unidirectionally reinforced composites-I." *J. Reinforced Plastics and Composites* **1,** 153–164.

BATDORF, S. B., and GHAFFANIAN, R. (1982). "Tensile strength of unidirectionally reinforced composites-II." *J. Reinforced Plastics and Composites* **1,** 165–176.

BATTJES, J. A. (1977). Probabilistic aspects of ocean waves. *Proc. Seminar on Safety of Structures under Dynamic Loading.* University of Trondheim. Norway.

BEARD, L. R. (1962). *Statistical Methods in Hydrology.* U.S. Army Corps of Engineers, Sacramento, California.

BENJAMIN, J. R. and CORNELL, C. A. (1970). *Probability, Statistics and Decision for Civil Engineers.* Mc-Graw Hill, New York.

BENSON, M. A. (1968). "Uniform flood-frequency estimating methods for federal agencies." *Water Resour. Res.* **4,** 891–908.

BERMAN, S. M. (1961). "Convergence to bivariate extreme value distributions." *Ann. Inst. Statist. Math.* **13,** 217–223.

BERMAN, S. M. (1962a). "A law of large numbers for the maximum in a stationary Gaussian sequence." *Ann. Math. Statist.* **33,** 93–97.

BERMAN, S. M. (1962b). "Limiting distribution of the maximum term in a sequence of dependent random variables." *Ann. Math. Statist.* **33,** 894–908.

BERMAN, S. M. (1962c). "Equally correlated random variables." *Sankhya* **A 24,** 155–156.

BERMAN, S. M. (1964). "Limit theorems for the maximum term in stationary sequences." *Ann. Math. Statist.* **35,** 502–516.

BERMAN, S. M. (1971). "Asymptotic independence of the numbers of high and low level crossings of stationary Gaussian processes." *Ann. Math. Statist.* **42,** 927–945.

BERMAN, S. M. (1982). "Sojourns and extremes of stationary processes." *Ann. Probab.* **10,** 1–46.

BERMAN, S. M. (1985). "An asymptotic formula for the distribution of the maximum of a Gaussian process with stationary increments." *J. Appl. Prob.* **22,** 454–460.

BERRY, G. (1975). "Design of carcinogenesis experiments using the Weibull distribution." *Biometrika* **62**, 321–328.

BILLMANN, B. R., ANTLE, C. E. and BAIN, L. J. (1972). "Estatistical inference from censored Weibull samples." *Technometrics* **14**, 831–840.

BIONDINI, R. and SIDDIQUI, M. M. (1975). Record values in Markov sequences." In: *Statistical Inference and Related Topics,* **2** (edited by Puri). Academic Press, New York, 291–352.

BIRKENMAIER, M. and NARAYANAN, R. (1982). Fatigue resistance of large high tensile steel stay tendons. IABSE Colloquium, Fatigue of Steel and Concrete Structures, Lausanne, 663–672.

BIRNBAUM, Z. W. and SAUNDERS, S. C. (1958). "A Statistical Model for Life-Length of Materials." *J. Amer. Statist. Assoc.* **53**, 151–159.

BLOM, G. (1958). *Statistical Estimates and Tranformed Beta-Variables.* Almqvist and Wiksell, Uppsala, Sweden. John Wiley, New York.

BLOM, G. (1962). "Nearly Best Linear Estimates of Location and Scale Parameters." *SG,* 34–46.

BLOOMER, N. T. and ROYLANCE, T. F. (1965). "A large scale fatigue test of aluminium specimens. *Aeronautical Quarterly* **16**, 307–322.

BOFINGER, E. and BOFINGER, V. J. (1965). "The correlation of maxima in samples drawn from a bivariate normal distribution." *Austr. J. Statist.* **7**, 57–61.

BOFINGER, V. J. (1970). "The correlation of maxima in several bivariate nonnormal distributions. *Austr. J. Statist.* **12**, 1–7.

BOGDANOFF, J. L. and SCHIFF, A. (1972). *Earthquake Effects in the Safety and Reliability Analysis of Engineering Structures.* International Conference on Structural Safety and Reliability (edited by A. M. Freudenthal), Washington D.C., Pergamon Press, 147–148.

BOGDANOFF, J. L. and KOZIN, F. (1985). *Probabilistic Models of Cumulative damage.* John Wiley and Sons, New York.

BOLOTIN, V. V. (1961). *Statistical Methods in Structural Mechanics.* Holden Day, Inc.

BOLOTIN, V. V. (1971). "Mathematical and experimental models of fracture." *Strength of Materials* **3**, 133–139.

BOLOTIN, V. V. (1981). *Wahrscheinlichkeitsmethoden zur Berechnung von Konstruktionen.* Veb Verlag für Bauwesen, Berlin.

BOLOTIN, V. V. (1984). *Random Vibration of Elastic Systems.* Ed. Nijhoff.

BOOS, D. D. (1984). "Using extreme value theory to estimate large percentiles." *Technometrics* **26**, 33–39.

BORGMAN, L. E. (1963). "Risk criteria." J. Waterw. Harb. Eng. Div., *ASCE,* **89, WW3,** 1–35.

BORGMAN, L. E. (1970). "Maximum wave height probabilities for a random number of random intensity storms." Proceedings 12th Conference on Coastal Engineering, Washington, D.C.

BORGMAN, L. E. (1973). "Probabilities of highest wave in hurricane." J. Waterw. Harb. Coast. Eng. Div., *ASCE* **99, WW2,** 185–207.

BRANGER, J. (1972). *Life Estimation and Prediction of Fighter Aircraft.* International Conference on Structural Safety and Reliability (edited by A. M. Freudenthal), Washington, D.C., Pergamon Press, 341–349.

BRAVINSKI, V. G. and OSIPOV, M. V. (1961). "Effect of the scale factor on time dependence of the cermaic materials." *Soviet Physics-Doklady* **6**, 250–252.

BRETSCHNEIDER, C. C. (1959). Wave variability and wave spectra for wind generated gravity waves. Tech. Memo. **118**, U.S. Beach Erosion Board, Washington, D.C.

BROWN, B. M. and RESNICK, S. I. (1977). "Extreme values of independent stochastic processes." *J. Appl. Probab.* **14**, 732–739.

BROWNLEE, K. A. (1965). *Statistical Theory and Methodology in Science and Engineering.* John Wiley and Sons. New York.

BUCKLAND, W. R. (1964). *Statistical Assessment of the Life Characteristic: a Bibliographic Guide.* Charles Griffin & Company Limited, London; Hafner Publishing Company, New York.

BÜHLER, H. and SCHREIBER, W. (1957). "Lösung eineger aufgaben der Dauerfestigkeit mit dem treppenstufen-Verfahren." *Archiv für eisenhüttenwessen* **28**, 153–156.

BURY, K. V. (1974). "Distribution of smallest lognormal and gamma extremes." *Statistische Hefte* **15**, 105–114.

BUTLER, J. P. (1927a). *Reliability Analysis in the Estimation of Transport-Type Aircraft Fatigue Performance.* International Conference on Structural Safety and Reliability (edited by A. M. Freudenthal), Washington, D.C.

BUTLER, J. P. (1972b). "The development of reliability analysis methods for the assurance of aircraft structural fatigue performance." Proceedings of the Colloquium on Structural Reliability: The Impact of Advanced Materials on Engineering Design (edited by J. L. Swedlow, T. A. Cruse and J. C. Halpin), Carnegy Mellon University, Pittsburg, Pa.

BUXBAUM, O. and SVENSON, O. (1972). Extreme value analysis of flight load measurements. *Aicraft Fatigue: Design Operational and Economic Aspects,* Proceedings of Symposium, Melbourne, Australia (edited by J. Y. Mann and I. S. Milligan). Pergamon Press, Rushcutters Bay, Australia, 297–322.

CABAÑA, E. M. and WSCHEBOR, M. (1981). "An estimate for the tails of the distribution of the supremum for a class of stationary multiparameter Gaussian processes." *J. Appl. Probab.* **18**, 536–541.

CACOULLOS, T. and DECICCO, H. (1967). "On the distribution of the bivariate range." *Technometric* **9**, 476–480.

CAMPBELL, J. W. and TSOKOS, C. P. (1973). "The asymptotic distribution of maxima in bivariate samples." *J. Amer. Statist. Assoc.* **68**, 734–739.

CARTWRIGHT, D. E. and LONGUET-HIGGINS, M. S. (1956). "The statistical distribution of the maxima of a random function." *Proc. Roy. Soc. Ser. A,* **237**, 212–232.

CASTILLO, E., LOSADA, M. and PUIG-PEY, J. (1977). *Análisis probabilista del número de olas y su influencia en la altura de cálculo de obras marítimas.* Revista de Obras Públicas, Madrid, 639–648.

CASTILLO, E., MORENO, E. and PUIG-PEY, J. (1980). "Nuevos modelos de distribución de extremos basados en aproximaciones en las ramas." *Trabajos de Estadística e Investigón Operativa* **34**, 6–23.

CASTILLO, E., MORENO, E. and PUIG-PEY, J. (1982). *Criterios mínimo cuadráticos de ajuste de distribuciones de probabilidad a datos experimentales.* Revista de Obras Públicas, 433–439.

CASTILLO, E., ASCORBE, A. and FERNANDEZ-CANTELI, A. (1983). Static progressive failure in multiple tendons. A statistical approach. 44th session of ISI. Madrid, 45.1.

CASTILLO, E., FERNANDEZ-CANTELI, A., ASCORBE, A. and MORA, E. (1983). The Box-Jenkins model and the progressive fatigue failure of large parallel elements stay-tendons. ASI-NATO, Statistical extremes and applications. Lisbon.

CASTILLO, E., FERNANDEZ-CANTELI, A., ASCORBE, A. and MORA, E. (1984). "Aplicación de los modelos de series temporales al análisis estadístico de la resistencia de tendones de puentes atirantados." *Anales de Ingeniería Mecánica* **2**, 379–382.

CASTILLO, E., FERNANDEZ-CANTELI, A., MORA, E. and ASCORBE, A. (1984). "Influencia de la longitud en la resistencia a fatiga de tendones en puentes atirantados." *Anales de Ingeniería Mecánica* **2**, 383–389.

CASTILLO, E., FERNANDEZ-CANTELI, A., ESSLINGER, V. and THRLIMANN, B. (1985). "Statistical models for fatigue analysis of wires, strands and cables." *IABSE Proceedings,* 82, 85, 1–40.

CASTILLO, E. and GALAMBOS, J. (1985a). The characterization of a regression model associated with fatigue problems. Conference on weighted distributions. Penn. State Univ.

CASTILLO, E. and GALAMBOS, J. (1985b). Bivariate models with normal conditionals. Conference on weighted distributions. Penn. State Univ.

CASTILLO, E. and GALAMBOS, J. (1986a). Characterization of bivariate densities via conditional densities. Technical Report, Temple University.

CASTILLO, E. and GALAMBOS, J. (1986b). Determining the domain of attraction of an extreme value distribution. Technical Report, Temple University.

CASTILLO, E. and GALAMBOS, J. (1987). "Lifetime regression models based on a functional equation of physical nature." *J. Appl. Prob.* 24, p 160–169.

CAVANIE, A., ARHAN, M. and EZRATY, R. (1976). A statistical relationship between individual heights and periods of storm waves. Proceedings BOSS'76, Trondheim, 354–360.

CHAKRABARTI, S. K. and COOLEY, R. P. (1977). "Statistical distributions of periods and heights of ocean waves." *J. Geophys. Res.* **82**, 1363–1368.

CHAN, L. K. and KABIR, A. B. M. (1969). "Optimum quantiles for the linear estimation of the parameters of the extreme value distribution in complete and censored samples." *Naval Res. Logist. Quart.* **16**, 381–404.

CHAN, L. K. and JARVIS, G. A. (1970). "Convergence of moments of some extreme order statistics." *J. Statist. Res.* **4**, 37–42.

CHAN, L. K. and MEAD, E. R. (1971). "Linear estimation of the parameters of the extreme-value distribution based on suitably chosen order statistics." *IEEE Transactions on Reliability,* **R-20**, 74–83.

CHANDLER, K. N. (1952). "The distribution and frequency of record values." *J. Royal Statist. Soc.* **B14**, 220–228.

CHANDRA, M., SINGPURWALLA, N. D. and STEPHENS, M. A. (1981). "Kolmogorov statistics for tests of fit for the extreme-value and Weibull distributions." *J. Amer. Statist. Assoc.* **76**, 729–731.

CHAPLIN, W. S. (1880). "The relation between the tensile strengths of long and short bars." *Van Nostrand's Engineering Magazine* **23**, 441–444.

CHAPLIN, W. S. (1882). "On the relative tensile strengths of long and short bars." *Proceedings of the Engineer's Club,* Philadelphia **3**, 15–28.

CHECHULIN, B. B. (1954). "On the statistical theory of brittle strength." *Zhurnal Teknicheskoi Fiziki* **24**, 292–298.

CHENG, S. (1985). "On limiting distributions of order statistics with variable ranks from stationary sequences." *Ann. Probab.* **13**, 1326–1340.

CHERNICK, M. R. (1980). "A limit theorem for the maximum term in a particular EARMA (1, 1) sequence." *J. Appl. Probab.* **17**, 869–873.

CHERNICK, M. R. (1981a). "A limit theorem for the maximum of autoregressive processes with uniform marginal distributions." *Ann. Probab.* **9**, 145–149.

CHERNICK, M. R. (1981b). "On strong mixing and Leadbetter's D condition." *J. Appl. Probab.* **18**, 746–769.

CHERNOFF, H. and LIEBERMAN, G. J. (1954). "Use of normal probability paper." *J. Amer. Statist. Assoc.* **49**, 778–785.

CHERNOFF, H. and LIEBERMAN, G. J. (1956). "The use of generalized probability paper for continuous distributions." *Ann. Math. Statist.* **27**, 806–818.

CHOBISOC, D. M. (1964). "On limit distributions for order statistics." *Theory Probab. Appl.* **9**, 142–148.

CHIZHOV, V. M. (1972). "Application of the method of extreme values to the distribution of maximum operating loads." *Uchenye Zapiski* **3**, 45–50.

CHOW, V. T. (1951). "A general formula for hydrologic frequency analysis." *Eos Trans. AGU* **32**, 231–237.

CHOW, V. T. (1964). *Handbook of Applied Hydrology.* McGraw Hill, New York.

CLOUGH, D. J. and KOTZ, S. (1965). "Extreme-value distributions with a special queuing model application." *J. Canad. Op. Res. Soc.* **3**, 96–109.

CLOUGH, D. J. (1969). "An asymptotic extreme-value sampling theory for estimation of a global maximum." *J. Canad. Op. Res. Soc.* **7**, 102–115.

COHEN, A. C. (1965). "Maximum likelihood estimation in the Weibull distribution based on complete and on censored samples." *Technometrics* **7**, 579–588.

COHEN, A. C. (1973). "Multi-censored sampling in the three parameter Weibull distribution." *Bull. Intern. Statist. Inst.* **45**, 277–282.

COHEN, J. P. (1982). "The penultimate form of approximation to normal extremes." *Adv. Appl. Prob.* **14**, 324–339.

COHEN, J. P. (1982). "Convergence rates for the ultimate and penultimate approximations in extreme-value theory." *Adv. Appl. Prob.* **14**, 833–854.

COLEMAN, B. D. (1956). "Time dependence of mechanical breakdown phenomena." *J. Appl. Physics* **27**, 862–866.

COLEMAN, B. D. (1957a). "Time dependence of mechanical breakdown in bundles of fibers I: Constant total load." *J. Appl. Phys.* **28**, 1058–1064.

COLEMAN, B. D. (1957b). "A stochastic process model for mechanical breakdown." *Trans. Soc. Rheol.* **1**, 153–168.

COLEMAN, B. D. and MARQUARDT, D. W. (1957). "Time dependence of mechanical breakdown in bundles of fibers II. The infinite ideal bundle under linearly increasing loads." *J. Appl. Phys.* **28**, 1065–1067.

COLEMAN, B. D. (1958a). "On the strength of classical fibres and fibre bundles." *J. Mech. Solids* **7**, 60–70.

COLEMAN, B. D. (1958b). "Time dependence of mechanical breakdown in bundles of fibers III: the power law breakdown rule." *Trans. Soc. Rheol.* **2**, 195–218.

COLEMAN, B. D. (1958c). "Statistics and time dependence of mechanical breakdown in fibers." *J. Appl. Phys.* **29**, 968–983.

COLEMAN, B. D. and MARQUARDT, D. W. (1958). "Time dependence of mechanical breakdown in bundles of fibers IV: Infinite ideal bundle under oscillating loads." *J. Appl. Phys.* **29**, 1091–1099.

CONOVER, W. J. (1965). "A *k*-sample model in order statistics. *Ann. Math. Statist.*" **36**, 1223–1235.

COOIL, B. (1985). "Limiting multivariate distribution of intermediate order statistics." *Ann. Prob.* **13**, 469–477.

CORNIERO, M. A. (1981). Análisis estadístico de fenómenos naturales cuasi-estacionarios mediante simulación numérica. Distribución de máximas alturas de ola. Doctoral dissertation. University of Santander, Spain.

CORTEN, H. T. (1964). Reinforced plastics. *Engineering Design for Plastics* (edited by E. Baer). Reinhold Publishing Corporation, New York; Chapman and Hall, Ltd., London, 869–994.

COURT, A. (1953). "Wind extremes as design factors." *J. of the Franklin Institute* **256**, 39–55.

CRAMÉR, H. (1962). "On the maximum of a normal stationary stochastic process." *Bull. Amer. Math. Soc.* **68**, 512–516.

CRAMÉR, H. (1965). "A limit theorem for the maximum values of certain stochastic processes." *Theory Probab. Appl.* **10**, 126–128.

D'AGOSTINO, R. B. (1971). "Linear estimation of the Weibull parameters." *Technometrics* **13**, 171–182.

DALEY, D. J. and HALL, P. (1984). "Limit laws for the maximum of weighted and shifted i.i.d. random variables." *Ann. Probab.* **12**, 571–587.

DALRYMPLE, T. (1960). Flood-frequency analysis. U. S. Geol. Surv., Water Supply Pap. 1543-A.

DANIELS, H. E. (1945). "The statistical theory of the strength of bundles of threads." *Proc. Royal Soc.* **A183**, 405–435.

DATTATRI, J. (1973). "Waves off Mangalore Harbor-west coast of India." *J. Waterw. Harb. Coast. Eng. Div.*, ASCE, **WW1**, 39–58.

DAVENPORT, A. G. (1968a). The relationship of wind structure to wind loading.

Proceedings International Seminar on Wind Effects on Building and Structures. University of Toronto Press, Toronto, Canada.

DAVENPORT, A. G. (1968b). The dependence of wind loads upon meteorological parameters. *Proceedings Wind Effects on Buildings and Structures,* Ottawa, **1**, University of Toronto Press, Toronto, Canada.

DAVENPORT, A. G. (1972). Structural safety and reliability under wind action. *International Conference on Structural Safety and Reliability* (edited by A. M. Freudenthal), Washington, D. C., Pergamon Press, 131–145.

DAVENPORT, A. G. (1978). Wind structure and wind climate. *Int. Res. Seminar on Safety of Structures Under Dynamic Loading,* Trondheim., 238–256.

DAVID, F. N. and JOHNSON, N. L. (1954). "Statistical treatment of censored data I. Fundamental formulae." *Biometrika* **41**, 228–240.

DAVID, H. A. (1973). "Concomitants of order statistics." *Bull. Inst. Internat.* Statist. **45**, 295–300.

DAVID, H. A. and GALAMBOS, J. (1974). "The asymptotic theory of concomitants of order statistics." *J. Appl. Probab.* **11**, 762–770.

DAVID, H. A., O'CONNELL, M. J. and YANG, S. S. (1977). "Distribution and expected value of the rank of a concomitant of an order statistic." *Ann. Statist.* **5**, 216–223.

DAVID, H. A. (1981). *Order Statistics.* John Wiley and Sons, New York.

DAVIDENKOV, N., SHEVANDIN, E. and WITTMANN, F. (1947). "The influence of size on the brittle strength of steel." *J. Appl. Mech.* **14**, 63–67.

DAVIS, R. A. (1979). "Maxima and minima of stationary sequences." *Ann. Probab.* 7, 453–460.

DAVIS, R. A. (1981). Limit laws for upper and lower extremes from stationary mixing sequences. Techn. Report, Massachussetts Institute of Technology.

DAVIS, R. A. (1982). "Limit laws for the maximum and minimum of stationary sequences." *Z. Wahrsch. verw. Geb.* **61**, 31–42.

DAVIS, R. and RESNICK, S. (1984). "Tail estimates motivated by extreme value theory." *Ann. Statist.* **12**, 1467–1487.

DAVIS, R. and RESNICK, S. (1985). "Limit theory for moving averages of random variables with regulary varying tail probabilities." *Ann. Probab.* **13**, 179–195.

DAWSON, D. A. and SANKOFF, D. (1967). "An inequality for probabilities." *Proc. Amer. Math. Soc.* **18**, 504–507.

DEHEUVELS, P. (1973). "Sur la convergence de sommes de minimums de variables aléatoires." *Comptes Rendus de l'Académie des Sciences,* Paris A **276**, 309–312.

DEHEUVELS, P. (1974). "Valeurs extrémales d'échantillons croissants d'une variable aléatoire réelle." *Ann. Inst. H. Poincaré* B **10**, 89–114.

DEHEUVELS, P. (1978). "Caracterisation complete des lois extremes multivariées et de la convergence aux types extremes." *Publ. Inst. Statist. Univ.* Paris **23**.

DEHEUVELS, P. (1981). Univariate extreme values. Theory and applications. *Proceedings of the ISI meeting,* Buenos Aires, 837–857.

DEHEUVELS, P. (1984). "The characterizations of distributions by order statistics and record values: A unified approach." *J. Appl. Probab.* **21**, 326–334.

DENGEL, D. (1971). "Einige Grundlegende Gesichtspunkte für die Planung und Auswertung von Dauerschwingversuchen." *Materialprüfung* **13**, No. 5, 145–151.

DENZEL, G. E. and O'BRIEN, G. L. (1975). "Limit theorems for extreme values of chain-dependent processes." *Ann. Probab.* **3**, 773–779.

DEO, C. M. (1971). "On maxima of Gaussian sequences. Abstract." *Ann. Math. Statist.* **42**, 2176.

DEO, C. M. (1973a). "A note on strong mixing Gaussian sequences." *Ann. Probab.* **1**, 186–187.

DEO, C. M. (1973b). "A weak convergence theorem for Gaussian sequences." *Ann. Probab.* **1**, 1061–1064.

DIXON, W. J. (1951). "Ratios involving extreme values." *Ann. Math. Statist.* **22**, 68–78.

DODD, E. L. (1923). "The greatest and the least variate under general laws of error." *Trans. Amer. Math. Soc.* **25**, 525–539.

DOWNTON, F. (1966a). "Linear estimates of parameters in the extreme value distribution." *Technometrics* **8**, 3–17.

DOWNTON, F. (1966b). "Linear estimates with polynomial coefficients." *Biometrika* **53**, 129–141.

DRAPER, L. (1963). Derivation of a design wave from instrumental records of sea waves. *Proceedings of the institution of Civil Engineers* (London) **26**, 291–303.

DUBEY, S. D. (1966a). "Some test functions for the parameters of the Weibull distributions. "*Naval Research Logistics Quarterly* **13**, 113–128.

DUBEY, S D. (1966b). "Transformations for estimation of parameters." *J. Ind. Statist. Assoc.* **4**, 109–124.

DUBEY, S. D. (1966c). "On some statistical inferences for Weibull laws." *Naval Res. Logist. Quart.* **13**, 227–251.

DUBEY, S. D. (1966d). "Hyper-efficient estimator of the location parameter of the Weibull laws." *Naval Res. Logist. Quart.* **13**, 253–264.

DUBEY, S. D. (1966e). "Asymptotic efficiencies of the moment estimators for the parameters of the Weibull laws." *Naval Res. Logist. Quart.* **13**, 265–288.

DUBEY, S. D. (1966f). "Characterization theorems for several distributions and their applications." *J. Industrial Math. Soc.* **16**, 1–22.

DUBEY, S. D. (1967a). "Some percentile estimators for Weibull parameters." *Technometrics* **9**, 19–129.

DUBEY, S. D. (1967b). "Normal and Weibull distributions." *Naval Res. Logist.* Quart. **14**, 69–79.

DUBEY, S. D. (1967c). "On some permissible estimators of the location parameter of the Weibull and certain other distributions." *Technometrics* **9**, 293–307.

DUBEY, S. D. (1967d). "Some simple estimators for the shape parameter of the Weibull law." *Naval Res. Logist. Quart.* **14**, 489–512.

DUBEY, S. D. (1967e). "Monte Carlo study of the moment and maximum likelihood estimators of Weibull parameters." *Trabajos de Estadística* **18**, 131–141.

DUCHENE-MARULLAZ, P. (1972). "Etude des vitesses maximales du vent." *Cahiers du Centre Scientifique et Technique du Batiment*, Cahier **1118**, No. 131.

DUEBELBEISS, E. (1979). "Dauerfestigkeitsversuche mit einem modifizierten Treppenstufenverfahren." *Materialprüfung* 16, Nr. 8, 240–244.

DUFOUR, R. and MAAG, U. R. (1978). "Distribution results for modified Kolmogorov-Smirnov statistics for truncated or censored samples." *Technometrics* 20, 29–32.

DUMOUCHEL, W. H. (1983). "Estimating the stable index α in order to measure tail thickness: A critque." *Ann. Statist.* 11, 1019–1031.

DWASS, M. (1960). Some k-sample rank order tests. *Contributions to probability and Statistics*. Stanford, Calif., Stanford Univ. Press, 198–202.

DWASS, M. (1964). "Extremal processes." *Ann. Math. Statist.* 35, 1718–1725.

DWASS, M. (1966). "Extremal processes II." *III. J. Math.,* 10, 381–391.

DWASS, M. (1973). Extremal processes III. Discussion Paper 41, Northwestern Univ.

DZIUBDZIELA, W. (1972). "Limit distributions of extreme order statistics in a sequence of random size." *Applicationes Math.* 13, 199–205.

DZIUBDZIELA, W. and KOPOCINSKI, B. (1976). "Limiting properties of the kth record values." *Applicationes Math.* 15, 187–190.

DZIUBDZIELA, W. (1984). "Limit laws for k^{th} order statistics from strong-mixing processes." *J. Appl. Probab.* 21, 720–729.

EARLE, M. D., EFFERMEYER, C. C. and EVANS, D. J. (1974). "Height-period joint probabilities in hurricane Camille." *J. Waterw. Harb. Coast. Eng. Div.,* ASCE 3, 257–264.

EATON, M. L. (1982). "A review of selected topics in multivariate probability inequalities." *Ann. Statist.* 10, 11–43.

EICKER, F. (1979). "The asymptotic distribution of the suprema of the standardized empirical processes." *Ann. Statist.* 7, 116–138.

ELDREDGE, G. G. (1957). "Analysis of corrosion pitting by extreme-value statistics and its application to oil well tubing caliper surveys." *Corrosion* 13, 67–76.

EMBRECHTS, P. and GOLDIE, C. H. (1981). "Comparing the tail of an infinitely divisible distribution with integrals of its Lévy measure." *Ann. Probab.* 9, 468–481.

ENDICOTT, H. S. and WEBER, K. H. (1956). Extremal nature of dielectric breakdown— effect of sample size. Symposium on Minimum Property Values of Electrical Insulating Materials, *ASTM Special Technical Publication No. 188,* ASTM, Philadelphia, Pa., 5–11.

ENDICOTT, H. S. and WEBER, K. H. (1957). Electrode area effect for the impulse breakdown of transformer oil. AIEE Transactions. Part III: Power Apparatus and Systems 76, 393–398.

ENGELHARDT, M. (1975). "On simple estimation of the parameters of the Weibull or extreme-value distribution." *Technometrics* 17, 369–374.

ENGELHARDT, M., BAIN, L. J. and WRIGHT, F. T. (1981). "Inferences on the parameters of the Birnbaum-Saunders fatigue life distribution based on maximum likelihood estimation." *Technometrics* 23, 251–256.

EPSTEIN, B. (1948a). "Statistical aspects of fracture problems." *J. Appl. Phys.* 19, 140–147.

EPSTEIN, B. (1948b). "Applications of the theory of extreme values in fracture problems." *J. Amer. Statist. Assoc.* **43**, 403–412.

EPSTEIN, B. and HAMILTON, B. (1948). "The theory of extreme values and its implications in the study of the dielectric strength of paper capacitors." *J. Appl. Phys.* **19**, 544–550.

EPSTEIN, B. (1954). "Truncated life tests in the exponential case." *Ann. Math. Statist.* **25**, 555–564.

EPSTEIN, B. and SOBEL, M. (1954). "Some theorems relevant to life testing from an exponential distribution." *Ann. Math. Statist.* **25**, 373–381.

EPSTEIN, B. (1960). "Elements of the theory of extreme values." *Technometrics* **2**, 27–41.

ESARY, J. D. and MARSHALL, A. W. (1974). "Multivariate distributions with exponential minimums." *Ann. Statist.* **2**, 84–98.

ETTINGER, P. (1966a). "Sur l'attraction de la somme de deux variables aléatoires par l'une des lois limites de la théorie des valeurs extremes, dans certains cas de dépendance." *Comptes Rendus de l'Academie des Sciences,* Paris, 262, 138–140.

ETTINGER, P. (1966b). "Conditions nécessaires et suffisantes pour qu'une variable aléatoire X soit attirée par l'une des trois classes de la théorie des valeurs extremes quand X est à valeurs dans un espace euclidien à p dimensions; classe que attire la somme de deux variables aléatoires indépendantes à valeurs dans un espace euclidien à p dimensions." *Comptes Rendus de l'Académie des Sciences* (Paris) A **263**, 620–623.

ETTINGER, P. (1967). "Etude d'une condition nécessaire et suffisante pour qu'une variable aléatoire à valeurs dans un espace euclidien à m dimensions et dont les marges sont attirées par des lois limites differentes, sont attirée par une des lois limites de la théorie des valeurs extremes." *Comptes Rendus de l'Académie des Sciences* (Paris) A **265**, 138–141.

EVANS, U. R. (1955). "Statistical methods in metallurgy." *Metallurgia* **52**, 107–111.

FABENS, A. J. and NEUTS, M. F. (1970). "The limiting distribution of the maximum term in a sequence of random variables defined on a Markov chain." *J. Appl. Probab.* **7**, 754–760.

FERNANDEZ-CANTELI, A. (1982). Statistical interpretation of the Miner-number using an index of probability of total damage. IABSE Colloquium Fatigue of Steel and Concrete Structures, Lausanne.

FERNANDEZ-CANTELI, A., ESSLINGER, V. and THURLIMANN, B. (1984). Ermüdungsfestigkeit von bewehrungs und spannstählen. Bericht Nr. 8002-1. Institut für Baustatik und Konstruktion, ETH, Zürich.

FINETTI, B. De (1932). "Sulla legge di probabilità degli estremi." *Metron* **9**, 127–138.

FINKELSTEIN, B. V. (1953). "On the limiting distribution of the extreme terms of a variational series of a two-dimensional random quantity." *Dokl. Akad.* SSSR **91**, 209–211.

FISHER, J. C. and HOLLOMON, J. H. (1947). "A statistical theory of fracture." *Transactions of the American Institute of Mining and Metallurgical Engineers* **171**, 546–61.

FISHER, R. A. and TIPPETT, L. H. C. (1928). "Limiting forms of the frequency distributions of the largest or smallest member of a sample. " *Proc. Cambridge Philos. Soc.* **24**, 180–190.

FORNEY, D. M. (1972). Reliability analysis methods for metallic structures. Proceedings of the Colloquium on Structural Reliability: The Impact of Advanced Materials on Engineering Design, 237–243.

FOWLER, F. H. (1945). "On fatigue testing under triaxial static and fluctuating stresses and a statistical explanation of size effect." *Trans. Amer. Soc. Mech. Eng.* **67**, 213–216.

FRANKEL, J. P. (1948). "Relative strength of Portland cement mortar in bending under various loading conditions." *Proc. Amer. Concr. Inst.* **45**, 21–32.

FRÉCHET, M. (1927). "Sur la loi de probabilité de l'écart maximum." *Ann. Soc. Polon. Math,* Cracow, **6**, 93.

FRÉCHET, M. (1951). "Sur les tableaux de corrélation dont les marges sont données." *Ann. Univ. Lyon,* **A**, Ser. 3, **14**, 53–77.

FREUDENBERG, W. and SZYNAL, D. (1976). "Limit laws for a random number of record values." *Bull. Acad. Pol. Sci. Ser. Math. Astron. et Phys.* **24**, 193–199.

FREUDENTHAL, A. M. (1946). "The statistical aspect of fatigue of materials." *Proc. Roy. Soc. Lond.* **A187**, 416–429.

FREUDENTHAL, A. M. (1951). Planning and interpretation of fatigue tests. Symposium on Statistical Aspects of Fatigue Tests, Special Technical Publication No. 121, ASTM, Philadelphia, Pa.

FREUDENTHAL, A. M., GUMBEL E. J. and DERMAN, C. (1953). Minimum life in fatigue failure. Technical report T-1A, Department of Industrial Engineering, Columbia University, New York

FREUDENTHAL, A. M. and GUMBEL, E. J. (1953). "On the statistical interpretation of fatigue tests." *Proc. Roy. Soc. Lond.* **A216**, 319–322.

FREUDENTHAL, A. M. and GUMBEL E. J. (1954a). "Minimum life in fatigue." *J. Amer. Statist. Assoc.* **49**, 575–597.

FREUDENTHAL, A. M. and GUMBEL E. J. (1954b). "Failure and survival in fatigue." *J. Appl. Phys.* **25**, 1435.

FREUDENTHAL, A. M. (1954). "Safety and the probability of structural failure." *Proc. ASCE* **80**, separate 468.

FREUDELTHAL, A. M. (1956). Physical and statistical aspects of cumulative damage. *Colloquium on Fatigue,* Stockholm (edited by W. Weibull and F. K. G. Odquist). Springer-Verlag, 53–62.

FREUDENTHAL, A. M. and GUMBEL E. J. (1956a). "Physical and statistical aspects of fatigue." *Adv. Appl. Mech.* **4**, 117–158.

FREUDENTHAL, A. M. and GUMBEL E. J. (1956b). Distribution functions for the prediction of fatigue life and fatigue strength. *Proceedings of the International Conference on Fatigue of Metals.* Institution of Mechanical Engineers, London.

FREUDENTHAL, A. M., ITAGAKI, H. and SHINOZIKA, M. (1966). Time to first failure for various distributions of time to failure, AFML-TR-66-241, Air Force Materials Laboratory, Wright-Patterson AFB, Ohio. AD807419.

FREUDENTHAL, A. M. (1972). Reliability analysis based on time to the first failure. *Proceedings of the Symposium on Aircraft Fatigue: Design, Operational and Economic Aspects,* Melbourne, Australia (edited by J. Y. Mann and I. S. Milligan). Pergamon Press.

FREUDENTHAL, A. M. (1974). "New aspects of fatigue and fracture mechanics." *Eng. Fract. Mech.* **6,** 775–793.

FREUDENTHAL, A. M. (1975). Reliability assessment of aircraft structures based on probabilistic interpretation of the sactter factor. AFML-TR-74-198, Air Force Materials Laboratory, Wright-Patterson AFB, Ohio. AD-A014359.

FRANSEN, A. (1971). "Estimation of location and scale parameters from ordered samples, with numerical applications to the Gumbel and Weibull distributions." *FOA Reports* **5,** 1–19.

FRANSEN, A. (1974). Table of coefficients for the best linear estimates of location and scale parameters for the Gumbel and Weibull distributions for samples sizes up to and including $n = 31$. Tables of coefficients for the Cramér-Rao lower bound and for the minimum variance and the efficiency for the Gumbel distribution." *FOA Reports* **8,** 1–31. N74-29033.

GABRIEL, K. (1979). Anwendung von statistischen methoden und wahrscheinlichkeitsbetrachtungen auf das verhalten von bündeln und seilen als zugglieder aus vielen und langen drähten. Weitgespannte flächentragwerke 2, Internat. symposium, Stuttgart.

GAEDE, K. (1942). "Anwendung statistischer Untersuchungen auf die Prüfung von Baustoffen," *Bauingenieur* **23,** 291–296.

GALAMBOS, J. and RÉNYI, A. (1968). "On quadratic inequalities in the theory of probability." *Studia Sci. Math. Hungar.* **3,** 351–358.

GALAMBOS, J. (1969). "Quadratic inequalities among probabilities." *Ann. Univ. Sci. Budapest, Sectio Math.* **12,** 11–16.

GALAMBOS, J. (1972). On the distribution of the maximum of random variables." *Ann. Math. Statist.* **43,** 516–521.

GALAMBOS, J. (1973). "The distribution of the maximum of a random number of random variables with applications." *J. Appl. Probab.* **10,** 122–129.

GALAMBOS, J. and SENETA, E. (1973). "Regularly varying sequences." *Proc. Amer. Math. Soc.* **41,** 110–116.

GALAMBOS, J. (1974). "A limit theorem with applications in order statistics." *J. Appl. Probab.* **11,** 219–222.

GALAMBOS, J. (1975a). "Methods for proving Bonferroni type inequalities." *J. London Math. Soc.* **9,** 561–564.

GALAMBOS, J. (1975b). "Characterizations of probability distributions by properties of order statistics. Statistical distributions in scientific work" (edited by G. P. Patil et al.). *D. Reidel Dordrecht,* **3,** 71–88.

GALAMBOS, J. (1975c). "Order statistics of samples from multivariate distributions." *J. Amer. Statist. Assoc.* **70,** 674–680.

GALAMBOS, J. (1975d). "Limit laws for mixtures with applications to asymptotic theory of extremes." *Zeitschrift fur Wahrsch. verw. Geb.* **32,** 197–207.

GALAMBOS, J. and SENETA, E. (1975). "Record times." *Proc. Amer. Math. Soc.* **50**, 383–387.

GALAMBOS, J. (1977a). *The Asymptotic Theory of Extreme Order Statistics. The Theory and Applications of Reliability* (edited by C. Tsokos and I. Shimi). Academic Press, New York, 151–164.

GALAMBOS, J. (1977b). "Bonferroni inequalities." *Ann. Probab.* **5**, 577–581.

GALAMBOS, J. and KOTZ, S. (1978). *Characterizations of Probability Distributions.* Lecture Notes in Mathematics, Springer Verlag, Heidelberg.

GALAMBOS, J. (1978). *The Asymptotic Theory of Extreme Order Statistics.* John Wiley and Sons, New York. (its russian translation was published in 1984 by NAUKA, Moscow).

GALAMBOS, J. (1980). "A statistical test for extreme value distributions." *Colloquia Mathematica Societatis János Bolyai,* 221–229.

GALAMBOS, J. and MUCCI, R. (1980). "Inequalities for linear combinations of binomial moments." *Publicationes Mathematicae* **27**, 263–268.

GALAMBOS, J. (1981). "Extreme value theory in applied probability." *Math Scientist* **6**, 13–26.

GALAMBOS, J. (1982a). *The Role of Exchangeability in the Theory of Order Statistics.* Exchangeability in Probability and Statistics. North-Holland Publishing Company.

GALAMBOS, J. (1982b). "The role of functional equations in stochastic model building," *Aequationes Mathematicae* **25**, 21–41.

GALAMBOS, J. (1984a). "Order statistics." *Handbook of statistics* **4**, 359–382.

GALAMBOS, J. (1984b). Introduction, order statistics, exceedances. Law of large numbers. In: Statistical Extremes and Applications. NATO ASI Series. D. Reidel Publishing Company.

GALAMBOS, J. (1984c). Asymptotics; Stable laws for extremes; tail properties. In: Statistical Extremes and Applications. NATO ASI Series. D. Reidel Publishing Company.

GALAMBOS, J. (1984d). Rates of convergence in extreme value theory. In: Statistical extremes and Applications. NATO ASI Series. D. Reidel Publishing Company.

GALAMBOS, J. (1985a). Multivariate order statistics. In: *Encyclopedia of Statistical Sciences* **6**, 100–104.

GALAMBOS, J. (1985b). "A new bound on multivariate extreme value distributions." *Annales Universitatis Rolando Eötvös* **27**, 37–40.

GEFFROY, J. (1958). "Contributions à la théorie des valeurs extrêmes." *Publ. Inst. Statist. Univ. Paris,* **7**, 37–185.

GERONTIDIS, I. and SMITH, R. L. (1982). "Monte Carlo generation of order statistics from general distributions." *Appl. Statist.* **31**, 238–243.

GILLET, H. W. (1940). "The size effect in fatigue." *Metals and Alloys* **A11**, 19–94.

GIMÉNEZ-CURTO, L. A. (1979). Comportamiento de los diques rompeolas bajo la acción del oleaje. Doctoral Dissertation, University of Santander, Spain.

GLAZ, J. and JOHNSON, B. M. (1984). "Probability inequalities for multivariate distributions with dependence structures." *J. Amer. Statist. Assoc.* **79**, 436–440.

GNEDENKO, B. V. (1941). "Limit theorems for the maximal term of a variational series." *Comptes Rendus de l'Académie des Sciences de l'URSS* **32**, 7–9.

GNEDENKO, B. V. (1943). "Sur la distribution limite du terme maximum d'une série aléatoire." *Ann. Math.* **44**, 423–453.

GOLDSTEIN, N. (1963). "Random numbers for the extreme value distribution." *Publ. Inst. Statist. Univ. Paris* **12**, 137–158.

GOMES, M. I. (1978). Some probabilistic and statistical problems in extreme value theory. Ph. D. Thesis. Sheffield, England.

GOMES, M. I. (1979). "Rates of convergence in extreme value theory." *Rev. Univ. Santander,* **2**, 1021–1023.

GOMES, M. I. (1984). "Penultimate limiting forms in extreme value theory." *Ann. Inst. Statist. Math.* **36**, 71–85.

GOODKNIGHT, R. G. and RUSSEL, T. L. (1963). "Investigation of statistics of wave heights." *J. of Waterw. Harb. Div.*, ASCE **89**, WW2, 29–54.

GOVINDARAJULU, Z. (1964). "A supplement to Mendenhall's bibliography on life testing and related topics." *J. Amer. Statist. Assoc* **59**, 1231–1291.

GOVINDARAJULU, Z. (1966). "Characterization of the exponential and power distributions." *Skand. Aktuar.* **49**, 132–136.

GOVINDARAJULU, Z. and JOSHI, M. (1968). "Best linear unbiased estimation of location and scale parameters of Weibull distribution using ordered observations." Reports of Statistical Application Research, *Union Japan. Scient. Eng.* **15**, 1–14.

GRAVEY, A. (1985). "A simple construction of an upper bound for the mean of the maximum of *n* idemtically distributed random variables." *J. Appl. Probab.* **22**, 844–851.

GREEN, R. F. (1967). "Partial attraction of maxima." *J. Appl. Probab* **13**, 159–163.

GRIFFITH, A. A. (1920). "The phenomena of rupture and flow in solids." *Philosophical Trans. Roy. Soc.* Lond. **A221**, 163–198.

GRIFFITH, A. A. (1924). The theory of rupture. *Proceedings of the First International Congress for Applied Mechanics,* Delft, 55–63.

GRIGORIU, M. (1984a). "Extreme of correlated translation series." *J. Struct. Eng. Div.*, ASCE, **110**, ST7.

GRIGORIU, M. (1984b). "Estimate of extreme winds from short records." *J. Struct. Eng. Div.*, ASCE, **110**, ST7, 1467–1484.

GRINGORTEN, I. I. (1963). "Envelopes of ordered observations applied to meteorelogical extremes." *J. Geophys. Res.*, 815–826.

GRINGORTEN, I. I. (1963). "A plotting rule for extreme probability paper." *J. Geophys. Res.* **68**, 813, 814.

GROSS, A. J. and HOSMER, D. W. (1978). "Approximating tail areas of probability distributions." *Ann. Statist.* **6**, 1352–59.

GROVER, H. J. (1966). Fatigue of aircraft structures, NAVAIR 01-1A-13. U.S. Government Office, Washington.

GUMBEL, E. J. (1934). "Les moments des distributions finales de la première et de la dernière valeur." *Comptes Rendus de l'Académie des Sciences,* Paris, **198**, 141–143.

GUMBEL, E. J. (1935). "Les valeurs extremes des distributions statistiques." *Annales de l'institut Henri Poincaré* **5**, 115–158.

GUMBEL, E. J. (1935). "La plus grande valeur." *Aktuárské Vedy* **5**, 83–39, 133–143 and 145–160.

GUMBEL, E. J. (1937). Der grösste Wert einer statistischen Veränderlichen. Comptes Rendus du Congrès International des Mathématiciens, Oslo 2,200–203.

GUMBEL, E. J. (1939). "Statistische theorie der grössten werte." *Zeitschrift für Schweizerische Statistik und Volkswirtschaft* **75**, 250–271.

GUMBEL, E. J. (1946). "Determination commune des constants dans les distributions des plus grandes valeurs." *Comptes rendus de l'Académie des Sciences*, Paris, **222**, 34–36.

GUMBEL, E. J. (1954a). Statistical theory of extreme values and some practical applications. National Bureau of Standards, Applied Mathematics Series, No. 33, U.S. Government Printing Office, Washington.

GUMBEL, E. J. (1954b). "The maxima of the mean largest value and of the range." *Ann. Math. Statist.* **25**, 76–84.

GUMBEL, E. J. and CARLSON, P. G. (1954). "Extreme values in aeronautics." *J. Aeron. Science.* **21**, 389–398.

GUMBEL, E. J. and LIEBLEIN, J. (1954). "Some applications of extreme value methods." *Amer. Statist'n* **8**, 14–17.

GUMBEL, E. J. (1955). *Statistical Estimation of the Endurance Limit.* Technical report T-3A, Department of Engineering, Columbia Univ. Press., New York.

GUMBEL, E. J. (1957a). *Extreme Values in Technical Problems.* Technical report T-7A, Department of Engineering, Columbia Univ., Press., New York, AD120920.

GUMBEL, E. J. (1957b). *Statistical Analysis of Fatigue.* Technical report T-8A, Department of Engineering, Columbia Univ., New York, AD126029.

GUMBEL, E. J. (1958). *Statistics of Extremes.* Columbia Univ. Press., New York.

GUMBEL, E. J. (1960). "Bivariate exponential distributions." *J. Amer. Statist. Assoc.* **55**, 698–707.

GUMBEL, E. J. (1962a). Statistical theory of extreme values. In: *Contributions to Order Statistics* (edited by A. S. Sarhan and B. G. Greenberg). John Wiley and Sons, New York, 56–93 and 406–431.

GUMBEL, E. J. (1962b). "Multivariate extremal distributions." *Bulletin de l'Institut International de Statistique* **39**, 471–475.

GUMBEL, E. J. (1962c). "Statistical theory of breaking strength and fatigue failure." *Bulletin de l'Institut International de Statistique* **39**, 375–393.

GUMBEL, E. J. (1963). "Parameters in the distribution of fatigue life." *J. Eng. Mech. Div.*, ASCE, **89**, EM5, 45–63.

GUMBEL, E. J. (1964). "Technische Anwendung der statistischen Theorie der Extremwerte." *Schweizer Archiv* **30**, 33–47.

GUMBEL, E. J. and GOLDSTEIN, N. (1964). "Analysis of empirical bivariate extremal distributions." *J. Amer. Statist. Assoc.* **59**, 794–816.

GUMBEL, E. J. (1965). "A quick estimation of the parameters in Fréchet's distribution." *Revue de l'Institut International de Statistique* **33**, 349–363.

GÜNBAK, A. R. (1978). Statistical analysis of 21 wave records off the danish white sand coast during one storm in January 1976. Division of Port and Ocean engineering. The Univ. of Trondheim, Norway.

GUPTA, V. K., DUCKSTEIN, L. and PEEBLES, R. W. (1976). "On the joint distribution of the largest flood and its time of occurrence," *Water Resources Res.* **12**, 295–304.

GUPTA, R. C. (1984). "Relationship between order statistics and record values and some characterization results." *J. Appl. Probab.* **21**, 425–430.

GUPTA, R. C. and LANGFORD, E. C. (1984). "On the determination of a distribution by its median residual life function: A functional equation." *J. Appl. Probab.* **21**, 120–128.

GURNEY, C. (1945). "Effect of length on tensile strength." *Nature* **155**, 273–274.

GURNEY, C. (1947). The statistical estimation of the effect of size on the breaking strength of rods. Aeronautical Research Council, London, Report Memorandum No. 2157.

GURNEY, C. and PEARSON, S. (1947). The effect of length on the strength of a steel wire. Aeronautical Research Council, London, Report Memorandum No. 2158.

GYIRES, B. (1975). "Linear order statistics in the case of samples with non-independent elements." *Publ. Math. Debrecen* **22**, 47–63.

HAAN, L. de (1970). "On regular variation and its application to the weak convergence of sample extremes." *Mathematical Centre Tracts* **32**, Amsterdam.

HAAN, L. de (1971). "A form of regular variation and its application to the domain of attraction of the double exponential distribution." *Zeitschrift für Wahrsch. werw. Geb.* **17**, 241–258.

HAAN, L. de, and HORDIJK, A. (1972). "The rate of growth of sample maxima." *Ann. Math. Statist.* **43**, 1185–1196.

HAAN, L. de, and RESNICK, S. I. (1973). "Almost sure limit points of record values." *J. Appl. Probab.* **10**, 528–542.

HAAN, L. de (1974a). "Equivalence classes of regularly varying functions." *Stochastic Processes Appl.* **2**, 243–259.

HAAN, L. de (1974b). "Weak limits of sample range." *J. Appl. Probab.* **11**, 836–841.

HAAN, L. de, and RESNICK, S. I. (1977). "Limit theory for multivariate sample extremes." *Zeitschrift für Wahrsch. verw. Geb.* **40**, 317–337.

HAAN, L. de (1981). "Estimation of the minimum of a function using order statistics." *J. Amer. Statist. Assoc.* **76**, 467–469.

HAAN, L. de, and RESNICK, S. I. (1984). "Asymptotically balanced functions and stochastic compactness of sample extremes." *Ann. Probab.* 588–608.

HAEUSLER, E. and TEUGELS, J. L. (1985). "On asymptotic normality of Hill's estimator for the exponent of regular variation." *Ann. Statist.* **13**, 743–756.

HAJDIN, N. (1976). "Vergleich zwischen den Paralleldrahtseilen und verschlossenen Seilen am Beispiel der Eisenbahnschrägseilbrücke über die Save in Belgrad." *IVBH-Vorbericht zum* **10**, 471–475.

HALL, P. (1978a). "On the extreme terms of a sample from the domain of attraction of a stable law." *J. London Math. Soc.* **18**, 181–191.

HALL, P. (1978b). "Representations and limit theorems for extreme value distributions." *J. Appl. Probab.* **15**, 639–644.

HALL, P. (1978c). "Some asymptotic expansions of moments of order statistics." *Stochastic Process. Appl.* **7**, 265–275.

HALL, P. (1979). "On the rate of convergence of normal extremes." *J. Appl. Prob.* **16**, 433–439.

HALL, P. (1982). "On estimating the endpoint of a distribution." *Ann. Statist* **10**, 556–568.

HALL, P. and WELSH, A. H. (1985). "Adaptive estimates of parameters of regular variation," *Ann. Statist.* **13**, 331–341.

HAMILTON, B. and RAWSON, H. (1970). "The determination of the flaw distributions on various glass surfaces from Hertz fracture experiments." *J. Mech. Phys. Solids* **18**, 127–147.

HANNAMAN, G. W., SPURGIN, A. J. and LUKIC, Y. (1985). A model for assessing human cognitive reliability in PRA studies. Third IEEE Conference on Human reliability, Monterey, California.

HARRIS, R. (1970). "An application of extreme value theory to reliability theory." *Ann. Math. Statist.* **41**, 1456–1465.

HARTER, H. L. (1964). Expected values of exponential, Weibull and Gamma order statistics. ARL Technical Report 64–31, Aerospace Research Laboratory, Wright-Patterson AFB. Ohio, AD436763.

HARTER, H. L. and MOORE, A. H. (1965a). "Point and interval estimators, based on *m* order statistics, for the scale parameter of a Weibull population with known shape parameter." *Technometrics* **7**, 405–422.

HARTER, H. L. and MOORE, A. H. (1965b). "Maximum likelihood estimation of the parameters of Gamma and Weibull populations from complete and from censored samples." *Technometrics* **7**, 639–643.

HARTER, H. L. and MOORE, A. H. (1965c). "Asymptotic variances and covariances of maximum-likelihood estimators, from censored samples, of the parameters of Weibull and Gamma populations." *Ann. Math. Statist.* **38**, 557–570.

HARTER, H. L. and MOORE, A. H. (1965d). "A note on estimation from a Type *I* extreme-value distribution." *Technometrics* **9**, 325–331.

HARTER, H. L. and MOORE, A. H. (1968). "Maximum-likelihood estimation, from doubly censored samples, of the parameters of the first asymptotic distribution of extreme values." *J. Amer. Statist. Assoc.* **63**, 889–901. Corrigenda: (1973), **68**, 757.

HARTER, H. L. (1977). A survey of the literature on the size effect on material strength. AFFDL-TR-77-11, Air Force Flight Dynamics Laboratory, Wright-Patterson AFB, Ohio, AD-A041535.

HARTER, H. L. (1978a). A Cronological Annotated Bibliography on Order Statistics, Volume: Pre-1950. U. S. Government Printing Office, Washington.

HARTER, H. L. (1978b). "A bibliography of extreme value theory." *Intern. Statist. Rev.* **46**, 279–306.

HARTLEY, H. O. and DAVID, H. A. (1954). "Universal bounds for mean range and extreme observation." *Ann. Math. Statist.* **25**, 85–99.

HASOFER, A. M. (1968). "A statistical theory of the brittle fracture." *Int. J. Fract. Mech.* **4**, 439–452.

HASOFER, A. M. and SHARPE, K. (1969). "The analysis of wind gusts." *Austral. Meteorol. Mag.* **17**, 198–214.

HASOFER, A. M. (1972). "Wind-load design based on statistics." *Proc. Inst. Civil Eng.* **51**, 69–82.

HASSANEIN, K. M. (1964). Estimation of the parameters of the extreme value distribution by order statistics. Univ. of North Carolina, Chapel Hill, N.C. AD622257.

HASSANEIN, K. M. (1968). "Analysis of extreme-value data by sample quantiles for very large samples." *J. Amer. Statist. Assoc.* **63**, 877–888.

HASSANEIN, K. M. (1969). "Estimation of the parameters of the extreme value distribution by use of two or three order statistics." *Biometrika* **56**, 429–436.

HASSANEIN, K. M. (1971). "Percentile estimators for the parameters of the Weibull distribution." *Biometrika* **58**, 673–676.

HASSANEIN, K. M. (1972). "Simultaneous estimation of the parameters of the extreme value distribution by sample quantiles." *Technometrics* **14**, 63–70.

HATANO, T. (1969). "Theory of failure of concrete and similar brittle solid on the basis of strain." *International J. Fract. Mech.* **5**, 73–79.

HELGASON, T. and HANSON, J. M. (1976). Fatigue strength of high-yield reinforcing bars. National Cooperative Highway Research Program, Report **164**.

HELLER, B. and PADERTAB. (1974). "Functional equations of some types of distribution functions used in statistics." *Acta Technica CSAV* **19**, 162–169.

HELMERS, R. (1980). "Edgeworth expansions for linear combinations of order statistics with smooth weight functions." *Ann. Statist.* **8**, 1361–1374.

HELMERS, R. (1981). "A Berry-Esseen theorem for linear combinations of order statistics." *Ann. Probab.* **9**, 342–347.

HENERY, R. J. (1981). "Place probabilities in normal order statistics models for horse races." *J. Appl. Probab.* **18**, 839–852.

HERBACH, L. H. (1963). "Estimation of the three Weibull parameters using an overlooked consequence of a well-known property of the Weibull distribution" (abstract). *Ann. Math. Statist.* **34**, 681–682.

HERBACH, L. H. (1984). Introduction, Gumbel model. In: *Statistical Extremes and Applications.* NATO ASI Series. D. Reidel Publishing Company.

I'HERMITE, R. (1939). "Étude statistique des bétons de chantier." *Annales de I'Institut Technique du Batiment et des Travaux Publics,* Paris, **4**, 66–75.

HERSHFIELD, D. M. (1962). "Extreme rainfall relationships." *J. Hydraul. Div.,* ASCE, **HY6**, 73–92.

HIGUCHI, T., LEEPER, H. M. and DAVIS, D. S. (1948). "Determination of tensile strength of natural rubber and GR-S. Effect of specimen size." *Analytical Chemistry,* **20**, 1029–1033.

HILL, L. R. and SCHMIDT, P. L. (1948). "Insulation breakdown as a function of area" (abstract). *Electrical Engineering, 67,* 76.

HILL, B. M. (1975). "A simple general approach to inference about the tail of a distribution." *Ann. Statist.* **3,** 1163–1174.

HILMES, J. B. and EKBERG, C. E. (1965). Statistical analysis of fatigue characteristics of under-reinforced prestressed concrete flexural members. Report IOWA Engineering Experiment Station, *Iowa State Univ.*

HIRT, M. A. (1977). "Neue Erkenntnisse auf dem Gebiet der ermüng und deren Berücksichtigung bei der Bemessung von Eisenbahnbrücken." *Der Bauingenieur* **52,** 255–262.

HOGLUND, T. (1972). "Asymptotic normality of sums of minima of random variables." *Ann. Math. Statist.* **43,** 351–353.

HOMMA, T. (1951). "On the asymptotic independence of order statistics." *Rep. Stat. Appl. Res.* JUSE **1,** 1–3.

HOROWITZ, J. (1980). "Extreme values from a non-stationary stochastic process: An application to air quality analysis." *Technometrics* **22,** 469–478.

HOSKING, J. R. M. (1984). "Testing whether the shape parameter is zero in the generalized extreme-value distribution." *Biometrika* **71,** 367–374.

HOSKING, J. R. M., WALLIS, J. R. and WOOD, E. F. (1985). "Estimation of the generalized extreme-value distribution by the method of probability weighted moments." *Technometrics* **27,** 251–261.

HOUMB, O. G. and OVERVIK, J. (1977). On the statistical properties of 115 wave records from the norwegian continental shelf. Div. of Ports and Ocean Engineering. The Univ. of Trondheim, Norway.

HUANG, J. S. (1974). "Characterizations of the exponential distribution by order statistics." *J. Appl. Probab.* **11,** 605–608.

HUANG, J. S. and GHOSH, M. (1982). "A note on strong unimodality of order statistics." *J. Amer. Statist. Assoc.* **77,** 929–930.

HÜCK, M. (1983). "Ein verbessertes Verfahren für die Auswertung von Treppenstufenversuchen." *Z. Werkstofftech.* **14,** 406–417.

HUNTER, D. (1976). "An upper bound for the probability of a union." *J. Appl. Probab.* **13,** 597–603.

IBRAGIMOV, I. A. and ROZANOV, Y. A. (1970). *Gaussian Stochastic Processes.* Izdatelstvo Nauka, Moscow.

IBRAGIMOV, I. A. and LINNIK, Y. V. (1971). *Independent and Stationary Sequences of Random Variables.* Wolters-Noorhoff, Groningen.

IKEDA, S. and MATSUNAWA, T. (1970). "On asymptotic independence of order statistics." *Ann. Inst. Statist. Math.* Tokyo **22,** 435–449.

JAESCHKE, D. (1979). "The asymptotic distribution of the supremum of the standardized empirical distribution function on subintervals." *Ann. Statist.* **7,** 108–115.

JASTRZEBSKI, P. (1961). "The influence of the length on the strength of steel bars subjected to tension." *Bull. Acad. Polon. Sciences* **9,** 129–137.

JELLINEK, H. H. G. (1958). "The influence of imperfections on the strength of ice." *Proc. Phys. Soc.,* London, **71,** 797–814.

JENSEN, A. (1969). A characteristic application of statistics in hydrology. 37th Session ISI, London.

JOHNSON, N. L. and KOTZ, S. (1972). *Distributions in Statistics.* John Wiley and Sons, New York.

JOHNS, M. V. and LIEBERMAN, G. J. (1966). "An asymptotically efficient confidence bound for reliability in the case of the Weibull distribution." *Technometrics* **8,** 135–175.

JOHNSON, A. I. (1953). *Strength, Safety and Economical Dimensions of Structures.* Victor Pettersons Bokindustri Aktiebolag, Stockholm.

JONES, B. H. (1975). *Probabilistic Design and Reliability.* Composite materials, **8:** Structural Design and Analysis (edited by C. C. Chamis). Academic Press.

JUDICKAJA, P. I. (1974). "On the maximum of a Gaussian sequence." *Theory Probab. Math. Statist.* **2,** 259–267.

JUNCOSA, M. L. (1949). "On the distribution of the minimum in a sequence of mutually independent random variables." *Duke Math. J.* **16,** 609–618.

KABIR, A. B. M. (1968). "Quantile estimators of the parameters of extreme value and Weibull distribution." *Bull. Inst. Statist. Res. and Training* **2,** 92–100.

KAGAN, A. M., LINNIK, Y. V. and RAO, C. R. (1973). *Characterization Problems in Mathematical Statistics.* John Wiley, New York.

KAO, J. H. K. (1956). "Weibull distribution in life-testing of electron tubes" (abstract). *J. Amer. Statist. Assoc.* **51,** 514

KARR, A. (1976). Two extreme value processes arising in hydrology. *J. Appl. Probab.* **13,** 190–194.

KAWATA, T. (1951). "Limit distributions of single order statistics." *Rep. Stat. Appl. Res.* JUSE **1,** 4–9.

KENDALL, D. G. (1967). "On finite and infinite sequences of exchangeable events." *Studia Sci. Math.* Hungar. **2,** 319–327.

KIMBALL, B. F. (1946). "Significant statistical estimation functions for the parameters of the distribution of maximum values." *Ann. Math. Statist.* **17,** 299–309.

KIMBALL, B. F. (1949). "An approximation to the sampling variance of an estimated maximum value of given frequency based on fit of doubly exponential distribution of maximum values." *Ann. Math. Statist.* **20,** 110, 113.

KIMBALL, B. F. (1960). "On the choice of plotting poitions on probability paper." *J. Amer. Satist. Assoc.* **55,** 546–60.

KIRBY, W. (1969). "On the random occurrences of major floods" *Water Resour. Res.* **5,** 778–784.

KOGAEV, V. P. (1964). Effect of stress concentration and scale factor on fatigue strength viewed from its statistical aspect. V. sb. Voprosy Mekhaniki Ustalosti, Mashinostroenie, Moscow, 67–100.

KONTOROVA, T. A. (1940). "Statistical theory of strength." *J. Techn. Phys.* **10,** 886–890.

KONTOROVA, T. A. (1943). "Concerning one of the applications of the statistical theory of the scale factor." *J. Techn. Physics* **13,** 296–308.

KONTOROVA, T. A. and FRENKEL, J. I. (1943). "Statistical theory of brittle strength of real crystals." *J. Techn. Phys.* **7,** 108–114.

KONTOROVA, T. A. and TIMOSHENKO, O. A. (1949). "Generalization of the statistical theory of strength for non-uniformly stressed bodies." *J. Techn. Phys.* **19,** 355–370.

KOTZ, S. and JOHNSON, N. L. (1982). *Encyclopedia of Statistical Sciences.* John Wiley and Sons, New York.

KOUNIAS, E. G. (1968). Bounds for the probability of a union, with application. Ann. Math. Statist. **39,** 2154–2158.

KOUNIAS, S. and MARIN, J. (1976). "Best linear Bonferroni bounds." *SIAM J. Appl. Math.* **30,** 307–323.

KRONMAL, R. A. and PETERSON, A. V. (1981). "A variant of the acceptance rejection method for computer generation of random variables," *J. Amer. Statist. Assoc.* **76,** 446–451.

KUBAT, P. and EPSTEIN, B. (1980). "Estimation of quantiles of location-scale distributions based on two or three order statistics." *Technometrics* **22,** 575–581.

KUCZMA, M. (1968). *Functional Equations in a Single Variable.* Polish Scientific Publishers, Warsaw.

KUDO, A. (1958). "On the distribution of the maximum value of an equally correlated sample from a normal population." *Sankhya* **A20,** 309–316.

KUNIO, T., SHIMIZU, M., YAMADA, K. and KIMURA, Y. (1974). "An interpretation of the scatter of fatigue limit on the basis of the theory of extreme value." *Trans. JSME* **40,** 2101–2109.

KWEREL, S. M. (1975a). "Most stringent bounds on aggregated probabilities of partially specified dependent systems." *J. Amer. Statist. Assoc.* **70,** 472–479.

KWEREL, S. M. (1975b). "Bounds on the probability of the union and intersection of m events." *Adv. Appl. Probability* **7,** 431–448.

KWEREL, S. M. (1975c). "Most stringent bounds on the probability of the union and intersection of m events for systems partially specified by $S_1, S_2, \ldots, S_k$ $2 \leq k \leq m$." *J. Appl. Probability* **12,** 612–619.

LAMPERTI, J. (1964). "On extreme order statistics." *Ann. Math. Statist.* **35,** 1726–1737.

LARDER, R. A. and BEADLE, C. W. (1975). "Strength distributions of single filaments." *J. Compos. Mater.* **9,** 241–243.

LARSEN, R. I. (1969). "A new mathematical model of air pollutant concentration averaging time and frequency." *J. Air Pollut. Contr. Assoc.* **19,** 24–30.

LAWLESS, J. F. (1978). "Confidence interval estimation for the Weibull and extreme value distribution." *Technometrics* **20,** 355–364.

LAWLESS, J. F. (1982). *Statistical Models and Methods for Lifetime Data* John Wiley and Sons, New York.

LEADBETTER, M. R. (1966). "On streams of events and mixtures of streams." *J. Roy. Statist. Soc.,* Ser. **B28,** 218–227.

LEADBETTER, M. R. (1971). Extreme value theory for stochastic processes. Proc 5th Princeton Systems Symposium.

LEADBETTER, M. R. (1974). "On extreme values in stationary sequences." *Zeitschrift für Wahrsch. verw. Geb.* **28,** 289–303.

LEADBETTER, M. R. (1975). Aspect of extreme value theory for stationary processes—A survey. In: *Stochastic Processes and Related Topics*, 1. (edited by Puri), New York.

LEADBETTER, M. R., LINDGREN, G. and ROOTZÉN, H. (1978). "Conditions for the convergence in distribution of maxima of stationary normal processes." *Stoch. Proc. Appl.* **8**, 131–139.

LEADBETTER, M. R. and ROOTZÉN, H. (1982). "Extreme value theory for continuous parameter stationary processes." *Z. Wahrsch. verw. Geb.* **60**, 1–20.

LEADBETTER, M. R., LINDGREN, G. and ROOTZÉN, H. (1983). *Extremes and Related Properties of Random Sequences and Processes.* Springer-Verlag, New York.

LEE, L. and THOMPSON, W. A. (1975). Reliability of multiple component systems. Report No. 75-1, Dept. Statist., Univ. of Missouri, Columbia, Mo. AD-AO13349.

LEHMAN, E. H. (1963). "Shapes, moments and estimators of the Weibull distribution," *IEEE Trans. Reliab.* **12**, 32–38.

LÉVI, R. (1949). "Calculs probabilistes de la securité des constructions." *Annales des Ponts et Chaussées* **119**, 493–539.

LIEBLEIN, J. (1951). "On certain estimators for the parameters of the distribution of largest values" (abstract). *Ann. Math. Statist.* **22**, 487.

LIEBLEIN, J. (1953). "On the exact evaluation of the variances and covariances of order statistics in samples from the extreme-value distribution." *Ann. Math. Statist.* **24**, 282–287.

LIEBLEIN, J. (1954). A new method of analyzing extreme value data. NACA Technical Note 3053, National Advisory committee for Aeronautics, Washington.

LIEBLEIN, J. and ZELEN, M. (1956). "Statistical investigation of the fatigue life of deep-groove ball bearings," *Journal of Research of the National Bureau of Standards* **57**, 273–316.

LIFSHITZ, J. M. and ROTEM, A. (1970). Longitudinal strength of unidirectional fibrous composites. AFML-TR-70-194, Air Force Materials Laboratory, Wright-Patterson AFB, Ohio, AD716629.

LILLIEFORS, H. W. (1969). "On the Kolmogorov-Smirnov test for the exponential distribution with mean unknown." *J. Amer. Statist. Assoc.* **64**, 387–389.

LINDGREN, G. (1971). "Extreme values of stationary normal processes." *Z. Wahrsch. verw. Geb.* **17**, 39–47.

LINDGREN, G. (1972a). "Wave-length and amplitude in Gaussian noise." *Adv. Appl. Probab* **4**, 81–108.

LINDGREN, G. (1972b). "Wave-length and amplitude for a stationary Gaussian process after a high maximum." *Z. Wahrsch. verw. Geb.* **23**, 293–326.

LINDGREN, G. (1974). "A note on the asymptotic independence of high level crossings for dependent Gaussian processes." *Ann. Probab.* **2**, 535–539.

LINDGREN, G. MARÉ, J. and ROOTZÉN, H. (1975). "Weak convergence of high level crossings and maxima for one or more Gaussian processes." *Ann. Probab.* **3**, 961–978.

LINDGREN, G. (1980). "Extreme values and crossings for the χ^2-process and other functions of multidimensional Gaussian processes, with reeliability applications." *Adv. Appl. Probab.* **12**, 746–774.

LINDGREN, G. and RYCHLIK, I. (1982). "Wave characteristic distributions for Gaussian waves-wave-length, amplitude and steepness." *Ocean Engineering* **9**, 411–432.

LITTELL, R. C. and RAO, P. V. (1978). "Confidence regions for location and scale parameters based on the Kolmogorov-Smirnov goodness of fit statistic." *Techno-metrics* **20**, 23–27.

LITTLE, R. E. (1972). "Estimating the median fatigue limit for very small up and down quantal response tests and for S-N data with runouts. Probabilistic Aspects of Fatigue." *ASTM Special Technical Publication* **511**, 29–42.

LITTLE, R. E. and JEBE, E. H. (1975). *Statistical Design of Fatigue Experiments.* Applied Science Publishers, London.

LOCAL climatological data annual and monthly summaries. U. S. Department of Commerce, National Environmental Data Service, National Climatic Center, Ashville, N.. C.

LOGAN, K. H. and GRODSKY, V. A. (1931). "Soil corrosion studies, 1930. Rates of corrosion and pitting of bare ferrous specimens." *Journal of Research of the National Bureau of Standards* **7**, 1–35.

LOGAN, K. H. (1936). "Soil corrosion studies. Rates of loss of weight and pitting of ferrous specimens." *Journal of Research of the National Bureau of Standards* **16**, 431–466.

LOH, W. (1984). "Estimating an endpoint of a distribution with resampling methods." *Ann. Statist.* **12**, 1543–50.

LONGUET-HIGGINS, M. S. (1952). On the statistical distribution of the heights of sea waves. J. Mar. Res. **9**, 245–266.

LONGUET-HIGGINS, M. S. (1975). "On the joint distribution of the periods and amplitudes of sea waves." *J. Geophys. Res.* 80, 2688–2694.

LOSADA, M. A., CASTILLO, E. and PUIG-PEY, J. (1978). "Incidencia de los datos disponibles en la fiabilidad de la estimación de la ola de cálculo en las obras marítimas." *Rev. Obras Públicas,* N. **3161**, 667–671 and 779–781.

LOYNES, R. M. (1965). "Extreme values in uniformly mixing stationary stochastic processes." *Ann. Math. Statist.* **36**, 993–999.

LURIE, D. and HARTLEY, H. O. (1972). "Machine generation of order statistics for Monte-Carlo computations." *Amer. Statist'n* **26**, 26–27.

MAENNIG, W. W. (1967). Untersuchungen zur Planung und Auswertung von Dauer-schwingversuchen an Stahl in Bereichen der Zeit und der Dauerfestigkeit. Fortschr. Ber. VDI-Z, Reihe 5, Nr. **5**, VDI-Verlag, Düsseldorf.

MAENNIG, W. W. (1970). "Bemerkungen zur Beurteilung des Dauerschwingverhaltens von Stahl und einige Untersuchungen zur Bestimmung des Dauerfestigkeits-bereichs." *Materialprüfung* **12**, Nr. 14, 124–131.

MAENNIG, W. W. (1971). "Vergleichende Untersuchung über die Eignung der Treppenstufen methode zur Berechnung der Dauerschwingfestigkeit." *Materialprü-fung* **13**, Nr. 1, 6–11.

MAHMOUD, M. W. and RAGAD, A. (1975). "On order statistics in samples drawn from the extreme value distribution." *Math. Operationsforschung u. Stat.* **6**, 800–816.

MALLOWS, C. L. (1973). "Bounds on distribution functions in terms of expectations of order statistics." *Ann. Probab.* **1**, 297–303.

MANN, N. R. (1965). Point and inteval estimates for reliability parameters when failure times have the two-parameter Weibull distribution. Ph. D. Dissertation, Univ. of California, Los Angeles. University Microfilms, Inc., Ann Arbor, Michigan.

MANN, N. R. (1967a). Results on location and scale parameter estimation with application to the extreme value distribution. ARL 67-0023, Aerospace Research Laboratory., Wright-Patterson AFB, Ohio, AD653575.

MANN, N. R. (1967b). "Tables for obtaining the best linear invariant estimates of parameters of the Weibull distribution." *Technometrics* **9**, 629–645.

MANN, N. R. (1968a). Results on statistical estimation and hypothesis testing with application to the Weibull and extreme value distributions. ARL 68-0068, Aerospace Research Laboratory, Wright-Patterson AFB, Ohio, AD672979.

MANN, N. R. (1968b). "Point and interval estimation procedures for the two-parameter Weibull and extreme-value distributions." *Technometrics* **10**, 231–256.

MANN, N. R. (1970a). Extension of results concerning parameter estimators, tolerance bounds and warranty periods for Weibull models. ARL 70-0010, Aerospace Research Laboratory, Wright-Patterson AFB, Ohio, AD708159.

MANN, N. R. (1970b). Estimation of location and scale parameters under various models of censoring and truncation. ARL 70-0026, Aerospace Research Laboratory, Wright-Patterson AFB, Ohio, 707868.

MANN, N. R. (1970c). "Estimators and exact confidence bounds for Weibull parameters based on a few ordered observations." *Technometrics* **12**, 345–361.

MANN, N. R. (1971). "Best linear invariant estimation for Weibull parameters under progressive censoring." *Technometrics* **13**, 521–533.

MANN, N. R., FERTIG, K. W. and SCHEUER, E. M. (1971). Confidence and tolerance bounds and a new goodness-of-fit test for two-parameter Weibull or extreme-value distributions. ARL 71-0077, Aerospace Research Laboratory, Wright-Patterson AFB, Ohio, AD727797.

MANN, N. R., SCHAEFER, R. E. and SINGPURWALLA, N. D. (1974). *Methods for Statistical Analysis of Reliability and Life Data.* John Wiley and Sons, New York.

MANSOUR, A. (1972). "Methods of computing the probability of failure under extreme values of bending moment," *J. Ship Res.* **6**, 113–123.

MARCUS M. B. and PINSKY, M. (1969). "On the domain of attraction of exp[−exp (−x)]." *J. Math. Anal. Appl.* **28**, 440–449.

MARCUS, M. B. (1974). "Asymptotic maxima of continuous Gaussian processes." *Ann. Probab.* **2**, 702–713.

MARDIA, K. V. (1964a). "Some results on the order statistics of the multivariate normal and Pareto type I populations." *Ann. Math. Statist.* **35**, 1815–1818.

MARDIA, K. V. (1964b). "Asymptotic independence of bivariate extremes." *Calcutta Stat. Assoc. Bull.* **13**, 172–178.

MARDIA, K. V. (1970). *Families of Bivariate Distributions*. Griffin, London.

MARKOVIC, R. D. (1965). Probability functions of best fit to distributions of annual precipitation and runoff, Hydrol. Pap. 8, Colorado State Univ., Fort Collins.

MARSAGLIA, G. (1984). "The exact-approximation method for generating random variables in a computer." *J. Amer. Statist. Assoc.* **79**, 218–221.

MARSHALL, A. W. and OLKIN, I. (1967a). "A multivariate exponential distribution." *J. Amer. Statist. Assoc.* **62**, 30–44.

MARSHALL, A. W. and OLKIN, I. (1967b). "A generalized bivariate exponential distribution." *J. Appl. Probab.* **4**, 291–302.

MARSHALL, A. W. and OLKIN, I. (1983). "Domains of attraction of multivariate extreme value distributions." *Ann. Probab.* **11**, 168–177.

MARSZAL, E. and SOJKA, J. (1974). "Limiting distributions and their application." *Wyz. Szkol. Ped. Krakow Rocznik Nauk* **7**, 59–76.

MASON, D. M. (1981). "Bounds for weighted empirical distribution functions." *Ann. Probability* **9**, 881–184.

MASON, D. M. (1982). "Laws of large numbers for sums of extreme values." *Ann. Probability* **10**, 754–764.

MASON, D. M. (1983). "The asymptotic distribution of weighted empirical distribution functions." *Stoch. Proc. Appl.* **15**, 99–109.

MASON, D. M. and SCHUENEMEYER, J. H. (1983). "A modified Kolmogorov-Smirnov test sensitive to tail alternatives." *Ann. Statist.* **11**, 933–946.

MATALAS, N. C. (1963). "Probability distribution of low flows." *U. S. Geol. Surv., Prof. Pap.* **434-A**, 1–27.

MATALAS, N. C. and WALLIS, J. R. (1973). "Eureka! it fits a Pearson type III distribution." *Water Resour. Res.* **9**, 281–291.

MATHAR, R. (1984). "The limit behaviour of the maximum of random variables with applications to outlier-resistance." *J. Appl. Probab.* **21**, 646–650.

MCADAM, D. J. GEIL, G. W., WOODWARD, D. H. and JENJINS, W. D. (1948). Influence of size and the stress system on the flow stress and fracture stress of metals. Metals Technology **15**, Technical Paper No. 2373.

MCCLINTOCK, F. A. (1955). "The statistical theory of size and shape effects in fatigue." *J. Appl. Mech.* **22**, 421–426.

MCCOOL, J. I. (1969). "Unbiased maximum-likelihood estimation of a Weibull percentile when the shape parameter is known." *IEEE Trans. Reliab.* **18**, 78–79.

MCCOOL, J. I. (1970a). "Inference on Weibull percentiles and shape parameter from maximum likelihood estimates." *IEEE Trans. Reliab.* **19**, 2–9.

MCCOOL, J. I. (1970b). "Evaluating Weibull endurance data by the method of maximum likelihood." *Trans. Amer. Soc. Lubric. Eng.* **13**, 189–202.

MCCORD, J. R. (1964). "On asymptotic moments of extreme statistics." *Ann. Math. Statist.* **35**, 1738–1745.

MCCOOL, J. I. (1966). "Inference from the third failure in a sample of 30 from a Weibull distribution." *Indust. Qual. Control* **23**, 109–114.

McElvaney, J. and Pell, P. S. (1974). "Fatigue damage of asphalt under compound loading." *J. Transp. Eng., ASCE,* **100,** TE3, 701–718.

Mandel, J. (1959). "The theory of extreme values." *ASTM Bulletin* No. 236, 29–30.

Maritz, J. S. and Munro, A. H. (1967). "On the use of the generalized extreme value distribution in estimating extreme percentiles." *Biometrics* **23,** 79–103.

Meeker, W. Q. and Nelson, W. B. (1975). Tables for the Weibull and smallest extreme value distribution. General Electric Co.

Meer, H. P., Van and Plantema, F. J. (1949). Vermoeiing van construties en constructiedelen. Rapport S. 357, Nationaal Luchtvaartlaboratorium, Amsterdam.

Mejzler, D. G. (1950). "On the limit distribution of the maximal term of a variational series." *Dopovidi Akad. Nauk Ukrain.* SSR **1,** 3–10.

Mejzler, D. G. (1953). "The study of the limit laws for the variational series." *Trudy Inst. Mat. Akad. Nauk Uzbek.* SSR **10,** 96–105.

Mejzler, D. G. (1956). "On the problem of the limit distribution for the maximal term of a variational series." *L'vov Politechn. Inst. Naucn. Zp.* **38,** 90–109.

Mejzler, D. G. (1978). "Limit distributions for the extreme order statistics." *Canad. Math. Bull.* **21,** 447–459.

Mendenhall, W. (1958). "A bibliography on life testing and related topics." *Biometrika* **45,** 521–543.

Menon, M. V. (1963). "Estimation of the shape and scale parameters of the Weibull distribution." *Technometrics* **5,** 175–182.

Metcalfe, A. G. and Smitz, G. K. (1964). "Effect of length on the strength of glass fibres." *Proc. Amer. Soc. Testing and Materials* **64,** 1075–1093.

Mexia, J. T. (1967). "Studies on the extreme double exponential distribution. Sequential estimation and testing for the location parameter of Gumbel distribution." *Revista da Faculdade de Ciencias, Universidade de Lisboa* **A12,** 5–14.

Michael, J. R. (1983). "The stabilized probability plot." *Biometrika* **70,** 11–17.

Mihram, G. A. (1969). "Complete sample estimation techniques for reparameterized Weibull distributions." *IEEE Trans. Reliab.* **18,** 190–195.

Mikhailov, V. G. (1964). "Asymptotic independence of vector components of multivariate extreme order statistics." *Teor. verojatnost. i primen.* **19,** 817–821.

Millan, J. and Yevjevich, V. (1971). Probabilities of observed draughts. Colorado State Univ. Hydrology papers, N. 50, Fort Collins, Colorado.

Mises, R. von (1936). "La distribution de la plus grande de n valeurs." *Revue Mathématique de l'Union Interbalkanique* (Athens) **1,** 141–160.

Mistéth, E. (1973). "Determination of the critical loads considering the anticipated durability of structures." *Acta Technica Academiae Scientiarum Hungaricae* **74,** 21–38.

Mistéth, E. (1974). "Dimensioning of structures for flood discharge according to the theory of probability." *Acta Technica Academiae Scientiarum Hungaricae* **76,** 107–127.

MITTAL, Y. (1974). "Limiting behaviour of maxima in stationary Gaussian sequences." *Ann. Probab* **2**, 231–242.

MITTAL, Y. and YLVISAKER, D. (1975). "Limit distributions for the maxima of stationary Gaussian processes." *Stochastic Processes Appl.* **3**, 1–18.

MITTAL, Y. (1976). Maxima of partial samples in Gaussian sequences. Technical Report No. 3, Stanford Univ.

MITTAL, Y. and YLVISAKER, D. (1976). "Strong laws for the maxima of stationary Gaussian processes." *Ann. Prob.* **4**, 357–371.

MITTAL, Y. (1979). "A new mixing condition for stationary Gaussian processes." *Ann. Probab.* **7**, 724–730.

MOGYORODI, J. (1967). "On the limit distribution of the largest term in the order statistics of a sample of random size." *Magyar Tud. Akad. Mat. Fiz. Oszt. Kozl.* **17**, 75–83.

MONFORT, M. A. J., VAN (1970). "On testing that the distribution of extremes is of type I when type II is the alternative." *J. Hydrol.* **11**, 421–427.

MONFORT, M. A. J., VAN (1973). "An asymmetric test on the type of the distribution of extremes." *Mededelingen Landbouwhogeschool* **73**, 1–15.

MOORE, A. H. and HARTER, H. L. (1967). "One-order-statistic conditional estimators of shape parameters of limited and Pareto distributions and scale parameters of type II asymptotic distributions of smallest and largest values." *IEEE Trans. Reliab.* **16**, 100–103.

MORGENSTERN, D. (1956). "Einfache Beispiele zweidimensionaler Verteilungen." *Mitteilungsblatt Math. Stat.* **8**, 234–235.

MORI, T. (1976). "Limit laws for maxima and second maxima for strong mixing processes." *Ann. Probab.* **4**, 122–126.

MORI, T. (1981). The relation of sums and extremes of random variables. Buenos Aires ISI meeting, 879–902.

MORI, T. and SZÉKELI, G. J. (1985). "A note on the background of several Bonferroni-Galambos type inequalities." *J. Appl. Probab.* **22**, 836–843.

MORIGUTI, S. (1951). "Extremal properties of extreme value distributions." *Ann. Math. Statist.* **22**, 523–536.

MOSES, F. (1974). "Reliability of structural systems." *J. of Structural Division*, ASCE, **100**, ST9, 1813–1820.

MOSZYNSKI, W. (1953). "Calculation of fatigue life of machine elements by aid of probability calculations." *Przeglad Mechaniczny* **2**, 271–274 and 325–331.

MOYER, C. A., BUSH, J. J., and RULEY, B. T. (1962). "The Weibull distribution function for fatigue life." *Research and Standards* **2**, 405–411.

MUCCI, R. (1977). Limit theorems for extremes. Doctoral dissertation, Temple University.

MURTHY, V. K. and SWARTZ, G. B. (1972). Annotated bibliography on cumulative fatigue damage and structural reliability models. ARL 72-0161, Aerospace Research Laboratory, Wright-Patterson AFB, Ohio, AD754062.

MURZEWSKI, J. (1972). "Optimization of structural safety for extreme load and strength distributions." *Archiwum Inzynierii Ladowej* **18**, 573–583.

NAGARAJA, H. N. (1977). "On a characterization based on record values." *Austral J. Statist.* **20**, 176–182.

NAGARAJA, H. N. (1985). Characterization of distributions based on linear regressions of adjacent order statistics. Conference on weighted distributions, Penn State Univ., Pa.

NAIR, K. A. (1976). "Bivariate extreme value distributions." *Comm. in Statist.* **5**, 575–581.

NELSON, W. and THOMPSON, V. C. (1971). "Weibull probability papers." *J. Qual. Technol.* **3**, 45–50.

NELSON, W. (1982). *Applied Life Data Analysis.* John Wiley and Sons, New York.

NEUTS, M. F. (1967). "Waiting times between record observations." *J. Appl. Probab.* **4**, 206–208.

NEWELL, G. F. (1964). "Asymptotic extremes for *m*-dependent random variables." *Ann. Math. Statist.* **35**, 1322–1325.

NISIO, M. (1967). "On the extreme values of Gaussian processes." *Osaka J. Math.* **4**, 313–326.

NORTH, M. (1980). "Time-dependent stochastic model of floods." *J. Hydrol. Div.* ASCE, HY5, 649–665.

NOWINSKI, J. (1956). "A series of experiments concerning the scale effect on the rupture of steel wires." *Rozprawy Inzynierskie* **59**, 565–572.

OBERT, L. (1972). *Brittle Fracture of Rock. Fracture: An Advanced Treatise,* VII: Fracture of Nonmetals and Composites. Academic Press, New York.

O'BRIEN, G. L. (1974a). "Limit theorems for the maximum term of a stationary process." *Ann. Probab.* **2**, 540–545.

O'BRIEN, G. L. (1974b). "The maximum term of uniformly mixing stationary processes." *Zeitschrift fur Wahrsch. verw. Geb.* **30**, 57–63.

OCHI, M. K. (1973). "On prediction of extreme values." *J. Ship Res.* **17**, 29–37.

OTTEN, A. and van MONFORT, M. A. J. (1978). "The power of two tests on the type of distributions of extremes." *J. Hydrol.* **37**, 195–199.

PARRAT, N. J. (1972). *Fibre-reinforced Materials Technology.* Van Nostrand Reinhold Company, New York.

PASSOS COELHO, D. and PINHO GIL, T. (1963). "Studies on extreme double exponential distribution I. The location parameter." *Revista da Faculdade de Ciencias de Lisboa* **A10**, 37–46.

PATIL, G. P., BOSWELL, M. T., JOSHI, S. W. and RATNAPARKHI, M. V. (1984). *Dictionary and Classified Bibliography of Statistical Distributions in Scientific Work.* **1**: Discrete Models. International Cooperative Publishing House.

PATIL, G. P., BOSWELL, M. T. and RATNAPARKHI, M. V. (1984). *Dictionary and Classified Bibliography of Statistical Distributions in Scientific Work.* Vol **2**: Continuous Univariate Models. International Cooperative Publishing House.

PATIL, G. P., BOSWELL, M. T., RATNAPARKHI, M. V. and ROUX, J. J. J.

(1984). *Dictionary and Classified Bibliography of Statistical Distributions in Scientific work.* **3**: Multivariate Models. International Cooperative Publishing House.

PAYNE, A. O. (1972). "A reliability approach to the fatigue of structures. Probabilistic Aspects of Fatigue." *ASTM Special Technical Publication* No. **511**, 106–149.

PEIRCE, F. T. (1926). "Tensile strength for cotton yarns. v.—The weakest link— theorems on the strength of long and of composite specimens." *J. Text. Inst.* **17**, T355–368.

PETERSON, R. E. (1930). "Fatigue tests of small specimens with particular reference to size effects." *Trans. Amer. Soc. Steel Treat.* **18**, 1041–1056.

PETERSON, R. E. (1949a). "Approximate statistical method for fatigue data." *ASTM Bull.* **156**, 50–52.

PETERSON, R. E. (1949b). Application of fatigue data to machine design. Fatigue Strength of Metals, Amer. Soc. Metals, Cleveland, Ohio, 60–78.

PHOENIX, S. L. and TAYLOR, H. M. (1973). "The asymptotic strength distribution of a general fiber bundle." *Adv. Appl. Probab.* **5**, 200–216.

PHOENIX, S. L. (1975). "Probabilistic inter-fiber dependence and the asymptotic strength distribution of classic fiber bundles." *Intern. J. Eng. Science* **13**, 287–304.

PHOENIX, S. L. (1978). "The asymptotic time to failure of a mechanical system of parallel members." *SIAM J. Appl. Math.* **34**, 227–246.

PHOENIX, S. L. and SMITH, R. L. (1983). "A comparison of probabilistic techniques for the strength of fibrous materials under local load-sharing among fibers." *Int. J. Solids Struct.* **19**, 479–496.

PHOENIX, S. L. and TIERNEY, L. J. (1983). "A statistical model for the time dependent failure of unidirectional composite materials under local elastic load-sharing among fibers." *Eng. Fract. Mech.* **18**, 193–215.

PHOENIX, S. L. and WU, E. M. (1983). "Statistics for the time dependent failure of Kevlar-49/epoxy composites: Micromechanical modeling and data interpretation." *Mech Comp. Mater.,* 135–215.

PICKANDS, J. III (1967a). "Sample sequences of maxima." *Ann. Math. Statist.* **38**, 1570–1574.

PICKANDS, J. III (1967b). "Maxima of stationary Gaussian processes." *Zeitschrift fur Wahrsch. verw. Geb.* **7**, 190–233.

PICKANDS, J. III (1968). "Moment convergence of sample extremes." *Ann. Math. Statist.* **39**, 881–889.

PICKANDS, J. III (1969a). "An iterated logarithm law for the maximum in a stationary Gaussian sequence." *Zeitschrift fur Wahrsch. verw. Geb.* **12**, 344–353.

PICKANDS, J. III (1969b). "Asymptotic properties of the maximum in a stationary Gaussian process." *Trans. Amer. Math. Soc.* **145**, 75–86.

PICKANDS, J. III (1971). "The two-dimensional Poisson process and extremal processes." *J. Appl. Probab.* **8**, 745–756.

PICKANDS, J. III (1975). "Statistical inference using extreme order statistics." *Ann. Statist.* **3**, 119–131.

PICKANDS, J. III (1981). Multivariate extreme value distributions. Proc. 43rd ISI Meeting. Buenos Aires, 859–877.

PLACKETT, R. L. (1947). "Limits of the ratio of mean range to standard deviation." *Biometrika* **34**, 120–122.

POSNER, E. C. (1965). "The application of extreme value theory to error free communication." *Technometrics* **7**, 517–529.

POSNER, E. C., RODENICH, E. R., ASHLOCK, J. C. and LURIE, S. (1969). "Application of an estimator of high efficiency in bivariate extreme value theory." *J. Amer. Statist. Assoc.* **64**, 1403–1415.

PROT, M. (1949a). "La securité. *Ann. Ponts Chauss.* **119**, 19–49.

PROT, M. (1949b). "Statistique et sécurité." *Revue de Métallurgie* **46**, 716–718.

PROT, M. (1949c). "Essais statistiques sur mortiers et bétons." *Ann. Inst. Techn. Bat. Trav. Publics* **2**, 1–40.

PROT, M. (1950). "Vues nouvelles sur la sécurité des constructions." *Mém. Soc Ing. Civils* France **103**, 50–57.

PUTTICK, K. E. and THRING, M. W. (1952). "Effect of specimen length on the strength of a material with random flaws." *J. Iron Steel Inst.* **172**, 56–61.

PUTZ, R. R. (1952). "Statistical distribution for ocean waves." *Trans. Amer. Geophys. Union* **33**, 685–692.

QUALLS, C. (1969). "On the joint distribution of crossings of high multiple levels by a stationary Gaussian process." *Ark. Mat.* **8**, 129–137.

QUALLS, C. and WATANABE, H. (1972). "Asymptotic properties of Gaussian processes." *Ann. Math. Statist.* **43**, 580–596.

RAO, C. V. S. K. and SRIDHAR, J. K. (1972). Statistical aspects of strength and fracture behaviour of concrete. Proceedings of the Mechanical Behaviour of Materials International Conference, Kyoto, Japan, IV: Concrete and Cement Paste, Glass and Ceramics. The Society of Materials Science, Kyoto, Japan, 53–62.

REEDER, H. A. (1972). "Machine generation of order statistics." *Amer. Statistn*, **26**, 56–57.

REICH, B. M. (1970). "Flood series compared to rainfall extremes." *Water Resour. Res.* **6**.

REISS, R. D. (1981). "Uniform approximation to distributions of extreme order statistics." *Adv. Appl. Probab.* **13**, 533–547.

RÉNYI, A. (1953). "On the theory of order statistics." *Acta Math. Acad. Sci. Hungar.* **4**, 191–231.

RÉNYI, A. (1958). "Quelques remarques sur les probabilités d'événements dépendantes." *J. Math. Pures Appl.* **37**, 393–398.

RÉNYI, A. (1962). On outstanding values of a sequence of observations. In: *Selected paper of A. Rényi*, **3**. Akadémiai Kiadó Budapest, 50–65.

RÉNYI, A. (1963), "On stable sequences of events." *Sankhya* **A25**, 293–302.

RESHETOV, D. N. and IVANOV, A. S. (1972). "How ageing affects failure properties." *Russ. Eng. J.* **52**, 6–9.

RESNICK, S. I. and NEUTS, M. F. (1970). "Limit laws for maxima of a sequence of random variables defined on a Markov chain." *Adv. Appl. Probab* **2**, 323–343.

RESNICK, S. I. (1971a). "Tail equivalence and its applications." *J. Appl. Probab.* **8**, 136–156.

RESNICK, S. I. (1971b). "Asymptotic location and recurrence properties of maxima of a sequence of random variables defined on a Markov chain." *Zeitschrift fur Wahrsch. verw Geb.* **18**, 197–217.

RESNICK, S. I. (1971c). "Products of distribution functions attracted to extreme value law." *J. Appl. Probab.* **8**, 781–793.

RESNICK, S. I. (1972). "Stability of maxima of random variables defined on a Markov chain." *Adv. Appl. Probab.* **4**, 285–295.

RESNICK, S. I. (1973a). "Limit laws for record values." *Stoch. Proc. Appl.* **1**, 67–82.

RESNICK, S. I. (1973b). "Record values and maxima." *Ann. Probab.* **1**, 650–662.

RESNICK, S. I. (1973c). "Extremal processes and record value times." *J. Appl. Prob.* **10**, 863–868.

RICE, S. O. (1939). "The distribution of the maxima of a random curve." *Amer. J. Math.* **61**, 409–416.

RICE, S. O. (1944). "Mathematical analysis of random noice." *Bell. Syst. Tech.* **23**, 282–332.

RICE, S. O. (1945). "Mathematical analysis of random noice." *Bell. Syst. Tech.* **24**, 46–156.

RINGER, L. J. and SPRINKLE, E. E. III (1972). "Estimation of the parameters of the Weibull distribution from multicensored samples." *IEEE Trans. Reliab.* **21**, 46–51.

ROBERTS, E. M. (1979a). "Review of statistics of extreme values with applications to air quality data, Part I: review." *J. Air Pollut. Control Assoc.* **29**, 632–637.

ROBERTS, E. M. (1979b). "Review of statistics of extreme values with applications to air quality data, Part II: Applications." *J. Air Pollut. Control Assoc.* **29**, 733–740.

ROOTZÉN, H. (1974). "Some properties of convergence in distribution of sums and maxima of dependent random variables." *Zeitschrift fur Wahrsch. verw. Geb.* **29**, 295–307.

ROOTZÉN, H. (1978). "Extremes of moving averages of stables processes." *Ann. Probab.* **6**, 847–869.

ROSEN, B. W. (1964). "Tensile failure of fibrous composites." *AIAA J.* **2**, 1985–1991.

ROSEN, B. W. and Dow, N. F. (1972). *Mechanics of Failure of Fibrous Composites.* Fracture: An Advanced Treatise, Vol. VII: Fracture of Nonmetals and Composites (edited by H. Liebowitz). Academic Press, New York, 611–674.

ROSEN, B. W. and ZWEBEN, C. H. (1972). Tensile failure criteria for fiber composite materials. NASA Contractor Report CR-2057, National Aeronautics and Space Administration, Washington, D.C.

RUSTAGI, J. S. (1957). "On minimizing and maximizing a certain integral with statistical applications." *Ann. Math. Statist.* **28**, 309–328.

SACHS, P. (1972). *Wind Forces in Engineering.* Pergamon Press, Oxford, England.

SARHAN, A. E. and GREENBERG, B. G. (1962). *Contributions to Order Statistics*. John Wiley and Sons, New York.

SARPHIE, C. S. (1972). Evaluation of a reliability analysis method for fatigue life of aircraft structures. Proc. Coll. Struct. Reliab: The Impact of Advanced Materials on Engineering Design, 246–264.

SATHE, Y. S., PRADHAN, M. and SHAM, S. P. (1980). "Inequality for the probability of the occurrence of at least m out of n events." *J. Appl. Probab.* **17,** 1127–1132.

SAUNDERS, S. C. (1972). The treatment of data in fatigue analysis. Proc. Coll. Struct. Reliab.: The impact of Advanced Materials on Engineering Design, 136–155.

SCHAFER, D. (1974). "Confidence bounds for the minimum fatigue life." *Technometrics* **16,** 113–123.

SCHUCANY, W. R. (1972). "Order statistics in simulation." *J. Statist. Comput. Simul.* **1,** 281–286.

SCHUELLER, G. I. (1984). *Application of Extreme Values in Structural Engineering*. In: Statistical extremes and Applications. NATO ASI Series. D. Reidel Publishing Company.

SCHUETTE, E. H. (1954). "A simplified statistical procedure for obtaining design-level fatigue curves." *Proc. ASTM* **54,** 853–874.

SCHÜPBACH, M. and HÜSLER, J. (1983). "Simple estimators for the parameters of the extreme-value distribution based on censored data." *Technometrics* **25,** 189–192.

SCHUSTER, E. F. (1984). "Classification of probability laws by tail behaviour." *J. Amer. Statist. Assoc.* **9,** 936–939.

SCOP, P. M. and ARGON, A. S. (1969). "Statistical theory of strength of laminated composites II." *J. Compos. Mater.,* 30–47.

SEDRAKYAN, L. G. (1958). Two problems of the statistical strength theory. Select. Transl. Math. Statist. Probab. **4,** Amer. Math. Soc., Providence, Rhode Island, 289–294.

SELLARS, F. (1975). "Maximum heights of ocean waves." *J. Geophys. Res.* **80,** 398–404.

SEN, P. K. (1959). "On the moments of the sample quantiles." *Calcutta Statist. Ass. Bull.* **9,** 1–19.

SEN, P. K. (1961). "A note on the large sample behaviour of extreme sample values from distributions with finite endpoints." *Bull Calcutta Statist. Assoc.* **10,** 106–115.

SEN, P. K. (1968). "Asymptotic normality of sample quantiles for m-dependent processes." *Ann. Math. Statist.* **39,** 1724–1730.

SEN, P. K. (1970). "A note on order statistics for heterogeneous distributions." *Ann. Math. Statist.* **41,** 2137–2139.

SEN, P. K. (1973). "On fixed size confidence bands for the bundle strength of filaments." *Ann. Statist.* **1,** 526–537.

SEN, P. K., BHATTACHARYYA, B. B. and SUH, M. W. (1973). "Limiting behaviour of the extremum of certain sample functions." *Ann. Statist.* **1,** 297–311.

SENETA, E. (1976). *Regularly Varying Functions*. Lecture Notes in Mathematics 508, Springer-Verlag, Heidelberg.

SERENSEN, S. V. and STRELYAEV, V. S. (1962). "The static structural strength of fibre glass materials." *Russian Eng. J.* **42**, 8–14.

SERFOZO, R. (1982). "Functional limit theorems for extreme values of arrays of independent random variables." *Ann. Probab.* **10**, 172–177.

SETHURAMAN, J. (1965). "On a characterization of the three limiting types of the extreme." *Sankhya* **A27**, 357–364.

SHANE, R. and LYNN, W. (1964). "Mathematical model for flood risk evaluation. *J. Hydraul. Div.,* ASCE, **90**, HY6, 1–20.

SHELLARD, H. C. (1958). "Extreme wind speeds over Great Britain and Northern Ireland." *Meteorol. Magaz.* **37**.

SHEPP, L. A. (1979). "The joint density of the maximum and its location for a Wiener process with drift." *J. Appl. Probab.* **16**, 423–427.

SHINOZUKA, M. (1972). "Probabilistic modeling of concrete structures." *J. Eng. Mech. Div.,* ASCE, **98**, EM6, 1433–1451.

SHORROCK, R. W. (1972a). "On record values and record times." *J. Appl. Probab.* **9**, 316–326.

SHORROCK, R. W. (1973). "Record values and inter-record times." *J. Appl. Probab.* **10**, 543–555.

SHUR, M. G. (1965). "On the maximum of a Gaussian stationary process." *Theory Probab. Appl.* **10**, 354–357.

SIA, Norm 161 (1979). Stahlbauten. Schweiz. Ingenieur und Architekten Verein, Zürich.

SIBUYA, M. (1960). "Bivariate extreme statistics." *Ann. Inst. Stat. Math.* **11**, 195–210.

SIEBKE, H. (1980). "Beschreibung einer Bezugsbasis zur Bemessung von Bauwerken auf Betriebsfestigkeit." *Schweissen und Schneiden* **32**, H. 8, 304–314.

SIMIU, E. and FILLIBEN, J. J. (1975). Statistical analysis of extreme winds. NBS TR-868. Nation. Bur. Stand., Washington, D.C.

SIMIU, E. and FILLIBEN, J. J. (1976). "Probability distribution of extreme wind speeds." *J. Struct. Div.,* ASCE, **102**, ST9, 1861–1877.

SIMIU, E. and SCANLAN, R. H. (1977). *Wind Effects on Structures: An Introduction to Wind Engineering.* John Wiley and Sons. New York.

SIMIU, E., BIÉTRY, J. and FILLIBEN, J. J. (1978). Sampling error in estimation of extreme winds. *J. Struct. Div.,* ASCE, ST3, 491–501.

SIMIU, E., CHANGERY, M. J. and FILLIBEN, J. J. (1979). Extreme wind speeds at 129 stations in the contiguous United States. Building Science Series, Nation. Bur. Stand., Washington, D.C.

SIMIU, E., FILLIBEN, J. J. and SHAVER, J. R. (1982). "Short term records and extreme wind speeds." *J. Struct. Div.,* ASCE, **108**, ST11, 2571–2577.

SINGPURWALLA, N. D. (1971). "Statistical fatigue models: a survey. *IEEE Trans. Reliab.* **20**, 185–189.

SINGPURWALLA, N. D. (1972a). Statistical failure models—a survey. Proc. Coll. Struct. Reliab.: The Impact of Advanced Materials on Engineering Design, 86–103.

SINGPURWALLA, N. D. (1972b). "Extreme values for a lognormal law with applications to air pollution problems." *Technometrics* **14,** 703–711.

SIRVANCI, M. and YANG, G. (1984). "Estimation of the Weibull parameters under type I censoring." *J. Amer. Statist. Assoc.* **79,** 183–187.

SLOCUM, C. E. and HANCOCK, E. L. (1906). *Text-book on the Strength of Materials.* Ginn and Co., Boston. 1911.

SMIRNOV, N. V. (1952). Limit distributions for the terms of a variational series. *Amer. Math. Soc. Transl.* **67,** 1–64.

SMITH, R. L. (1980). "A probabilistic model for fibrous composite with local load-sharing. *Proc. Roy. Soc.* Lond. **A372,** 539–553.

SMITH, R. L. (1981). "Asymptotic distribution for the failure of fibrous materials under series-parallel structure and equal load-sharing." *J. Appl. Mech.* **103,** 75–82.

SNEYERS, R. (1984). Extremes in meteorology. In: *Statistical Extremes and Applications.* NATO ASI Series. D. Reidel Publishing Company.

SONI, A. H. and LITTLE, R. E. (1965). Fatigue Limit Analyses and Design of Fatigue Experiments. Proc. Tenth Conf. Design Experim. Army Res. Developm. Test., ARO-D Report, No. 65-2, U.S. Army Research Office, Durham, N.C.

SPINDEL, J. E., BOARD, B. R. and HAIBACH, E. (1979). *The Statistical Analysis of Fatigue Test Results.* ORE Utrecht.

STAM, A. J. (1973). "Regular variation of the tail of a subordinated probability distribution." *Adv. Appl. Probab.* **5,** 308–327.

STEPNOV, M. N. (1964). "Estimating the probability of failure in fatigue testing." *Trudy Aviatsionyyi i Teknologicheskii Institut,* Moscow, Part 61, 38–44.

STULEN, F. B. (1951). On the statistical nature of fatigue. Symposium ASTM Statistical Aspects of Fatigue Tests. ASTM Special Technical Publication No. 121, Philadelphia, Pa., 23–40.

SUAREZ BORES, P. (1964). Estructura del oleaje. Laboratorio de Puertos, Centro de Estudios y Experimentación de Obras Públicas, Madrid.

SUGIURA, N. (1962). "On the orthogonal inverse expansion with an application to the moments of order statistics." *Osaka J. Math.* **14,** 253–263.

SUGIURA, N. (1964). "The bivariate orthogonal inverse expansion and the moments of order statistics." *Osaka J. Math.* **1,** 45–59.

SUH, M. W., BHATTACHARYYA, B. B. and GRANDAGE, A. (1970). "On the distribution and moments of the strength of a bundle of filaments." *J. Appl. Probab.* **7,** 712–720.

SUZUKI, E. (1961). "A new, procedure of statistical inference on extreme values." *Pap. Meterol. Geophys.* **12,** 1–17.

SWEETING, T. J. (1985). "On domains of uniform local attraction in extreme value theory." *Ann. Probab.* **13,** 196–205.

TANG, J. P. and YAO, J. T. P. (1972). "Fatigue damage factor in structural design." *J. Struct. Div.,* ASCE, **98,** ST1, 125–134.

TEUGELS, J. L. (1981). "Limit theorems on order statistics." *Ann. Probab.* **9,** 868–880.

THIRUVENGADAM, A. (1972). Corrosion Fatigue at High Frequencies and High

Hydrostatic Pressures. Stress Corrosion Cracking of Metals—A State of the Art. ASTM Special Technical Publication No. 518, Philadelphia, Pa.

THOMAN, D. R., BAIN, L. J. and ANTLE, C. E. (1969). "Inferences on the parameters of the Weibull distribution." *Technometrics* **11**, 445–460.

THOMAN, D. R., BAIN, L. J. and ANTLE, C. E. (1970). "Maximum likelihood estimation, exact confidence intervals for reliability, and tolerance limits in the Weibull distribution." *Technometrics* **12**, 363–371.

THOM, H. C. S. (1967). Asymptotic Extreme Value Distributions Applied to Wind and Waves. NATO Seminar on extreme value problems, Faro, Portugal.

THOM, H. C. S. (1968a). Toward a Universal Climatological Extreme Wind Distribution. Intern. Res. Seminar on Wind effects on Buildings and Structures, National Research Council Proceedings, Ottawa, Canada.

THOM, H. C. S. (1968b). "New distributions of extreme winds in the United States." *J. Struct. Div.*, ASCE, **94**, ST7, 1787–1801.

THOM, H. C. S. (1969). "Application of climatological analysis to engineering design data." *Rev. Belge Statist. Rech. Opérat.* **9**, 2–13.

THOM, H. C. S. (1971). "Asymptotic extreme value distributions of wave height in open ocean." *J. Mar. Res.* **29**, 19–27.

THOM, H. C. S. (1973). "Extreme wave height distributions over oceans." *J. Waterw. Harb. Coast. Div., ASCE*, **99**, WW3, 355–374.

THOMAS, D. I. (1972). "On limiting distributions of a random number of dependent random variables." *Ann. Math. Statist.* **43**, 1719–1726.

THOMAS, D. R. and WILSON, W. M. (1972). "Linear order statistic estimation for the two-parameter Weibull and extreme value distributions from type II progressively censored samples." *Technometrics* **14**, 679–691.

THRASHER, L. W. and AAGARD, P. M. (1970). "Measured wave force data on Offshore platforms." *J. Petrol. Techn.*, 339–346.

TIAGO DE OLIVEIRA, J. (1957). "Estimators and tests for continuous populations with location and dispersion parameters." *Rev. Fac. Cienc.* Lisboa **A6**, 121–146.

TIAGO DE OLIVEIRA, J. (1958). "Extremal distributions." *Rev. Fac. Cienc.* Lisboa **A7**, 215–227.

TIAGO DE OLIVEIRA, J. (1959). Distribuiçoes de extremos. Curso de Matemáticas Superiores Prof. Mira Fernandes, Vol. **2**, Instituto Superior de Ciencias Económicas e Financeiras, Lisboa, 133–159.

TIAGO DE OLIVEIRA, J. (1961). "The asymptotical independence of the sample mean and the extremes." *Rev. Fac. Cienc.* Lisboa **A8**, 299–310.

TIAGO DE OLIVEIRA, J. (1962). "Structure theory of bivariate extremes: extensions." Estudos de Math. Estat. Econom. **7**, 165–195.

TIAGO DE OLIVEIRA, J. (1968). "Extremal processes: Definition and properties." *Publ. Inst. Stat. Univ.* Paris **17**, 25–36.

TIAGO DE OLIVEIRA, J. (1963). "Decision results for the parameters of the extreme value (Gumbel) distribution based on the mean and the standard deviation." *Trab. Estad. Invest. Oper.* **14**, 61–81.

TIAGO DE OLIVEIRA, J. (1970). "Biextremal distributions: statistical decision." *Trab. Estad. Invest. Oper.* **21**, 107–117.

TIAGO DE OLIVEIRA, J. (1971). A new model of bivariate extremes: Statistical decision. In: *Studi Prob. Stat. ricerca oper. in onore di G. Pompilj.* Oderisi, Gubbio, 1–13.

TIAGO DE OLIVEIRA, J. (1972). Statistics for Gumbel and Fréchet distributions. In: *Structural Safety and Reliability* (edited by A. Freundenthal). Pergamon Press, New York, 91–105.

TIAGO DE OLIVEIRA, J. (1974). "Regression in the nondifferentiable bivariate extreme models." *J. Amer. Statist. Assoc.* **69**, 816–818.

TIAGO DE OLIVEIRA, J. (1975). Bivariate extremes. Extensions. Proc. 40th Session ISI, Warsaw.

TIAGO DE OLIVEIRA, J. (1980). Bivariate extremes: foundations and statistics. In: *Proc. Fifth Int. Symp. Multivar. Anal.* (edited by P. R. Krishnaia), North Holland, 349–366.

TIAGO DE OLIVEIRA, J. (1981). "Statistical choice of univariate extreme models." *Statistical distributions in Scientific Work* **6**, 367–387.

TIAGO DE OLIVEIRA, J. (1984). *Statistical Extremes and Applications.* NATO ASI Series. D. Reidel Publishing Company, Lisbon.

TIDE, R. H. R. and VAN HORN, D. (1966). A statistical study of the static and fatigue properties of high strength prestressing strand. Fritz Engineering Lab., Report No. 309.2, Lehigh Univ.

TIERNEY, L. (1982). "Asymptotic bounds on the time to fatigue failure of bundles of fibers under local load-sharing." *Adv. Appl. Probab.* **14**, 95–121.

TILLY, G. P. and MOSS, D. S. (1982). Long endurance of steel reinforcement. IABSE Colloquium on Fatigue of Steel and Concrete Structures, Lausanne, 229–238.

TIMOSHENKO, S. P. (1953). *History of Strength of Materials.* McGraw Hill, Inc., New York.

TIPPETT, L. H. C. (1925). "On the extreme individuals and the range of samples taken from a normal population." *Biometrika* **17**, 364–387.

TODOROVIC, P. and ZELENHASIC, E. (1970). "A stochastic model for flood analysis." *Water Resour. Res.* **6**, 1641–1648.

TODOROVIC, P. (1970). "On some problems involving random number of random variables." **i41**, 1059–1063.

TODOROVIC, P. (1971). On Extreme Problems in Hydrology. Joint satistics meeting, Amer. Statist. Assoc. and Inst. Math. Statist., Colorado State Univ., Fort Collins.

TODOROVIC, P. and ROUSSELLE, J. (1971). "Some problems of flood analysis." *Water Resour. Res.* **7**, 1144–1150.

TODOROVIC, P. and WOOLHISER, D. A. (1972). "On the time when the extreme flood occurs." *Water Resourc. Res.* **8**, 1433–1438.

TODOROVIC, P. and SHEN, H. W. (1976). *Some Remarks on the Statistical Theory of Extreme Values Stoch. Approach.* Water Resour. **2** (edited by H. W. Sen). Univ. of Colorado Press., Fort Collins, Colorado.

TODOROVIC, P. (1978). "Stochastic models of floods." *Water Resour. Res.,* 345–356.

TODOROVIC, P. (1979). A Probabilistic Approach to Analysis and Predistion of Floods. Proc. 43rd Session ISI, Buenos Aires.

TRADINIK, W., KROMP, K. and PABST, R. F. (1981). Ueber die universalität des Weibull-ansatzes. Materialprüfung 23, No. 2.

TSIREL'SON, V. S. (1975). "The density of the distribution of the maximum of a Gaussian process." Theory Probab. Appl. 20, 847–856.

TUCKER, J. (1927). "The compressive strength dispersion of materials with applications." Franklin Institute 204, 751–781.

TUCKER, J. (1941). "Statistical theory of the effect of dimensions and of method of loading upon the modulus of rupture of beams." Proc. ASTM 41, 1072–1094.

TUCKER, J. (1945a). "Effect of dimensions of specimens upon the precision of strength data." Proc. ASTM 45, 952–960.

TUCKER, J. (1945b). "The maximum stresses present at failure of brittle materials." Proc. ASTM 45, 961–975.

TUCKER, M. J. (1963). "Analysis of records of sea waves." Proc. Instit. Civil Eng., London, 26, 305–316.

TURKMAN, K. F. and WALKER, A. M. (1983). "Limit laws for the maxima of a class of quasi-stationary sequences." J. Appl. Probab. 20, 814–821.

UDOGUCHI, T. and MATSUMURA, M. (1972). Statistical investigation on low-cycle fatigue life of high tension steel. Proc. of the International Conference on Mechanical Behaviour of Materials, Kyoto, Japan, 5: Composites, Testing and Evaluation. Soc. Mater. Sci., Kyoto, Japan, 458–470.

U. S. GEOLOGICAL SURVEY. Magnitude and Frequency of Floods in the United States. U. S. Geol. Surv. Water Supply Pap., 1963–1968.

Van-ZWET, W. R. (1964). Convex transformations of random variables. Mathematical Center Tracts 7, Amsterdam.

VANDAR, Ö and FINNIE, I. (1975). "An analysis of the Brazilian disk fracture test using the Weibull probabilistic treatment of brittle strength. Intern." J. Fract. 11, 495–508.

VILLASEÑOR, J. (1976). On Univariate and Bivariate Extreme Value Theory. Doctoral Dissertation, Iowa State Univ.

VILLASEÑOR, J. (1981). Norming Constants for Maxima Attracted to $\exp[\exp(-x)]$. Proc. 43rd Session ISI, Buenos Aires.

VINSON, J. R. and CHOU, T. W. (1975). Composite Materials and Their Use in Structures. John Wiley and Sons, New York.

VORLICEK, M. (1963). "The effect of the extent of stressed zone upon the strength of material." Acta Technica CSAV 8, 149–176.

VORLICEK, M. (1971a). "Statistical theory of the stress distribution effect on the strength." Stavebnicky Casopis 19, 274–288.

VORLICEK, M. (1971b). "Fundamentals of the statistical theory of strength." Acta Technica CSAV 16, 239–268.

VORLICEK, M. (1972). "Fundamentals of the statistical theory of strain." Acta Technica CSAV 17, 84–112.

WADDOUPS, M. E. and HALPIN, J. C. (1974). "The fracture and fatigue of composite structures." *Computers and Structures* **4**, 659–673.

WADSWORTH, N. J. and SPILLING, I. (1968). "Load transfer from broken fibres in composite materials." *British J. Appl. Phis.* **1**, 1049–1058.

WALKER, A. M. (1981). "On the classical Bonferroni inequalities and the corresponding Galambos inequalities." *J. Appl. Probab.* **18**, 757–763.

WALSH, J. E. (1969a). "Asymptotic independence between largest and smallest of a set of independent observations." *Ann. Inst. Statist. Math.,* Tokyo, **21**, 287–289.

WALSH, J. E. (1969b). "Approximate distributions for largest and for smallest of a set of independent observations." *S. Afr. Statist. J.* **3**, 83–89.

WARNER, R. F. and HULSBOS, C. L. (1966). "Fatigue properties of prestressing strand." *PCI J.* **11**, No. 2, 25–46.

WARNER, R. F. (1982). Fatigue of partially prestressed concrete beams. IABSE Colloquium Fatigue of Steel and Concrete Structures, Lausanne, 431–438.

WATSON, G. S. (1954). "Extreme values in samples from *m*-dependent stationary stochastic processes." *Ann. Math. Statist.* **25**, 798–800.

WATTS, J. H. V. (1977). Limit Theorems and Representations for Order Statistics from Dependent Sequences. Doctoral Dissertation, Statistics Dept., Univ. of N. C.

WATTS, V. ROOTZÉN, H. and LEADBETTER, M. R. (1982). "On limiting distribution of intermediate order statistics from stationary sequences." *Ann. Probab.* **10**, 653–662.

WEBBER, M. (1979). Analyse asymptotique del processus Gaussienns stationaires. Inst. Res. Math. Avan. Strasbourg.

WEBER, K. H. and ENDICOTT, H. S. (1956). "Area effect and its extremal basis for the electric breakdown of transformer oil." AIEE Trans., Part III: *Power Apparatus and Systems* **75**, 371–381.

WEBER, K. H. and ENDICOTT, H. S. (1957). "Extremal area effect for large area electrodes for the electric breakdown of transformer oil." AIEE Trans., Part III: *Power Apparatus and Systems* **76**, 1091–1098.

WEIBULL, W. (1939a). A statistical theory of the strength of materials. Ingeniörs Vetenskaps Akademien Handlingar, No. 151. Generalstabens Litografiska Anstalts Förlag, Stockholm.

WEIBULL, W. (1939b). The phenomenon of rupture in solids. Ingeniörs Vetenskaps Akademien Handlingar, No. 153. Generalstabens Litografiska Anstalts Förlag, Stockholm.

WEIBULL, W. (1949). Statistical representation of fatigue failures in solids. Trans. Roy. Inst. Technol., Stockholm, Separate No. 27.

WEIBUL, W. (1951). "A statistical distribution function of wide applicability." *J. Appl. Mech.* **18**, 293–297.

WEIBULL, W. (1952a). "Statistical design of fatigue experiments." *J. Appl. Mech.* **19**, 109–113.

WEIBULL, W. (1952b). "A survey of 'statistical effects' in the field of material failure." *Appl. Mech. Rev.* **5**, 449–451.

WEIBULL, W. (1959). Zur Abhängigkeit der Festigkeit von der Probegrösse. Ingenieur Archiv **XXVIII**, 360–362.

WEIBULL, W. (1961). *Fatigue Testing and Analysis of Results.* Pergamon Press, New York.

WEIBULL, W. (1977). References on the Weibull distribution. FTL A-report **A20**: 23. Forvarets Teletekniska Laboratorium, Stockholm.

WEINSTEIN, S. B. (1973). "Theory and application of some classical and generalized asymptotic distributions of extreme values." *IEEE Trans. Inf. Theor.* **19**, 148–154.

WEISS, L. (1969). "The joint asymptotic distribution of the k-smallest sample spacings." *J. Appl. Probab.* **6**, 442–448.

WEISSMAN, I. (1975). "Multivariate extremal processes generated by independent non-identically distributed random variables." *J. Appl. Probab.* **12**, 477–487.

WEISSMAN, I. (1978). "Estimation of parameters and large quantiles based on the kth largest observations." *J. Amer. Statist. Assoc.* **63**, 812–815.

WEISSMAN, I. (1980). "Estimation of tail parameters under type I censoring." *Comm. Statist. Theor. Meth.* **A9**, 1165–1175.

WELSH, R. E. (1971). "A weak convergence theorem for order statistics from strong mixing processes." *Ann. Math. Statist.* **42**, 1637–1646.

WELSCH, R. E. (1972). "Limit laws for extreme order statistics from strong mixing processes." *Ann. Math. Statist.* **43**, 439–446.

WELSCH, R. E. (1973). "A convergence theorem for extreme values from Gaussian sequences." *Ann. Probab.* **1**, 398–404.

WIEGEL, R. L. (1964). *Oceanographical Engineering.* Prentice Hall, Inc. Englewood Cliffs, N. J.

WILKS, S. S. (1962). *Mathematical Statistics.* John Wiley and Sons, New York.

WILSON, B. W. (1966). Design Sea and Wind Conditions for Offshore Structures. Proc. OECON, Offshore Exploration Conference, 665–708.

WINKLER, F. (1954). "Relation between bundle strength and strength of individual fibers or yarns." *Faserforschung und Textiltechnik* **5**, 398–403.

WIRSCHING, P. H. and YAO, J. T. P. (1970). "Statistical methods in structural fatigue." *J. Struct. Div., ASCE,* **96**, ST6, 1201–1219.

YANG, J. N. and TRAPP, W. J. (1974). "Reliability analysis of aircraft structures under random loading and periodic inspection." *AIAA J.* **12**, 1623–1630.

YANG, C. Y. TAYFUN, M. A. and HSIAO, G. C. (1974). *Stochastic Prediction of Extreme Waves and Sediment Transport in Coastal Waters.* Stochastic Problems in Mechanics. Proc. Symp. Univ. Waterloo, Ontario, Canada, Waterloo Press, 431–448.

YANG, S. S. (1977). "General distribution theory of the concomitants of order statistics." *Ann. Statist.* **5**, 996–1002.

YANG, S. (1981). "Linear functions of concomitants of order statistics with applications to non-parametric estimations of a regression function." *J. Amer. Statist. Assoc.* **76**, 658–662.

YAO, J. T. P. (1974). "Fatigue reliability and design." *J. Struct. Div.*, ASCE, **100**, ST9, 1827–1836.

YEN, C. S. (1969). Fatigue statistical analysis. *Metal Fatigue: Theory and Design* (edited by A. F. Madayag), Chapter 5. John Wiley and Sons, 140–169.

YEVJEVICH, V. (1984). Extremes in hydrology. In: *Statistical Extremes and Applications*. NATO ASI Series. D. Reidel Publishing Company.

ZELENHASIC, E. (1970). Theoretical Probability Distribution for Flood Peaks. Hydrol. Pap. 42, Colo. State Univ., Fort Collins, Colorado.

ZIDEK, J. V., NAVIN, F. P. D. and LOCKHART, R. (1979). Statistics of extremes: an alternate method with application to bridge design codes. *Technometrics* **21**, 185–191.

ZWEBEN, C. (1968). "Tensile failure of fiber composites." *AIAA J.* **6**, 2325–2331.

Index

Printed and bound by CPI Group (UK) Ltd, Croydon, CR0 4YY

03/10/2024

01040410-0009